BIOFUELS

Meeting the energy and environmental challenges of the transportation sector

Cover illustration : Agricultural and forest wastes used as raw material
for the second generation biofuel production.

IFP Energies nouvelles Publications

D. BALLERINI
Expert in Biotechnology and Biomass

BIOFUELS

Meeting the energy and environmental challenges of the transportation sector

Preface by
Olivier APPERT
IFP Energies nouvelles Chairman and CEO

Translated from the French
by Trevor Jones, Jill Rupnow (Lionbridge)

2012
Editions TECHNIP 25 rue Ginoux, 75015 PARIS, FRANCE

FROM THE SAME PUBLISHER

- Geomechanics applied to the petroleum industry
  J.F. NAUROY
- Heavy Crude Oils
  From Geology to Upgrading. An Overview
  A.Y. HUC
- CO_2 Capture
  Technologies to Reduce Greenhouse Gas Emissions
  F. LECOMTE, P. BROUTIN, E. LEBAS
- Corrosion and Degradation of Metallic Materials
  Understanding of the Phenomena and Applications in Petroleum and Process Industries
  F. ROPITAL
- Multiphase Production
  Pipeline Transport, Pumping and Metering
  J. FALCIMAIGNE, S. DECARRE
- A Geoscientist's Guide to Petrophysics
  B. ZINSZNER, F.M. PERRIN
- Acido-Basic Catalysis (2 vols.)
  Application to Refining and Petrochemistry
  C. MARCILLY
- Petroleum Microbiology (2 vols.)
  Concepts. Environmental Implications. Industrial Applications
  J.P. VANDECASTEELE
- Physico-Chemical Analysis of Industrial Catalysts
  A Practical Guide to Characterisation
  J. LYNCH
- Chemical Reactors
  From Design to Operation
  P. TRAMBOUZE, J.P. EUZEN
- Petrochemical Processes (2 vols.)
  Technical and Economic Characteristics
  A. CHAUVEL, G. LEFEBVRE
- The Technology of Catalytic Oxidations (2 vols.)
  P. ARPENTINIER, F. CAVANI, F. TRIFIRO
- Marine Oil Spills and Soils Contaminated by Hydrocarbons
  C. BOCARD

Printed in France

ISBN 978-2-7108-0983-8

Preface

Six years have already elapsed since the first edition of this book on biofuels was published*. The fundamental aspects concerning the interest in biofuels remain, but in view of the recent evolution and increasing awareness of the global energy and environmental stakes as well as the development of research actions conducted on second and third generation biofuels, it was essential to release a new edition.

Firstly, the growth in transport demand will remain strong over the coming decades, boosted by the development of the emerging countries, and will also remain highly dependent on oil.

Secondly, we have entered an era of energy transition forcing us to save energy, optimise its use by improving our energy intensity, but also to diversify our production sources, especially by developing non-carbon energies to reduce CO_2 emissions and fight against climate change.

In this context, biofuels contribute to lowering tension on the demand for oil and to reducing greenhouse gas emissions, even though the extent of this reduction has still to be specified in view of the numerous studies published on the subject.

Although highly useful, their development must also be compatible with harmonious and sustainable development of the planet. Whereas at the start of the century, the development of first-generation biofuels made use of excess agricultural areas, local situations of competition have since emerged between energy and food uses of some agricultural land. Although large agricultural areas are still available throughout the world, it is now essential to launch and continue ambitious research and development programs to develop biofuel production processes based on lignocellulosic materials (non-food materials obtained in particular from brushwood and forest waste to produce 2nd generation biofuels) or algae (3rd generation biofuels).

Combined with first generation biofuels and with no real threat to food, these technologies will make it possible to reach the target of 10% incorporation of biofuels in transport fuel set by the European Union by 2020.

* Les biocarburants. État des lieux, perspectives et enjeux du développement (in French). Éditions Technip, 2006.

Some of the authors of this book have been working for more than 25 years on the technological development of biofuels, in terms of both production and use. While the previous edition provided knowledge about the entire biofuel production chain, this edition draws up the new state of the art called for in view of recent changes. It arrives at exactly the right time to renew the interest in biofuels, including for air transport, and provide an insight on the technological research and development axes currently being investigated.

Olivier Appert
Chairman and CEO
IFP Energies nouvelles

Foreword

"Our grandchildren will remember us for not having thought about them" [1]

When preparing this book, I often thought about my grandson Alexandre and my granddaughter Elisa, to whom I dedicate this book, hoping that they will be happy despite the difficulties we can expect.

More sustainable energy technologies and systems must be developed, in particular in the transport sector, with the twofold objective of reducing oil dependence and limiting the environmental impacts, especially in terms of greenhouse gas emission.

Amongst the themes examined, that of biofuels offers real perspectives. They are obtained from plant raw materials such as oleaginous products for biodiesel, and cereals or sugar plants for ethanol. In the near future, they will be produced from lignocellulosic biomass such as agricultural and forest residues or energy crops. Consequently, the entire plant is used and the yields obtained in the conversion of solar and photosynthesis energy to bioenergy are significantly improved.

Both for the agricultural sector and the refining and automotive industries, biofuels represent an opportunity, in a context where reduction of greenhouse gas emissions (mainly carbon dioxide) must be increasingly taken into account to reduce the scale of global warming.

This book is a new edition of "Les biocarburants. État des lieux, perspectives et enjeux du développement", published in January 2006, by Éditions Technip, Paris. In this new edition, all the technical, economical and environmental aspects have been updated. It is divided into seven chapters, each of which can be read independently of the others.

---

1. The author of this quote highlights the unrestrained consumption of the planet's non-renewable resources by our civilisation mode which, without significant changes to our behaviour, will be difficult to sustain.

This book takes an exhaustive look at the strategic stakes, the global situation of biofuels and the availability of plant resources that can be converted into biofuels. Development in the production of biofuels has been exceptionally fast since 2003, with a downturn in 2008 and 2009 due to the economic crisis. This development can be explained by several reasons:

- rapid evolution of the international context with the rise in the price of the barrel of crude oil and the increasing importance placed on environmental issues by political leaders and managers of industrial companies,
- significant increase in the volumes of biofuels produced in countries such as the United States, Brazil and several European states, already big producers,
- emergence of new producing countries such as Canada, China, Thailand, Malaysia, etc.

A detailed state of the art of the various proven production processes is then given, with a description of the technologies, economic and environmental aspects and the characteristics of the products obtained.

Once again, this book focuses on the future and describes the new pathways using lignocellulosic biomass as raw material. The objective is to reduce competition between the food resource and the use of agricultural materials for energy purposes. Numerous research and development programmes are currently being conducted and are starting to lead to industrial applications.

Finally, over the last few years, numerous research groups have studied the use of algae to produce fuels. A technical update is provided on this subject, as well as on the production of methane and hydrogen by biochemical and biological pathways, respectively.

In conclusion, biofuels exhibit good properties when mixed with fossil fuels, and with respect to the environment. However, their economic competitiveness is still a problem.

Daniel Ballerini

Acknowledgements

This book is a new, significantly updated edition of "Les biocarburants. État des lieux, perspectives et enjeux du développement", published in January 2006, by Editions Technip, Paris.

My thanks go to the authors of the first edition who were prepared to dedicate some of their time to updating the technical, economic and environmental aspects, a task which involved much more work than I had originally expected. My thanks also go to the new authors who provided their skills and knowledge in several technical fields, in which they are recognised specialists.

I am grateful to Germain Martino (former director of the IFP Energies nouvelles Process Business Unit) and to Patrick Boisserpe (IFP Energies nouvelles) for having read the manuscript and whose comments, criticisms and corrections were invaluable.

I would also like to extend my thanks to Bernard Chaud (Tereos) and Georges Vermeersch (Prolea) for having shared with me their extensive experience and knowledge, highly useful when writing this book.

Numerous technical aspects were clarified with the help provided by a number of experts who I would like to thank: Michel Bloch and Jean-Alain Chodorge (Axens) for the production of ETBE and biofuels from vegetable oils, Mélissa Clodic (PhD student at Enerbio) on the energy stakes, Nicolas Le Bigot (CCFA) for the use of biofuels in engines, Rémy Marchal (IFP Energies nouvelles) on the aspects concerning the manufacture of methane and acetone-butanol, Jean-Christophe Pouet (ADEME) for the environmental balances and Julien Rousseau (Prolea), Damien Sagot (Diester Industrie) and Pierre Villeneuve (CIRAD) for the production of biodiesel.

I also benefited from the help of Dominique Allinquant (Graphidoc) to produce the figures and by Mireille Darthenay (IFP Energies nouvelles) with the book layout.

I am most grateful to IFP Energies nouvelles for having asked me to coordinate and write this new edition on biofuels.

I express my gratitude to Mr Olivier Appert, IFP Energies nouvelles Chairman and CEO, who did me the honour of writing the preface of this book.

Daniel Ballerini

Table of contents

Preface . . . V
Foreword . . . VII
Acknowledgements . . . IX
List of authors . . . XXI
List of abbreviations . . . XXIII
List of units . . . XXVII

Chapter 1
BIOFUELS: A PARTIAL SOLUTION WITH SEVERAL CHALLENGES

1.1 Retrospective . . . 3
1.1.1 Until the 1960s . . . 3
1.1.2 From the 1970s . . . 4

1.2 2000-2030: strategic stakes . . . 6
1.2.1 Energy supply and demand by 2030 . . . 6
1.2.2 Oil: the importance of OPEC . . . 8
1.2.3 Road transport in the world . . . 8
1.2.4 Environmental impacts (see Appendix 1) . . . 10
1.2.5 The safety stakes . . . 12
1.2.5.1 Risks related to bioethanol production . . . 14
1.2.5.2 Risks related to biodiesel production . . . 15
1.2.5.3 Safety stakes identified for 2nd generation biofuels . . . 16
1.2.6 Biofuels: a promising future . . . 17
1.2.6.1 Synthetic fuels derived from lignocellulosic biomass and/or organic waste . . . 20
1.2.6.2 Ethanol derived from lignocellulosic biomass . . . 20
1.2.6.3 Alternatives: heat, electricity, fuels, chemistry . . . 21
1.2.6.4 Conclusions . . . 22

1.3 The first step in biofuel production: production of biomass . . . 23
1.3.1 Competition with food agricultural crops . . . 23
1.3.2 Biodiversity . . . 24
1.3.3 Biomass, biofuels and soil organic matter . . . 25

1.3.3.1 Biomass and CO_2 compatibility ... 25
1.3.3.2 Evolution of soil carbon stock ... 26
1.3.3.3 Bioenergy feedstock and land use changes ... 26
1.3.3.4 Effect of cultivation practices ... 27

1.4 Vision of the refining industry ... 28
1.4.1 A key point for the future: the impact of helping reduce CO_2 emissions ... 29
1.4.2 Comparison of biofuels with the other non-conventional fuels ... 30
1.4.2.1 Comparison criteria ... 30
1.4.2.2 Alternative fuels obtained from fossil energies ... 31
1.4.2.3 BTL *(Biomass-to-Liquids)* ... 32

1.5 The vision of the car manufacturers ... 33
1.5.1 Control of polluting emissions ... 33
1.5.1.1 European regulations ... 33
1.5.1.2 Progress made ... 35
1.5.1.3 Expected progress ... 35
1.5.1.4 Interest of biofuels ... 36
1.5.2 Reduction of CO_2 emissions ... 37
1.5.2.1 Progress made ... 37
1.5.2.2 Expected progress ... 38
1.5.2.3 Interest of biofuels ... 39
1.5.3 Future trends in fuels ... 40
1.5.3.1 Characteristics of conventional fuels ... 40
1.5.3.2 Gasolines ... 41
1.5.3.3 Diesels ... 42
1.5.3.4 Trend in fuel characteristics ... 43

1.6 Biofuels in the world ... 44
1.6.1 North America ... 44
1.6.2 Latin and Central America ... 45
1.6.3 Europe ... 47
1.6.4 The other biofuel producing regions ... 48

References ... 49

Chapter 2

FIRST GENERATION BIOFUELS FOR SPARK IGNITION ENGINES: ETHANOL AND ETBE

2.1 Raw materials of vegetable origin ... 51
2.1.1 Alcohol-producing potential of sugar plants ... 52
2.1.2 Alcohol-producing potential of amylaceous plants ... 52

2.2 Ethanol production processes ... 54
2.2.1 Ethanol fermentation ... 54
2.2.1.1 Principle ... 54
2.2.1.2 Microorganisms ... 54
2.2.1.3 Ethanol fermentation parameters ... 55
2.2.1.4 Fermentation step ... 56

2.2.2 Ethanol extraction 59
2.2.2.1 Distillation 59
2.2.2.2 Other technologies 60
2.2.3 Effluents and co-products 60
2.2.4 Process diagrams 61
2.2.4.1 Production of ethanol from sugar plants 61
2.2.4.2 Production of ethanol from amylaceous plants 62
2.2.5 Energy balance 64
2.2.5.1 Production of ethanol from sugar beet 64
2.2.5.2 Production of ethanol from amylaceous plants 65
2.2.6 Potential ethanol production improvements 66
2.2.6.1 Pretreatment of amylaceous products 66
2.2.6.2 Ethanol fermentation 66
2.2.6.3 Improved energy performance 66

2.3 Ether fuels 67
2.3.1 Sources of raw materials for the production of ether fuels 69
2.3.2 Etherification reaction 71
2.3.2.1 Secondary reactions 73
2.3.2.2 Etherification catalysts 74
2.3.3 Etherification processes 77

2.4 Main industrial players 83

2.5 World ethanol production 83

2.6 Ethanol and ETBE, fuels 84
2.6.1 Ethanol fuel 84
2.6.1.1 Physical properties 86
2.6.1.2 Octane numbers 87
2.6.1.3 Volatility 88
2.6.1.4 Stability in the presence of water 89
2.6.1.5 Cold starting 90
2.6.1.6 Compatibility of materials 90
2.6.1.7 Ethanol in spark ignition engines 91
2.6.1.8 Ethanol in Diesel engines 93
2.6.2 ETBE-fuel 94

2.7 Chemical conversion of ethanol 96
2.7.1 Production of hydrogen 96
2.7.1.1 Production of hydrogen by ethanol steam reforming 96
2.7.1.2 Purification of the hydrogen produced 99
2.7.1.3 Production of hydrogen by partial oxidation 100
2.7.1.4 Production of hydrogen by autothermal reforming 100
2.7.1.5 Hydrogen production systems and technologies 101
2.7.1.6 Conclusion 101
2.7.2 Other ways for the conversion of ethanol into fuel base 102
2.7.2.1 Direct oligomerisation of ethanol into hydrocarbons 102
2.7.2.2 Ethanol conversion in two steps via ethylene 103
2.7.2.3 Conversion of ethanol via butadiene 104
2.7.2.4 Conversion of ethanol into butanol-1 105
2.7.2.5 Conversion of ethanol by reaction with another molecule 106
2.7.2.6 Conclusion 106

2.8 Environmental aspects ... 106
2.9 Economic aspects ... 109
2.9.1 Ethanol production cost in the United States ... 109
2.9.2 Ethanol production cost in Brazil ... 110
2.9.3 Ethanol production cost in Europe ... 112
2.9.4 ETBE production cost ... 112
References ... 113

Chapter 3
FIRST GENERATION BIOFUELS FOR DIESEL ENGINES

3.1 Raw materials of vegetable origin ... 118
3.1.1 World production of vegetable oils ... 118
3.1.2 Production of oilseeds outside Europe ... 119
3.1.3 European production of vegetable oils ... 119
3.1.4 Agricultural yields (seeds, oils, cakes) ... 121
3.1.5 Composition of vegetable oils ... 122
3.1.6 Manufacture of vegetable oils ... 122
3.1.6.1 The example of rapeseed ... 122
3.1.6.2 Other oils ... 125
3.1.7 Applications of vegetable oils ... 126
3.2 Methanol ... 128
3.2.1 Methanol synthesis ... 129
3.2.2 Methanol production processes ... 129
3.3 VOME production processes ... 129
3.3.1 The transesterification reaction ... 129
3.3.2 Homogeneous catalysis processes ... 131
3.3.2.1 Catalysts ... 131
3.3.2.2 Raw material specifications ... 131
3.3.2.3 Discontinuous processes ... 132
3.3.2.4 Continuous processes ... 133
3.3.2.5 Energy consumptions ... 135
3.3.3 Continuous heterogeneous catalysis processes ... 135
3.3.3.1 The catalyst ... 135
3.3.3.2 Raw material specifications ... 135
3.3.3.3 Continuous process ... 136
3.4 Production of Vegetable Oil Ethyl Esters (VOEE) ... 138
3.5 Other raw materials ... 139
3.6 Other processes, including hydrotreatment and hydroisomerisation of renewable fats ... 140
3.6.1 Hydrotreatment ... 140
3.6.2 Hydroisomerisation ... 143
3.6.3 Processes ... 144
3.6.3.1 Co-treatment with fossil feedstocks ... 144
3.6.3.2 Treatment in dedicated unit ... 145

3.7 Main industrial players 145

3.8 Global production of biodiesel 146

3.9 The glycerine market 147

3.10 Direct use of vegetable oils in Diesel engines 151

3.10.1 Characteristics of vegetable oils 151

3.10.2 Impact on combustion 151

3.10.3 Impact on emissions 153

3.10.4 Special case of agricultural tractors 154

3.10.5 Conclusion 155

3.11 Use of vome in Diesel engines 155

3.11.1 Characteristics of the products manufactured 156

3.11.2 VOME properties 161

3.11.3 Performance of VOME as fuel 163

3.11.4 Impact of VOME on emissions 165

3.12 Use of biofuels in air transport 166

3.13 Environmental aspects 166

3.14 Economic aspects 168

References 171

Chapter 4
LIGNOCELLULOSIC BIOMASS RESOURCES

4.1 Lignocellulosic biomass: source and current uses 173

4.1.1 Two major families of lignocellulosic resources 173

4.1.1.1 The co-product resource: agricultural and forest residues or surpluses 173

4.1.1.2 The dedicated resource: lignocellulosic crops dedicated to energy 174

4.1.2 Main figures of current sources 176

4.1.2.1 Distribution of main types of biomass 176

4.1.2.2 The forest resource 177

4.1.2.3 Agricultural residues 181

4.1.2.4 Dedicated crops 182

4.2 Raw material/conversion process adequacy 182

4.2.1 Composition of the raw material 182

4.2.1.1 Residues 182

4.2.1.2 Dedicated crops 184

4.2.2 Characteristics of the raw material likely to affect energy conversion 185

4.2.2.1 Thermochemical conversion 185

4.2.2.2 Biochemical conversion 185

4.3 Lignocellulosic biomass production and mobilisation conditions 185

4.3.1 Resource access constraints 186

4.3.2 Technical routes and conditioning 187

4.3.3 Competing uses 187

4.3.4 Environmental constraints 188

4.4 Potential for the production of lignocellulosic biomass, new means, new spaces 190
4.4.1 Availability of additional resources 190
4.4.1.1 Management of under-exploited forests 190
4.4.1.2 Development of production means and improvement in yields 191
4.4.2 New land availabilities 191
4.4.2.1 Under-exploited arable land 191
4.4.2.2 Release of new areas: improvement in food crop productivities 194
4.4.2.3 Protected or regulated buffer strips 194
4.4.3 New land exploitation balance 195

4.5 lignocellulosic biomass for energy purposes, conclusion and perspectives 195

References 197

Chapter 5
CONVERSION OF LIGNOCELLULOSIC BIOMASS BY THERMOCHEMICAL PATHWAY

5.1 Pyrolysis 199
5.1.1 Definition – Principle 199
5.1.1.1 Slow pyrolysis 200
5.1.1.2 Fast pyrolysis 201
5.1.1.3 Interest of pyrolysis 201
5.1.2 Basic pyrolysis mechanisms 202
5.1.2.1 Pyrolysis of cellulose 202
5.1.2.2 Pyrolysis of hemicelluloses 202
5.1.2.3 Pyrolysis of lignins 203
5.1.2.4 Pyrolysis of lignocellulosic biomass 203
5.1.2.5 Heat and mass transfers 204
5.1.3 Main pyrolysis reaction parameters 204
5.1.3.1 Heat fluxes imposed on the solid 205
5.1.3.2 Granulometry and humidity of the raw material 205
5.1.3.3 Temperature and residence time in gas phase 205
5.1.3.4 Pressure 206
5.1.4 Fast pyrolysis processes 206
5.1.4.1 Operating conditions required 206
5.1.4.2 General principle of the process 206
5.1.4.3 Main reactor technologies 207
5.1.4.4 Separation and condensation techniques 214
5.1.4.5 Mass and energy balances 215
5.1.5 Pyrolysis oils (bio-oils) 215
5.1.5.1 Chemical composition 215
5.1.5.2 Physico-chemical characteristics and qualities of pyrolysis oils 216
5.1.5.3 Environment and safety 217
5.1.6 Potential applications for pyrolysis oils 218
5.1.6.1 Combustion in boilers 218
5.1.6.2 Direct use in Diesel and Stirling engines 219

5.1.6.3 Combustion in gas turbines 219
5.1.6.4 Use as raw material in the biofuel sector 220

5.2 Torrefaction 222
5.2.1 Torrefaction stakes 222
5.2.2 Basic torrefaction mechanisms 223
5.2.3 Processes 225

5.3 Gasification 230
5.3.1 Chemical reactions and thermodynamic constraints for the production of syngas 232
5.3.1.1 Main reactions 232
5.3.1.2 Biomass representation and ideal steam gasification reaction 233
5.3.1.3 Steam gasification reactions at thermodynamic equilibrium 233
5.3.1.4 Reaction equilibrium considerations 234
5.3.1.5 Autothermal and allothermal processes 235
5.3.1.6 Gasification reaction in the presence of hydrogen 235
5.3.2 Industrial processes 236
5.3.2.1 Overview 236
5.3.2.2 Main types of reactor 236
5.3.2.3 Syngas characteristics 240
5.3.2.4 Generic problems related to the biomass gasification technologies . 240
5.3.3 Yields 244
5.3.3.1 General approach to the problem of yields 244
5.3.3.2 Main points of the process used to increase the fuel production yield 244
5.3.3.3 Yield calculations for various process options 245
5.3.4 Technico-economic aspects 247
5.3.5 Environmental aspects 249
5.3.6 Existing demonstration units and R&D requirements 249
5.3.7 Conclusion 252

5.4 Direct liquefaction or hydrothermal conversion 253
5.4.1 Principle and operating conditions 253
5.4.2 Processes 254

5.5 Liquid fuels and hydrogen produced from biomass syngas 255
5.5.1 Methanol synthesis 255
5.5.2 Dimethyl ether synthesis 257
5.5.3 Fuel methanol and MTBE 258
5.5.3.1 Methanol characteristics 259
5.5.3.2 MTBE characteristics 260
5.5.4 Fuel-DME 260
5.5.4.1 DME characteristics 260
5.5.4.2 Use of DME in Diesel engines 261
5.5.5 Fischer-Tropsch synthesis 262
5.5.6 BTL fuel 263
5.5.7 Hydrogen production 263
5.5.8 Fuel/electricity polygeneration 266

References 266

Chapter 6

CONVERSION OF LIGNOCELLULOSIC BIOMASS BY BIOCHEMICAL PATHWAY. PRODUCTION OF ETHANOL AND ABE

6.1 Ethanol production . . . 275

6.1.1 Substrate composition . . . 275

6.1.2 The process . . . 277

6.1.2.1 Pretreatment . . . 278

6.1.2.2 Hydrolysis . . . 284

6.1.2.3 Production of enzymes and their action mode . . . 286

6.1.2.4 Ethanol fermentation . . . 293

6.1.2.5 Use of lignin . . . 300

6.1.2.6 The various production diagrams . . . 301

6.1.2.7 Conclusion . . . 301

6.1.3 Environmental and energy aspects . . . 303

6.1.4 Economic aspects . . . 303

6.1.5 First and second generation ethanol production on the same industrial site . . . 304

6.2 Production of ethanol from CO + H_2 . . . 304

6.3 Acetonobutylic fermentation . . . 307

6.3.1 Principle . . . 307

6.3.2 History . . . 308

6.3.3 Raw materials . . . 309

6.3.4 The process . . . 309

References . . . 312

Chapter 7

OTHER BIOMASS-TO-ENERGY BIOCHEMICAL PATHWAYS

7.1 Methane fermentation . . . 319

7.1.1 History . . . 319

7.1.2 Principle . . . 320

7.1.3 Raw materials . . . 321

7.1.4 Processes . . . 322

7.1.4.1 Operating parameters . . . 322

7.1.4.2 Technologies . . . 322

7.1.4.3 Performance . . . 323

7.1.4.4 Purification of biogas . . . 323

7.1.4.5 Financial aspects . . . 324

7.1.4.6 Industrial sites in Europe . . . 324

7.1.4.7 Advantages – Disadvantages – Improvement axes . . . 325

7.2 Production of algae for energy uses . . . 326

7.2.1 Marine macroalgae . . . 326

7.2.2 Microalgae . . . 327

7.2.2.1 Algal strains . . . 328

7.2.2.2 Yield and productivity . . . 328

7.2.2.3 Culture of microalgae . . . 328

7.2.2.4 The overall process for production of algae and oils 330
7.2.2.5 Financial aspects 331
7.2.2.6 Pilot operations 331
7.2.2.7 Main companies involved in the production of microalgae for energy purposes 332
7.2.2.8 R&D programmes 332
7.2.2.9 Conclusions. 333

7.3 Hydrogen Production 334
7.3.1 Production by fermentation 334
7.3.2 Photobiological production 336
7.3.2.1 Photoautotrophic production 336
7.3.2.2 Photoheterotrophic production. 336
7.3.3 Combined productions by fermentative and photobiological pathways. 337
7.3.4 Conclusions and perspectives. 338

References. 338

APPENDIX 1

Environmental balance of biofuels 341

References. 346

APPENDIX 2

Crude oil refining 347

Index 351

List of authors

Nathalie Alazard-Toux
Director of Economic Studies at IFP Energies nouvelles
92852 Rueil-Malmaison Cedex (nathalie.alazard-toux@ifpen.fr)

Daniel Ballerini
Expert in Biotechnology and Biomass Chemistry
Former Head of Department at IFP Energies nouvelles
(daniel.ballerini@wanadoo.fr)

Laurent Bournay
Engineer at IFP Energies nouvelles, 69390 Vernaison
(laurent.bournay@ifpen.fr)

François Broust
Engineer of the Forest Department at CIRAD, 73 avenue J.-F. Breton,
34398 Montpellier (francois.broust@cirad.fr)

Dominique Casanave
Research Engineer, Process Division at IFP Energies nouvelles
69390 Vernaison (dominique.casanave@ifpen.fr)

Thierry Chapus
Engineer at IFP Energies nouvelles, 69390 Vernaison
(thierry.chapus@ifpen.fr)

Maurice Dohy
Former AGRICE Director at ADEME (maurice@dohy.fr)

Alain Forestière
Former Chemist Engineer and Director Expert at IFP Energies nouvelles
(alain.forestiere@neuf.fr)

Benoît Gabrielle
Environmental Physics Professor, AgroParisTech,
78850 Thiverval-Grignon (benoit.gabrielle@agroparistech.fr)

Gérard Hillion
Former Senior Research Engineer, Applied Chemistry and Physical Chemistry Division at IFP Energies nouvelles (ghillion@wanadoo.fr)

Stéphane His
Engineer at Technip, 92973 Paris La Défense (shis@technip.com)

Sylvie Lacombe
Research Engineer, Process Division at IFP Energies nouvelles (sylvie.lacombe@ifpen.fr)

Daphné Lorne
Engineer, Economics Department, Economics and Information Watch and Management Division, IFP Energies nouvelles, 92852 Rueil-Malmaison (daphne.lorne@ifpen.fr)

Guy Marlair
Technical Advisor, Accidental Risks Division,
INERIS, 60550 Verneuil-en-Halatte (guy.marlair@ineris.fr)

Eric Marty
Director of Emertec Gestion; Emertec Energy Environment Fund
Emertec Gestion, 4 bis rue Pasquier, 75008 Paris (emarty@emertec.fr)

Frédéric Monot
Head of the Biotechnology and Biomass Chemistry Department at IFP Energies nouvelles
92852 Rueil-Malmaison Cedex (frederic.monot@ifpen.fr)

Xavier Montagne
Deputy Scientific Director of IFP Energies nouvelles
92852 Rueil-Malmaison Cedex (xavier.montagne@ifpen.fr)

Anne Prieur
Engineer at GDF Suez, Saint-Denis La Plaine (anne.prieur-vernat@gdfsuez.com)

Patrick Rousset
Engineer of the Forest Department at CIRAD, 73 avenue J.-F. Breton,
34398 Montpellier (patrick.rousset@cirad.fr)

Jean-Marie Seiler
Research Director at the French Alternative Energies and Atomic Energy Commission (CEA)
CEA Grenoble, DTN/SETT, 17 rue des Martyrs, 38054 Grenoble Cedex 9
(jean-marie.seiler@cea.fr)

Jean-Bernard Sigaud
Former Refining-Petrochemistry-Gas Director at the IFP School
232 avenue Napoléon Bonaparte, 92500 Rueil-Malmaison (jean-bernard.sigaud@neuf.fr)

Karine Surla
Engineer at IFP Energies nouvelles, 69390 Vernaison
(karine.surla@ifpen.fr)

List of abbreviations

ABE Acetone-Butanol-Ethanol
ACEA European Automobile Manufacturers Association
ADEME Agence de l'Environnement et de la Maîtrise de l'Énergie (*French Environment and Energy Management Agency)*
ADM Archer Daniel Midlands
AEC Aid for Energy Crops
AFEX Ammonia Fiber Explosion
ALLIANCE Alliance of Automobile Manufacturers
ASTM American Society for Testing and Materials
ATP Adenosine triphosphate
BASF Badisch Anilin und Soda Fabrik
BCI BC International Corporation
BDP Bundesverband Deutscher Planzenzüchter
BOD Biochemical Oxygen Demand
BTG Biomass Technology Group BV
BTL Biomass-to-Liquids
BTU British Thermal Unit
CANMET Canadian Mineral and Energy Technology Centre
CAP Common Agricultural Policy
CBH Cellobiohydrolase
CBM Carbohydrate Binding Module
CCFA Comité des Constructeurs Français d'Automobiles (*French Automobile Manufacturers Committee*)
CCR Carbon Conradson Residue
CEA Commissariat à l'Énergie Atomique et aux Énergies Alternatives (*French Alternative Energies and Atomic Energy Commission*)
CEH Chemical Economics Handbook
CIS Commonwealth of Independent States
CEN Comité Européen de Normalisation (*European Committee for Standardisation*)
CEREOPA Centre d'Études et de Recherche sur l'Économie et l'Organisation des Productions Animales (*Centre for Research on the Economics and Management of Animal Production*)

CETIOM Centre Technique Interprofessionnel des Oléagineux Métropolitains (*Technical Centre for Oilseed Crops and Industrial Hemp*)

CFC Chlorofluorocarbons

CFPP Cold Filter Plugging Point

CIEMAT Centro de Investigaciones Energéticas, Medio Ambientales y Tecnológicas

CIRAD Centre de Coopération Internationale en Recherche Agronomique pour le Développement (*Centre for International Cooperation in Agronomic Research for Development*)

CITEPA Centre Interprofessionnel Technique d'Études de la Pollution Atmosphérique (*Interprofessional Technical Centre for Studies on Atmospheric Pollution*)

CM Catalytic Module

CMC Carboxymethylcellulose

CNRS Centre National de la Recherche Scientifique (*National Centre for Scientific Research*)

COD Chemical Oxygen Demand

CONCAWE The oil companies' European association for environment, health and safety in refining and distribution

CPDP Comité Professionnel du Pétrole (*Professional Committee for Petroleum*)

CRES Centre for Renewable Energy Sources

CTL Coal-to-Liquids

DBV Deutscher Bauernverband

DDGS Distiller's Dried Grain with Solubles

DFO Domestic Fuel Oil

DGEC Direction Générale de l'Énergie et du Climat (*General Directorate for Energy and Climate*) (Ministry of Ecology, Sustainable Development, Transportation and Housing)

DIPE Diisopropylether

DIREM Direction des Ressources Énergétiques et Minières (*Directorate for Energy and Mineral Resources*) (Ministry of Industry)

DM Dry Matter

DMC Direct Microbial Conversion

DME Dimethylether

DNA Deoxyribonucleic acid

DOE Department Of Energy (United States)

DOM Département d'Outre-Mer (*French Overseas Department*)

DP Degree of Polymerisation

DPE Dipentylether

DTU Danmarks Tekniske Universitet

EAMA European Automobile Manufacturers Association

EC European Commission

EEA European Energy Agency

EEC European Economic Community (now the European Union EU)

EG Endoglucanase

EIA Energy Information Administration (United States)

EMA Engine Manufacturers Association

EN European Norm

ENEL Ente Nazionale per l'Energia Elettrica

EPA Environment Protection Agency (United States)

ETAE Ethylteramylmethylether

ETBE Ethyltertiobutylether

EU European Union
EU 15 European Union with 15 Countries
EU 25 European Union with 25 Countries
EUCAR European Council for Automotive Research and Development
FAO Food and Agriculture Organization of the United Nations
FCC Fluid Catalytic Cracking
FFV Fuel Flexible Vehicle
FICFB Fast Internally Circulating Fluidised Bed
FPU Filter Paper Unit
FT Fischer-Tropsch
GCV Gross Calorific Value
GDP Gross Domestic Product
GHG GreenHouse Gas
GT Gas Turbine
GTBE Glycerol-terbutyl ether
GTL Gas-to-Liquids
HCCI Homogeneous Charge-Compression Ignition
HFC Hydrofluorocarbons
HMF Hydroxymethylfurfural
HTU Hydrothermal Upgrading
ICI Imperial Chemical Industries PLC
IEA International Energy Agency
IFN Inventaire Forestier National (*National Forest Inventory*)
IFPEN IFP Energies nouvelles
INRA Institut National de la Recherche Agronomique (*National Institute for Agricultural Research)*
IPCC Intergovernmental Panel on Climate Change
IPTBE Isopropylterbutylether
ITERG Institut des corps gras (*French Institute for Fats and Oils*)
JAMA Japan Automobile Manufacturers Association
JRC Joint Research Centre, European Commission
LCA Life Cycle Assessment
LCB Lignocellulosic biomass
LDC Least Developed Countries
LHV Lower Heating Value
LPG Liquefied Petroleum Gas
MAE Methylarylether
MEDD Ministère de l'Écologie, du Développement Durable, des Transports et du Logement (*Ministry of Ecology, Sustainable Development, Transportation and Housing*)
MGC Mitsubishi Gas Chemical
MON Motor Octane Number
MTBE Methyltertiobutylether
MTG Methanol To Gasoline
MW Molecular Weight
NAD Nicotinamide Adenine Dinucleotide
NADP Nicotinamide Adenine Dinucleotide Phosphate
NCV Net Calorific Value
NEDC New European Driving Cycle
NF Norme Française (*French standard*)
NREL National Renewable Energy Laboratory
OECD Organisation for Economic Cooperation and Development
OFP Ozone-Forming Potential
ONIDOL Organisation Nationale Interprofessionnelle des Oléagineux (*French National Oil-producing Interprofessional Organisation*)
OPEC Organization of Petroleum Exporting Countries
PAHs Polycyclic Aromatic Hydrocarbons

PEM	Proton Exchange Membrane
PFC	Perfluorocarbons
PPP	Pentose Phosphate Pathway
PSA	Pressure Swing Adsorption
PTFE	Polytetrafluoroethylene
PTT	Polytrimethylene terephthalate
PWC	Price Waterhouse Coopers
R&D	Research & Development
RFA	Renewable Fuels Association
RME	Rapeseed Methyl Esters
RON	Research Octane Number
SCEES	Service Central des Enquêtes et Etudes Statistiques (*Central Office for Statistical Surveys and Studies*) (Ministry of Agriculture)
SCOP	Surfaces en Céréales, Oléagineux et Protéagineux (*Cereal, Oilseed and Protein Crop Area*)
SDHF	Sucreries Distilleries des Hauts de France
SF6	Sulphur hexafluoride
SNPAA	Syndicat National des Producteurs d'Alcool Agricole (*National Union of Agricultural Alcohol Producers*)
SOFC	Solid Oxide Fuel Cell
SRC	Short Rotation Coppice
SRI	Stanford Research Institute
SSCF	Simultaneous Saccharification and Co-Fermentation
SSF	Simultaneous Saccharification and Fermentation
TAME	TerAmylMethylEther
TBA	Tertiobutylalcohol
TIPP	Taxe Intérieure sur les Produits Pétroliers (*French Domestic Tax on Petroleum Products*)
TNO	Netherlands Organisation for Applied Scientific Research
TOE	Tonne Oil Equivalent
TOM	Territoire d'Outre-Mer (*French Overseas Territory*)
UFOP	Union zur Förderung von Öl- und Proteinpflanzen
UNGDA	Union Nationale des Groupements de Distillateurs d'Alcool (*French National Federation of Alcohol Distillers Associations*)
UOP	Universal Oil Products
USDA	Union des Sucreries et Distilleries Agricoles (*Union of Agricultural Sugar Refineries and Distilleries*)
US DA	US Department of Agriculture
VDÖ	Verband Deutscher Öhlmühlen
VEETC	Volumetric Ethanol Excise Tax Credit
VLK	Verband der Landwirtschaftskammern
VNG	Vehicle Natural Gas
VOC	Volatile Organic Compounds
VOEE	Vegetable Oil Ethyl Esters
VOME	Vegetable Oil Methyl Esters
VSRC	Very Short Rotation Coppice
VTT	Technical Research Centre of Finland
WEO	Washington Engineering Office
WTO	World Trade Organisation
WWFC	Worldwide Fuel Charter
XDH	Xylose Dehydrogenase
XI	Xylose Isomerase
XR	Xylose Reductase

List of units

Area

- 1 hectare (ha) = 10 000 m^2
- 1 acre = 4047 m^2 (i.e. approximately 0.4 ha)
- 1 square inch = 6.45×10^{-4} m^2
- 1 square foot = 9.3×10^{-2} m^2

Volume

- 1 US gallon = 3.785 l
- 1 US barrel, bbl = 158.984 l
- 1 US bushel = 35.24 l
- 1 bushel corresponds to 25.4 kg of maize
- 1 stere = 1 m^3

Mass

- 1 pound (lb) = 0.4536 kg
- 1 Dalton (d) = 1.66×10^{-27} kg

Pressure

- 1 Pascal (Pa) = 1 Newton/m^2 (N/m^2)
- 1 bar = 10^5 Pa = 14.5 psi
- 1 atmosphere (atm) = 1.015 bar = 1015 mbar = 1.015×10^5 Pa
- 1 pound/square inch (psi) = 6.89×10^3 Pa

Energy

- 1 Joule (J) = 1 N × m
- 1 Gigajoule (GJ) = 10^9 J
- 1 Megajoule (MJ) = 10^6 J
- 1 calorie (cal) = 4.185 J
- 1 thermie = 4.185 MJ

- 1 Wh (Watt hour) = 3600 J
- 1 toe (tonne oil equivalent) = 41.86 GJ = 11.63 MWh
- 1 BTU (British Thermal Unit) = 1055 J = 252 cal

Electrical conductance

- 1 Siemens (S) = 1 Ohm^{-1}

Kinematic viscosity

- 1 Stoke (St) = 10^{-4} $m^2 \times s^{-1}$

Concentration

- 1 ppm = 1 part per million (1×10^{-6}), e.g.: 1 mg/kg
- 1 ppb = 1 part per billion (1×10^{-9}), e.g.: 1 µg/kg

1

Biofuels: a partial solution with several challenges

N. Alazard-Toux, D. Ballerini, M. Dohy, B. Gabrielle, G. Marlair, X. Montagne, J-.B. Sigaud

Biofuels are derived from plant materials, the biomass. For their growth and energy requirements, plants use the carbon in the carbon dioxide present in the atmosphere, via the process of photosynthesis. Unlike the depleting fossil energies such as oil, natural gas and coal, biomass produced in this way is a renewable energy source which offers a high growth potential.

The current first-generation biofuel processes (Figure 1.1) lead to two types of products:

- biodiesel, mixed with Diesel oil for use in Diesel engines; it consists of Vegetable Oil Methyl Esters (VOME), themselves manufactured from vegetables oils (rapeseed, sunflower, soya, palm),
- ethanol, obtained by fermenting the sugars contained in sugar plants (sugar beet and sugar cane) and in cereals such as wheat, corn and barley. It can be used directly, pure or mixed in gasolines, or after conversion into ETBE (ethyltertiobutylether).

First-generation biofuels offer good medium term development prospects, although they will be faced with limitations in terms of supplies of raw materials. The area of land dedicated to this type of crop is in fact steadily decreasing. In view of the predicted growth in world population and despite the continued improvement in agricultural efficiency, we can reasonably expect that the competition between the various uses of agricultural resources for food, industrial and energy purposes will become increasingly fierce.

Consequently, if biofuels are to represent a serious complement to energy supply in the transport sectors, the sources of plant raw materials must be diversified.

Other resources exist, in the form of lignocellulosic biomass: agricultural residues (straw, corn stalks), forest residues and wood conversion by-products. It will eventually be possible to extend this lignocellulosic biomass potential by growing dedicated perennial crops harvested annually, such as high yield herbaceous species, multiannual perennial crops such as Short Rotation Coppices (SRC): poplar and eucalyptus amongst others. These are only a few examples. Organic wastes from paper and agribusiness industries represent extensive resources and also offer a high potential for extending the range of carbon-containing materials that can be converted into biofuels.

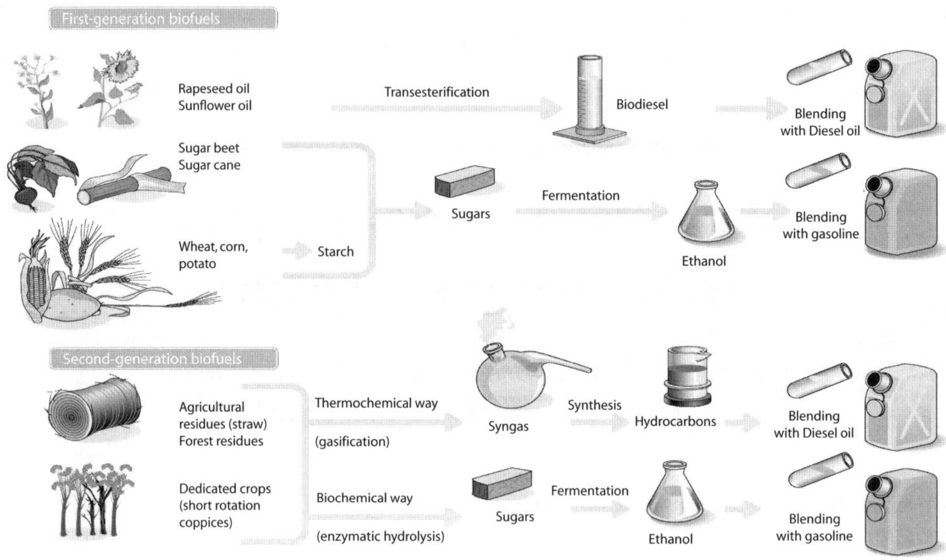

Figure 1.1

Biofuel production diagram.

Several technological barriers still remain to be removed, however, before the conversion processes of lignocellulosic biomass to biofuels (second generation biofuels) are operational at industrial scale.

This chapter provides the reader with several updates on biofuels.

Firstly, a retrospective recalls that biofuels were used from the end of the 19th century, before being rediscovered in the 1970s.

An update on the major strategic stakes then illustrates how biofuels can help to ease the energy and environmental constraints, by playing a role in the supply of fuels for transport and in limiting greenhouse gas emissions. The stakes related to biofuel production and distribution safety are also examined.

A discussion on the availability of agricultural land introduces the necessary production of so-called 2nd generation biofuels.

Three paragraphs are dedicated to the impact of biofuels on, respectively, the agricultural world, the refining industry and the automotive construction.

Lastly, biofuel productions at world and regional levels are evaluated: North America, Latin and Central America, Europe and other regions.

1.1 RETROSPECTIVE

1.1.1 Until the 1960s

The utilisation of biofuels in engines is not a recent practice. As early as the end of the 19th century, partial or total substitution of gasoline by denatured alcohol was already being considered by France.

On reading the newspapers from this period [1], it is striking to find a subject which is so topical today in an otherwise highly nationalistic line of thought. Apart from aspects concerning sustainable development, which was not a concern at the time, we discover the same issues as today regarding the substitution of petroleum products by compounds derived from plants.

There was a strong desire at the time to reduce dependence on foreign oil supplies and to make more effective use of agricultural surpluses, in particular sugar beet. Disposal of agricultural surpluses was also an isssue and solutions such as the creation of "export subsidies" were proposed.

The technical advantages and disadvantages related to the use of ethanol in engines were also discussed (denaturation of alcohol, lower energy content than gasolines, adaptation of carburettors, etc.). The problems of costs as well as direct and indirect taxes imposed by the state and the town (*régie* and *octroi*) were also analysed.

Lastly, significant marketing efforts (even if the term did not exist at the time) were made in exhibitions and car races where denatured alcohol, designated under the trade name of "electrine", was used as fuel.

In 1902, four cars powered with alcohol were entered in the French *Circuit du Nord* by Serpollet, who held the speed record with 121 km/h. The cars finished the race in the third, fourth, fifth and sixth positions. Despite the dreadful condition of the roads, the cars completed the 900 km circuit with an average speed of 71 km/h.

Before World War I, Parisian buses were running on a blend of gasoline and denatured alcohol. To increase the calorific value of the mixture, denatured alcohol was mixed with up to 50% coal benzine, at the time still known as benzol (a mixture of benzene-rich aromatics), before being added to gasoline in proportions of about 30%.

Between 1920 and 1950, ethanol was widely used as a fuel for vehicles. Consumption was 2 Mhl for example. Due to the desire for energy independence and in particular the political commitment to reduce the French trade-balance deficit, substitution of fossil fuels with sugar-beet-based ethanol was maintained at high rates throughout this period. The obligation to blend ethanol with regular gasoline was even a condition for imports of crude oil.

In the 1950s, Antar distributed "Tri-Super-Azur", a blend composed of 75% gasoline, 15% ethanol and 10% benzol [2].

In the 1960s, ethanol disappeared from the fuel market for at least two main reasons:

- oil was plentiful and prices were attractive, making biofuels less competitive,
- sugar-beet surpluses disappeared because of the strong demand from the agribusiness and chemical industries.

1.1.2 From the 1970s

It was only after the 1973 and 1978 oil shocks that there was a renewed interest in biofuels. Brazil launched a program called "Proalcool" (Brazilian Alcohol Program) whose objective was to produce ethanol from sugar cane and the United States tried to promote the use of "gasohol" (a contraction of gasoline and alcohol), a blend of 10% corn-ethanol and gasoline.

From the start of the 1920s, these two countries had set up small-scale regional ethanol production and distribution structures, mainly in states with abundant agricultural resources.

In Brazil, the world's leading sugar producer, as in the United States, the world's leading corn producer, development of the ethanol production industry represents a good way of regulating the export prices of these two raw materials while partially reducing the energy dependence of the two countries.

In the 1970s, development of the ethanol production industry in the United States was justified from an agricultural standpoint since it offered new outlets for raw materials such as corn, whose markets were saturating. In 1978, the Energy Tax Act introduced the excise tax exemption on fuels blended with at least 10% ethanol.

In the 1980s, the production of alcohol was driven by environmental concerns. From 1985, MTBE (methyltertiobutylether) was added to lead-free reformulated gasolines.

In 1990, the Clean Air Act Amendments required the use of oxygenated gasoline in areas where the atmospheric CO and ozone contents exceed the standards. This was the main reason for the development of reformulated gasolines with ether-fuels, mainly MTBE. In 1992, the Energy Policy Act extended the tax exemption to fuels containing less than 10% ethanol.

Since 2003-2004, the phasing out of MTBE in gasolines (see insert) in some states boosted the production of fuel ethanol as oxygenate substitute.

It was primarily the American determination to limit energy dependence (in 2002, the United States imported 9 M barrels/day of crude oil for a consumption of 15 M barrels/day, i.e. 60% of their requirements), ahead of other objectives such as reducing greenhouse gas emissions due to transport and optimising use of their agricultural land, which supported the trend to develop the use of biofuels, mainly ethanol, biodiesel production still being marginal.

From the 1970s in Brazil, two types of formulation were imposed and supported by tax incentives on the manufacture of alcohol and the purchase of cars:

- a blend of between 20% and 25% of anhydrous ethanol in gasoline,
- 92% ethanol (8% water) with no petroleum product.

Less emphasis was placed on these programs at the end of the 1980s, due to lower world oil prices, before picking up again in the early 2000s, in a context marked by a drop, even a complete cut, in government subsidies.

In France, the "Carburol plan", a national biofuel R&D program, was launched at the end of the 1970s. Its aim was to develop new technologies for the production of biofuels from lignocellulosic biomass: acetone-butanol-ethanol by the biochemical pathway, methanol by the thermochemical pathway. This plan to the construction of two demonstration

MTBE, which has a high octane number (RON =118), was introduced in the United States in lead-free gasolines from 1985. Its addition reduces the levels of some pollutants in engine emissions (figures vary depending on the type of vehicle fleet tested). The US Auto-Oil Program concluded that adding 15% MTBE reduced CO emissions by 10%, unburnt hydrocarbons by 5%, benzene and 1,3-butadiene discharges by 15%, but increased formaldehyde emissions by 25%. In all cases, no effect is observed on the nitrogen oxides and little effect on the OFP (Ozone-Forming Potential). These characteristics lead to an increase in the use of MTBE as fuel, especially in the United States, and in particular further to the enactment of the 1990 Clean Air Act Amendments, which required gasoline reformulation with an oxygen content of between 2% and 2.7% by weight in geographic areas which exceeded the maximum rates of atmospheric CO and ozone.

This massive use of MTBE progressively led to environmental problems, eventually creating unacceptable situations regarding the quality of drinking water resources polluted by accidental leaks occurring during storage and transport of MTBE and fuels reformulated with this ether.

Its high mobility and solubility (48g/l) in water and its resistance to biodegradation explain the persistence and possible extension over the next few years of groundwater pollution by MTBE. In California, the percentages of contaminated drinking water catchments, estimated between 0.3% and 1.2% in 1998, probably reach between 1% and 5% by 2010. The taste and odour of MTBE are perceptible from contents as low as 2µg/l. MTBE concentrations in water of between 20-40µg/l are accepted by the US EPA.

Concerning the potential risks of its use on human health, MTBE is considered as "probably carcinogenic" (assimilation to group 2A substances). On 25 March 1999, the Governor of California ruled that MTBE should be completely removed from the formulation of gasolines used within the State by the end of 2002. Further to this decision, which became law, MTBE was banned in most American states. A first study estimated the costs of depolluting MTBE-contaminated soils at between 20 and 30 billion dollars.

units: the ASCAB platform at Clamecy, in the Nièvre department (thermochemical pathway) and the ASCAF platform at Soustons, in the Landes department (biological pathway). The plan was stopped in the early 1990s.

In 1985, at a time when European directive 85/536/CEE dated 5 December of the same year defined the conditions for adding ethanol and ETBE (Ethyl-Tertio-Butyl-Ether) in gasolines, a consultative Commission was created in France for the production of substitution fuels. It included the ministries of industry, environment and agriculture, as well as research organisations and representatives of the professions concerned.

In 1987, the French government lowered taxes on ethanol and set up an optional fallow land scheme.

In 1992, as part of Common Agricultural Policy (CAP), the European Community imposed a 15% freeze on land set aside for cereal, oilseed and protein crops, from the 1992-1993 campaign. This measure aimed to solve the problems of agricultural surpluses which were difficult to export. Farmers were allowed to cultivate these frozen areas for non-food purposes – e.g. to produce biofuels – while benefiting from the compensation payment for uncultivated fallow land.

At the same time, the tax reductions were extended to biofuels developed in France:

- ethanol, produced from fermentation of sugar beet or cereal starch, to be used as ETBE after reaction with isobutene, blended with gasolines,
- Vegetable Oil Methyl Esters (VOME) obtained by reaction of rapeseed or sunflower oil with methanol, blended with Diesel oil.

The appearance of this new biofuel deserves mentioning. For many years in fact, ethanol was the only biofuel used since, at the time, cars were fuelled almost exclusively on gasoline. While as far back as the 1900 Paris World Fair, Rudolf Diesel [1] exhibited his new engine running on peanut oil, it was only in the early 80s that research was finally dedicated to synthesis of Vegetable Oil Methyl Esters (VOME) [3], resulting in the construction of the first industrial biodiesel production in 1992 at Compiègne (France).

Since then, programs for the development and industrialisation of these fuels have been launched in Europe (Germany, France, etc.) and throughout the world (Brazil, south east Asian countries, United States, etc.).

1.2 2000-2030: STRATEGIC STAKES

1.2.1 Energy supply and demand by 2030

In terms of energy prospective, the reference scenarios drawn up by several major organisations (US Department of Energy (DOE) – [4], the International Energy Agency (IEA) [5] and the European Union (EU) [6]) all agree that the world demand for primary energy will remain strong until 2030. It is expected to increase by 60 to 75% between 2000 and 2030, a figure confirmed by the new IEA reference scenario published in 2009 [7] which predicts that the demand will grow by 65 to 70% between 2000 and 2030 (Figure 1.2).

Fossil energies will continue to play a major role in the world energy balance (excluding non commercial energies) with a share which will remain in the region of 85%.

According to the IEA [5] and EU [6] reference scenarios, nuclear power, which currently caters for nearly 7% of world energy consumption, should record an average growth of less than 1% per year over the first 30 years of the millennium. According to the 2009 IEA scenario, the share of nuclear power in world primary energy consumption is not expected to change significantly.

With a high average annual growth rate, renewable energies (wind, solar, small-scale hydropower and biomass) could represent slightly less than 15% of the global energy balance by 2030.

1. In 1912, Rudolf Diesel wrote: “The use of vegetable oils for Diesel engines may seem insignificant today. But such oils may become in the course of time as important as petroleum or coal-based products of the present time”.

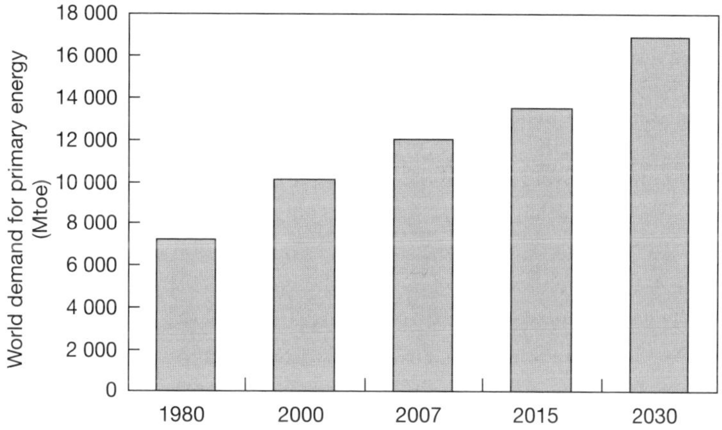

Figure 1.2

World demand for primary energy (Mtoe) according to the 2009 IEA scenario.
Source: IEA

Oil will remain the dominant energy with a 30% share, ahead of coal (28%) and gas (21%). Over the 2000-2030 timeframe, the global crude oil demand is expected to rise from 73 Mb/d (3.7 Gt/year) to 100 Mb/d (5 Gt/year) (Figure 1.3). This growth will nevertheless be accompanied by a shift of oil towards its specific applications such as petrochemistry and in particular transportation, for which substitutes are not readily available. In the next few decades, three quarters of the increase in demand for petroleum products will come from the transportation sector.

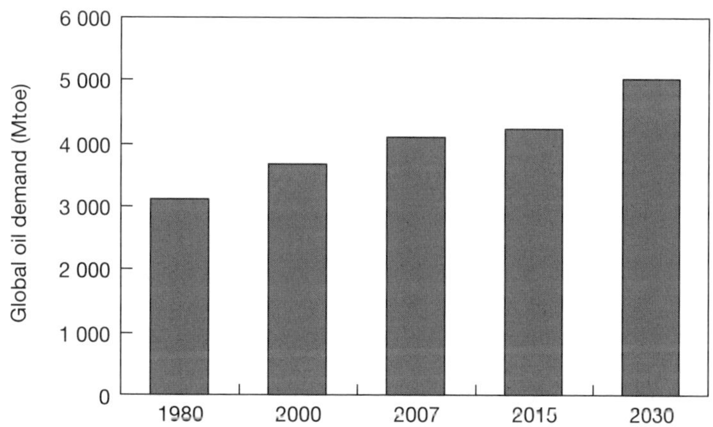

Figure 1.3

Global oil demand (Mtoe).
Source: IEA

1.2.2 Oil: the importance of OPEC

In the first oil crisis in 1973, the Organization of Petroleum Exporting Countries (OPEC) decided to cut oil exports, a trend confirmed with the second crisis in 1979. Since the end of the 1980s, however, an increase has been observed in the cartel's market share. With 40% of world production, OPEC has not yet regained the position it held in the early 70s (more than half of world supplies), but trends seem to suggest that it is only a matter of one or two decades.

If world oil demand continues to rise from 3.7 Gtoe in 2000 to 5.6 Gtoe in 2030 as predicted by the IEA reference scenario [7], with the current state of knowledge most of the major production areas outside OPEC offer only very limited development perspectives.

The North Sea currently represents nearly 6% of world supply; but its output has already started to decline. The same applies to the main production areas of the Asia-Pacific region. US production is not expected to increase significantly in the coming years. The main region outside OPEC to offer a high development potential for the next few years, the Commonwealth of Independent States (CIS) nevertheless seems unable to make up alone for the production shortfall in the long term. In the long term, only the OPEC countries, which concentrate most of the world's identified black gold reserves, in particular the Middle East oil countries, seem capable of satisfying the increased demand.

1.2.3 Road transport in the world

The demand for mobility is closely related to income levels, the distance separating home and workplace, leisure opportunities, etc. Over the last decades, higher incomes, technological progress, improved infrastructures and more leisure time have resulted in higher demand for mobility: more trips over generally longer distances.

This had led to a sharp rise in road transport. Over the last 20 years, while the Gross National Product (GNP) of the industrialised countries rose by more than 50%, the vehicle fleet in the OECD (Organisation for Economic Cooperation and Development) countries increased by 60% and distances travelled by 70%. In the industrialised countries, roads currently represent 90% of personal transport and three quarters of goods transport. In the rest of the world, the rise in road transport has been even more significant over the last two decades: the vehicle fleet has more than doubled. At the present time, however, the industrialised countries continue to account for most of the world fleet: four vehicles out of five.

Road transport remains highly dependent on oil. At world scale, petroleum-based fuels represent 98% of the energy used in this sector. Although solutions based on alternative energies exist (and have existed for many years in some cases), they remain marginal. They include Vehicle Natural Gas (VNG), Liquefied Petroleum Gas (LPG), oxygenated fuels, whether of chemical or agricultural origin, generally mixed with petroleum fuels. Transport even seems to be taking up an increasing position as end-use-sector consuming petroleum products.

In the next few years, the global mobility requirements can therefore be expected to increase and road transport should continue to develop, mainly boosted by the demand from the developing countries. During the next two decades, while the vehicle fleet is only expected to increase by a third in the OECD zone, it will rise by a factor of three or even four in the rest of the world. The same should be true for the distances travelled. In addition, air transport will continue to grow.

By 2030, consumption of petroleum products for transportation will increase even further, from 2.3 Gtoe in 2007 to 3.5 Gtoe in 2030 (Figure 1.4). Amongst the other energy consumptions, the trend for industrial requirements is comparable to that of transport, although the distribution of energy sources is quite different: the mains sources are electricity and coal, followed by gas, then a large share of renewable energies. The third highest consumer is the residential sector, with a large proportion of biomass.

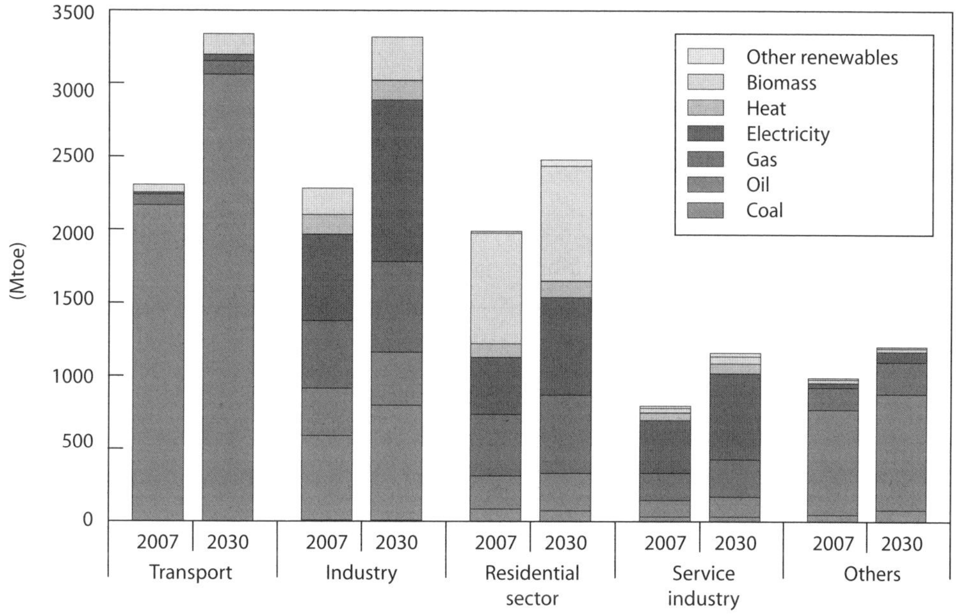

Figure 1.4

Energy consumption in various sectors, including transport.

Significant technical and regulatory progress has been made to reduce the emissions of local pollutants responsible for health problems, acid rain, etc. due to the increase in road traffic. Over the next 15 to 20 years, these efforts, observed mainly in the OECD countries, should substantially reduce a certain number of emissions (nitrogen oxides, particulates, carbon monoxide, etc.), depending on the pollutants. The same efforts have not been made in the rest of the world where these emissions are globally expected to increase in the future.

In the industrialised countries, releases of carbon dioxide due to transport have dropped significantly on new vehicles, for several years. Since the start of the 70s, however, global annual emissions of CO_2 for this sector have more than doubled and are expected to be virtually double again over the next 30 years.

1.2.4 Environmental impacts (see Appendix 1)

Although divided for many years on the subject, the international scientific community now agrees on the fact that human activity, especially and above all through its use of fossil fuels, has increased the concentration of atmospheric greenhouse gas, resulting in climate change on the surface of the planet (see insert). Studies conducted by the IPCC (Intergovernmental Panel on Climate Change) [8], an organisation including nearly 2000 researchers and experts throughout the world, created in 1988 to study the phenomenon, demonstrate that fast changes are taking place. The average atmospheric temperature has increased by 1.0°C since 1910. The results of the studies conducted under the supervision of the IPCC, referring to a set of 35 scenarios, predict that the world temperature should increase by another 1.4 to 5.8°C between 1990 and 2100. Other changes would also occur: mean sea-level rise of 0.1 to 0.8 m, with already an estimated rise in sea level of 16 cm in 100 years, faster desertification of some regions, reduction in snow cover (– 9% between 1950 and 2000) and glacier extents and, lastly, an increase in the frequency and intensity of extreme climatic phenomena.

Greenhouse effect and greenhouse gases (GHG)

About 50% of solar radiation is absorbed by the soil, which then retransmits the heat as infrared radiation. Some of this radiation is captured by the gases present naturally in low concentrations in the atmosphere (water vapour, CO_2, CH_4, N_2O). The atmospheric layer therefore retains the heat, like the windows of a greenhouse.

Without this "natural" greenhouse effect, the mean temperature on the Earth would be -18°C. As a comparison, the mean temperature on Mars is just-50°C. This is not because Mars is farther away from the Sun than the Earth, but because it lacks sufficient atmosphere to cause the greenhouse effect. In contrast, on Venus whose atmosphere consists mainly of CO_2, the temperature is over 400°C.

The United Nations Framework Convention on Climate Change was signed at Rio de Janeiro in June 1992, during the Earth Summit. While accepting that human activity had modified the world climate, the objective of the treaty was to stabilise "greenhouse gas concentrations in the atmosphere at a level that would prevent dangerous anthropogenic interference with the climate system".

Signed in 1997, the objective of the Kyoto Protocol [9] was to increase this international response by assigning enforceable targets to each signatory country to reduce greenhouse gas emissions. Between 2008 and 2012, mean annual emissions must be reduced by 5.2% compared with the 1990 figure, chosen as reference value.

The scientists working on this study also found that greenhouse gas emissions have a long-term impact on the composition of the atmosphere. For example, the lifetime of atmospheric CO_2 is over one century (Table 1.1). Relative stabilisation of CO_2 emissions over the 21st century would nevertheless increase the concentration of this gas in the atmosphere.

Table 1.1

Type of GHG	Global-warming potential (GWP) compared with CO_2	Percentage contribution to the greenhouse effect	Atmospheric lifetime (years)
CO_2	1	69.6	125
CH_4	21	12.4	12
N_2O	310	15.8	150
Fluorinated gases [a]	16 000	2.2	120 (mean)

[a] *PFC: Perfluorocarbon; HFC: Hydrofluorocarbons; SF6: Sulphur hexafluorides.*
Source: IPCC

The GHG emissions by source, in 2005, are shown on Figure 1.5.

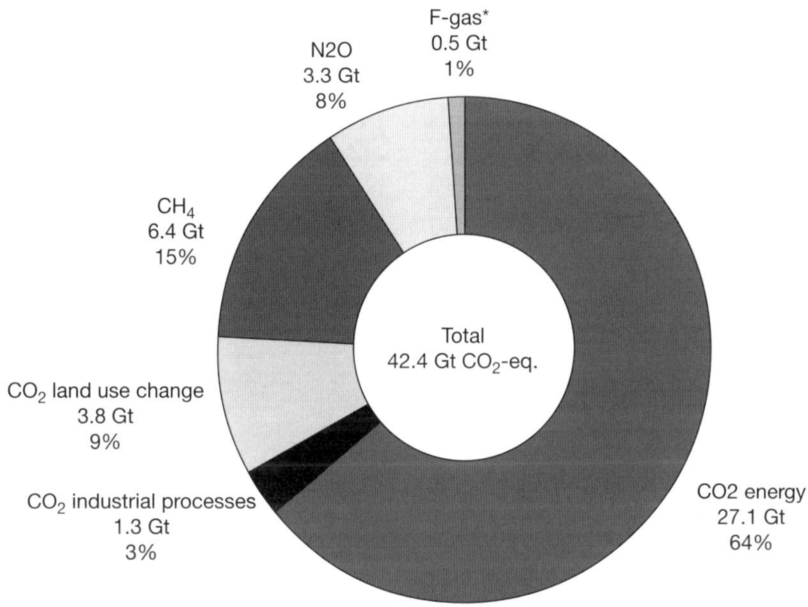

* *F-gas: hydrofluorocarbons (HFCs), perfluorocarbons (PFCs) and sulfur hexafluorides (SF6)*

Figure 1.5

GHG emissions by source, in 2005.
Source: IEA

Over the last century, the atmospheric content of the main greenhouse gas in terms of global contribution, carbon dioxide, has increased considerably due to the exploitation and use of fossil fuels. The increase was continuous. Over the last thirty years, the annual CO_2 emissions induced by energy consumption have increased by 65% on the Earth's surface. In the coming years, this trend is expected to continue at a similar pace.

At world level, two sectors in particular will contribute to this increase: electricity production and transport. In the developing countries, electricity production should be the main contributor to the increase in CO_2 emissions over the new few decades, and the transport sector to a lesser extent. In the industrialised countries, the two sectors, transport and electricity production, are expected to be equally responsible for the increasing carbon dioxide releases.

CO_2 emissions due to transport, according to various scenarios, are represented on Figure 1.6.

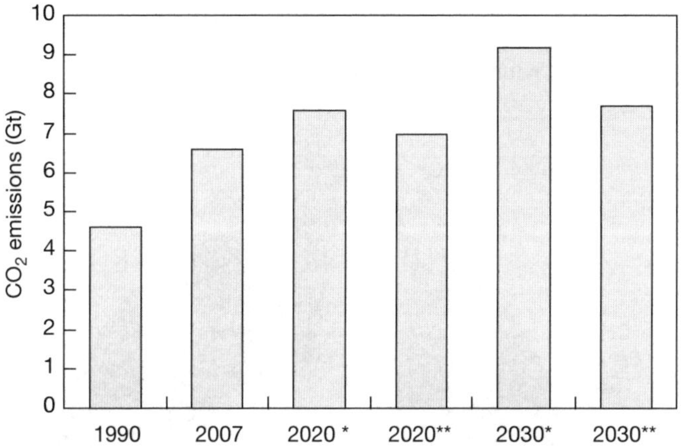

Figure 1.6

CO_2 emissions due to transport.
Source: IEA (* reference scenario. ** scenario 450)

1.2.5 The safety stakes

Faced with the development of biofuels, the knowledge scientifically validated on the safety aspects potentially associated with the use and production of biofuels must be updated and completed. This need was identified at a relatively late stage, with the incidents that accompanied the industrial expansion of first generation processes. A description of their typology has been recently obtained by analysing these incidents [10]. Most of the risks are relatively conventional, of variable type and scale depending on the processes and the places where the biofuels are produced and used. Their processing deserves careful consideration and involves a detailed analysis of the factors likely to generate accidents. For example, in December 2009, an accident was reported in a biodiesel production plant in the State of Washington: an operation in which sulphuric acid was used led to destruction of a mixing tank (explosion), causing extensive material damage.

The following table lists a few serious accidents involving biofuels:

Table 1.2 Serious accidents.

In January 2004 at Port Kembla (Australia), a tank containing 7000 m^3 of ethanol exploded and caught fire. Welding was being carried out on the tank at the time of the explosion. Two people were injured in the accident.
In September 2005 at West Burlington, Iowa (United States), a grain dryer caught fire in an ethanol plant. The roads around the plant were closed for one hour. One plant employee and two firemen were slightly injured.
In October 2006 at New Brighton (United States), a train carrying ethanol over the Beaver river was derailed, resulting in an explosion followed by a fire. No-one was injured, but there was a risk of polluting the river and the local residents were evacuated.
In September 2007 at Sao Paulo (Brazil), a 9-million litre ethanol tank exploded in a factory closed for reconstruction work. The explosion could have been caused by lightning.
In January 2008 in Ohio (United States) in a biodiesel plant, an operator checked the level of glycerine in a tank. On leaving, he inadvertently left the tank manhole open, allowing methanol vapours to escape from the 10 000-gallon tank. An employee noticed vapours in the building and, to ventilate, pressed a switch to open the door. The spark caused by pressing the switch set fire to the methanol vapours, causing an explosion. 4 plant employees were injured.
In April 2008 at Calgary (Canada), a man working in a new biodiesel plant was killed by an explosion. The man was welding above a biodiesel tank when it exploded. The explosion was followed by a fire which lasted several hours.
In May 2008 at New-York (United States), an explosion, followed by a fire, occurred in a garage. The owner was manufacturing biodiesel. The accident destroyed his garage, damaged the house and another garage.
In June 2008 in Minnesota (United States), a train transporting ethanol was derailed, resulting in spillage of the product. According to the investigation, the accident could have been caused by faulty rails. No-one was injured in the accident.
In June 2009 in Illinois (United States, a train transporting ethanol was derailed, causing an explosion followed by a fire. 6 people were injured and one person waiting in his car for the train to go past was killed. About 600 local residents were evacuated.
In October 2009 at Savannah (United States), a biodiesel tank on the site of a recently opened plant exploded. One person was injured.

To avoid delaying the development of this sector, it seems essential to consolidate the knowledge on safety issues by identifying and characterising the hazards presented by all products involved in the biofuel life cycle: agri-resources, reactants, catalysts, products used in the agri-resource pre-treatment operations or in the refining of finished products, various formulation additives, co-products and other materials involved in their upgrading. An overview of the risks related to biofuels and their typology, especially for bioethanol and biodiesel, is provided. These risks are related to the combustibility or flammability properties of the main products involved in the biofuel industry.

1.2.5.1 Risks related to bioethanol production

A certain number of technical problems concerning the safety of ethanol-blended fuels soon came to light: compatibility of ethanol with polymer materials (fuel supply circuits) and its corrosiveness with some metals (especially aluminium). However, careful review of the literature and exchanges with foreign researchers (SP, National Swedish Rescue Service) demonstrate that further studies are required so that the experts can reach a consensus of opinion, especially as regards E85 (gasoline mixed with 85% ethanol), before being able to perfectly control the danger of fire and explosion associated with the addition of ethanol in the gasoline pool. Evaluating the true danger of fire and explosion associated with the life cycle of fuel ethanol (in its various formulations) is in fact far more complex than simply observing the ethanol extended flammability limits (compared with gasoline) shifted towards higher temperatures (Table 1.3).

Table 1.3 Main characteristics of ethanol biofuels.

	ETBE	Ethanol	Unleaded gasoline 95	E10
Molar mass (g/mol)	102	46.07	102.5	–
Boiling point (°C)	72.8	78.4	< 35	–
Stoichiometric ratio	12.1	8.95	14.4	–
Reid vapour pressure RVP (kPa)	10	15.9	60/70 [a]	60(+ x)/70(+ x)
Flash point (°C)	– 20	12	– 40	~ – 40
Autoignition temperature (°C)	470	400	280-350	> 300 [b] (Total) > 250 [c] (Shell)
Lower/Upper Explosive Limits (vol%) (LEL/UEL)	1.2-8.8	3.3-19	1.4-8.7	–
Upper/Lower Flammability Limits (°C) (UFL, LFL)	Not available	9-44	(– 45)/(– 20)	< – 5 [d](UFL) [11]

[a] *Values calculated using EN 228 and applicable in France.*
[b] *Safety data sheet (Total).*
[c] *Safety data sheet (Shell).*
[d] *Estimation worthy of being confirmed experimentally.*

The other parameters to be taken into account to assess and control the accident scenarios resulting from the (direct or indirect) danger of fuel ethanol explosion and fire, include in particular:

- the various characteristics for each gasoline/ethanol mixture considered,
- the property of ethanol to form various azeotropes with the light gasoline compounds, causing "over-volatility" of the ethanol/gasoline mixtures, which varies depending on the proportions of the mixture, when gasoline remains the majority compound,
- in contrast, lower volatility of ethanol and E85 (compared with gasoline) making adaptations on the vehicles necessary,

- the risk of separation of the gasoline and ethanol phases depending on the water and ethanol contents,
- the risk of leaks due to the corrosiveness of ethanol with respect to some metals and the incompatibility with numerous plastics,
- the considerably higher electrical conductivity of ethanol and E85 compared with gasoline,
- the faster flame velocity,
- the polar nature of ethanol as solvent.

Safety issues arise, especially as regards logistics: installation of premium-grade gasoline refuelling depots, storage and distribution of ethanol blended fuels (unleaded 95 – E10) and superethanol (E85) in petrol stations, positioning of equipment with respect to the risk of explosible atmosphere and the associated environmental aspects, for which there is very little hindsight. The experience acquired and the globalisation of biofuel markets must not mask the fact that the technical solutions to be set up in order to develop the technologies necessary to implement biofuels in transport cannot be systematically generalised, without adaptation to the local context. The solution developed in Brazil to solve problems of cold starting in spark ignition engines due to the low volatility of ethanol was to add an auxiliary gasoline tank under the bonnet, periodically filled by local service station attendants. Obviously, this solution cannot be considered in Europe since it fails to meet basic safety principles in view of the European automotive fuel distribution context [12].

Lastly, when planning how to manage the risk of fire in the distribution and utilisation circuits, it is worth pointing out that fires of ethanol (and therefore of gasolines blended with large quantities of ethanol) behave quite differently from hydrocarbon fires and gasoline fires in particular. Apart from the advantages (shared with combustion in engines) of a reduced risk of pollution compared with gasolines, special extinguishing means are required due to ethanol's behaviour as polar solvent [13]. Gasohol fires (E10 United States) can be extinguished with low expansion foams such as AFFF or AR-AFFF (resistant to alcohols), but the application rates may have to be increased, especially in case of prolonged fire. Lastly, note that the development of ethanol fuels has justified the allocation of a new UN number to identify the transport of fuels composed of ethanol/gasoline mixtures for ethanol concentrations greater than 10% (entry UN 3475 added to RID/ADR 2009).

1.2.5.2 Risks related to biodiesel production

The "non flammable" nature of biodiesel often mentioned in general public documents (due to the high flash point of this biofuel – about 180°C) must not mask the reality of the dangers of explosion and fire observed when analysing the life cycle of biofuels, from the production of upgradeable agri-resources through to use in Diesel engine vehicles.

To illustrate this, we simply need to remember that biodiesel remains a fuel, and is therefore combustible, and above all to consider the dangers intrinsic to products involved in large quantities in the "transesterification" reaction, at the centre of the current biodiesel production technology.

This reaction, which consists in converting the fatty acids of triglycerides into methyl or ethyl esters that are directly compatible with Diesel oils of fossil origin:

– is generally carried out by using methanol or ethanol, which are toxic and flammable products,
– is a catalytic reaction. The homogeneous catalysts, either alkaline or acid, create a variable risk of fire and/or explosion, especially with the sodium or potassium alkoxides, which are amongst the most widely used [14].

Other products, either flammable or presenting fire or explosion risks, are involved in the life cycle of biodiesels, for example in the seed trituration plants (final extraction), or in the warehouses used to store co-products of plant origin (risks of self-heating, also present in ethanol supply chains).

Combustion tests were conducted on various oils and glycerine during a first orientation campaign at INERIS for various incident thermal fluxes imposed during Tewarson tests, simulating a given level of thermal attack. The oil burning rates are low. In spite of these figures, however, we must not forget that the true burning rates remain highly dependent on the actual conditions of an accident scenario. Tests, as well as feedback [15], show that once in steady state regime, these types of fire release large quantities of energy and substantial volumes of smoke.

In view of the technological progress made and the more widespread use of these biofuels, support will be provided on control of safety issues. Extensive research studies are being conducted, especially in the United States, to ensure the safety of pipeline transport, in view of its specific corrosion problems [16].

1.2.5.3 Safety stakes identified for 2nd generation biofuels

Second generation biofuels represent “disruptive” solutions compared with the first generation production technologies. They require new processes demanding a rigorous approach in terms of safety management, due in particular to the more severe processing conditions (higher temperature and pressure) especially for thermochemical pathways. These new pathways may also involve different types of formulated fuels (gasoline/butanol blends, stabilised Diesel/ethanol blends).

Safety stakes are obviously associated with the introduction of biofuels as an alternative to kerosene for civil and military air transport. These stakes are considered and examined in particular in specific research programs, not yet finished, such as SWAFEA and ALAFA-BIRD [17] in Europe and by the CAAFI alliance program [18] developed in the United States.

Lastly, biofuel production will gradually shift towards a new industrial production model called biorefining, which will also produce other material (chemical building-blocks, bio-products, biomaterials) or energy flows with the aim of optimising the use of biomass. This multiform production model will result in an unprecedented level of integration for the processes involved and high expectations in terms of flexibility, both upstream (range of biomasses to be upgraded) and downstream (energy vectors, biomaterials, biomolecules). This configuration will have major repercussions on risk control. To ensure an adequate level of control, all operating modes (and predicted malfunctions) of such a complex system must be correctly anticipated and evaluated right from the design phase. In view of the technologies which will emerge in this type of system, but also features such as recourse to various micro-organisms, nanotechnologies (separative techniques, catalysts, etc.) and new solvent families (ionic liquids, etc.), it is likely that the experience gained in safety management in traditional refineries (based on petroleum chemistry) will have to be adapted to this new field.

1.2.6 Biofuels: a promising future

The development of more sustainable energy technologies or systems is not something new for the transport sector. Research and development studies have been, and still are, conducted on numerous solutions, the objective being to combine lower environmental impact with reduced oil dependence and optimisation of the use of agricultural and forest land.

These studies aim at improving existing technologies, internal combustion engines (spark ignition or Diesel) in order to significantly reduce the consumption of these vehicles, as well as the search for alternative solutions in terms of engines (fuel cell, electric vehicle, hybrid vehicle) and fuels (natural gas, biofuels, hydrogen).

Amongst the solutions studied, biofuels offer real perspectives. Blending them with conventional fuels should reduce greenhouse gas emissions in the transport sector without substantially modifying the distribution infrastructures or the vehicles.

Although not the only solution to the problems of greenhouse gas emissions in the transport sector and the need to reduce the demand for oil, in the medium term the biofuel option represents one of the best cost-efficiency ratios [19].

The development of biofuel supply chains offers numerous advantages, even though biofuels currently cost more than fossil fuels, at equivalent energy content.

Biofuel supply chains may have a positive impact on the trade balance of a country or a community area (e.g. the European Union), if imported products are replaced by products manufactured within the country or within the community area (substitutes for imported fossil origin fuels or animal feeds).

It must be pointed out that domestic production of biofuels will only have a positive effect on the trade balance if the agricultural raw material also comes from the same country or community area. To maintain this positive impact on the trade balance, protective measures (customs duties, etc.) may have to be implemented to avoid or limit massive imports of biofuels or agricultural raw materials produced at cheaper cost in other countries.

The production of biofuels also offers new outlets for agriculture on the domestic market. The possibility of using some crops for two outlets, food and energy, will increase the potential sources of income for farmers as well as the total income, thereby helping to maintain or develop employment in rural areas.

In addition, even imported, biofuels help to secure supplies as they offer an alternative to imports of crudes and petroleum products, while diversity of sources is expected to diminish with time. Biofuels will not be imported from the same regions as oil and petroleum products

However, it is presently very difficult to assess the volumes of vegetable biomass which could eventually be used to produce energy, or to accurately estimate the quantities of energy which could be produced. Numerous studies have been conducted on the subject, in an attempt to assess the potential that bioenergies as a whole (production of heat and electricity, transport) could represent in the long term. The results obtained are far from being convergent, considerable differences remain between the most optimistic and the most pessimistic. On a global scale, the potential of biomass produced from crops dedicated to energy production (excluding in this case agricultural, forest or food industry waste) could vary from 1 to over 5.5 Gtoe per year by 2050 [19].

Numerous uncertainties remain. Assessing the areas that will be dedicated to these crops depends on the planet's food requirements which depend in turn on the increase in world population, diets and yields per hectare: it will be difficult to divert land required for food production to the production of energy crops. The global energy volume which could be obtained from these crops will also depend on the types of crop, their location and the evolution of yields. Lastly, the land and plant resources set aside for energy production will obviously not be dedicated solely to the production of biofuels.

Detailed studies have been conducted to evaluate the future world and regional demands for biofuels, especially by the IEA [7]. The demands increase systematically up to 2015 and 2030, possibly reaching slightly above 2.5 M barrels/day (Figure 1.7) for the United States and Europe by 2030. Replacing fuels by first generation biofuels, produced from amylaceous plants (wheat, corn) or sugar plants (sugar beet, sugar cane) and derived of vegetable fats (rapeseed, sunflower, soya, etc.) raises the problem of farmland allocation. We are faced with a problem of land and resource availability. Their use for the production of biofuels competes with their use for food production.

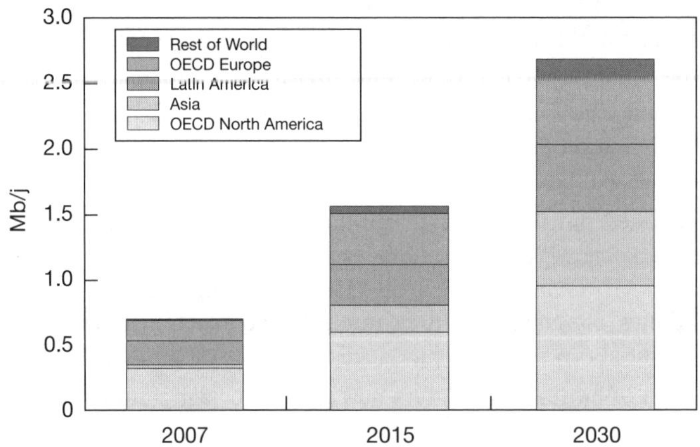

Figure 1.7

Future biofuels demand.

The pressure on land cultivated with these plants will therefore become increasingly problematic over the next 20 years. We may reasonably assume that, with the rise in world population and despite the projected increase in crop yields, the competition between the various uses of traditional resources will increase, production of food remaining a vital priority. Large areas of uncultivated arable land are available, however, for example in the developing countries, especially Africa.

At the same time, the development of the so-called alternative fuels (natural gas, liquefied petroleum gas, hydrogen) will remain limited. The projected 20% substitution of gasolines and Diesel by these alternative fuels by 2020 will be difficult to achieve, for at least three reasons:

- the marked lack of enthusiasm displayed by car drivers to use gaseous fuels,
- the unsolved technical difficulties (storage, transport) related to the use of hydrogen which, according to most experts, will not be solved for another 30 years,
- the limited effect of these alternative biofuels to lower emissions of greenhouse gas (CO_2), apart from hydrogen manufactured by electrolysis of water using nuclear, hydropower, solar or wind electricity.

Given this context of pressure on farmland and the limited development of alternative fuels, other ways of producing fuels from biomass will have to be found. Possibilities do exist, for example the production of fuels from lignocellulosic resources (straw, wood, etc.).

Besides cereals, beet and oilseeds, agricultural residues, for example cereal straw, forest waste and by-products of the wood conversion industries are other potential biomass resources. It will eventually be possible to extend these resources by growing dedicated annual crops (e.g. triticale), herbaceous perennial crops (e.g. miscanthus) and woody perennial crops (e.g. SRC, short rotation coppices such as poplar and eucalyptus).

Some organic wastes available in large quantities such as urban wastewater treatment plant sludges and effluents from the paper and agribusiness industries, represent a non-negligible potential for widening the range of carbon substrates that can be converted into biofuels.

Two biofuel synthesis pathways, using these new feedstocks, are under development:

- the thermochemical or dry pathway for the production of synthesis gas ($CO + H_2$) then liquid fuels (see Chapter 5),
- the biochemical or wet pathway for the manufacture of low-molecular weight alcohols, mainly ethanol (see Chapter 6).

The successful combination of new raw materials and new conversion processes should offer several advantages:

- no direct competition with food uses. We should not forget, however, that these new resources already have, and will continue to have, outlets other than that of conversion into biofuels, for example in the sectors of heat and electricity production, paper mills and chemistry,
- as regards energy consumption, the conversion processes would no longer require any external energy, in particular of fossil origin, which is currently the case with the synthesis of biodiesel on one side, and that of bioethanol from wheat starch or beet sugar, on the other side,
- these new synthesis pathways provide the best answers to the problem of reducing CO_2 emissions, as shown recently by a joint study conducted by the CONCAWE (European oil companies) EUCAR (European car manufacturers) and the EU Joint Research Centre (European Commission) [20],

- at least as regards cereal straws, these new technologies would probably increase biofuel production yield by 50%, per hectare cultivated,
- these new processes do not generate any by-products or co-products, for which applications and new outlets would have to be found,
- lastly, they offer good potential for reducing the cost of biofuels.

However, although processes using the traditional agri-resources are technically mature, it is not the case with those using lignocellulosic biomass: a number of technological barriers must be overcome before they are really operational at industrial scale.

The points mentioned above clearly indicate the importance of validating these new raw material supply chains rapidly and of developing these new technologies.

1.2.6.1 Synthetic fuels derived from lignocellulosic biomass and/or organic waste

The thermochemical pathway, also called the dry or gasification pathway, converts biomass into a gaseous mixture rich in CO and H_2, syngas, which after purification can be used to produce liquid fuels such as methanol, dimethylether, or a mixture of hydrocarbons obtained from the Fischer-Tropsch synthesis. The quality of the liquid fuels obtained is excellent, they contain no sulphur or aromatic compounds and, having a high cetane number, they can be used in modern Diesel engines.

Syngas can also be used to produce hydrogen, via the gas shift reaction in which CO reacts with water vapour, increasing the hydrogen content.

In Europe, it is primarily the Nordic countries (Finland, Sweden, Denmark) together with Austria and Germany, which have tried to develop lignocellulosic biomass gasification technologies. France has been working on the development of these new processes for several years.

In these countries, the bulk of these studies, many supported by European funding, have focused for several years on cogeneration solutions (production of electricity and heat). The main examples are the Varnamö (Sweden) and Güssing (Austria) demonstration units.

The interest for gasification has only very recently turned towards the production of liquid fuels, and in the longer term hydrogen. The technical, financial and environmental aspects are detailed in Chapter 5.

1.2.6.2 Ethanol derived from lignocellulosic biomass

This pathway converts two of the three main fractions of lignocellulosic biomass, cellulose and hemicelluloses, into ethanol, while the third one lignin provides the energy required to implement the conversion process.

Optimisation of this pathway will consist in improving the ethanol yield, by overcoming two major technological barriers:

- poor efficiency of the enzymatic hydrolysis of lignocellulosic materials. The goal of current research and development studies throughout the world, aimed at improving the activity of cellulolytic enzymes, is to bring the level of efficiency and cost of this hydrolysis step in line with that of enzymatic hydrolysis of the amylaceous materials contained in wheat or corn,

- poor conversion into ethanol of the C5 sugars contained in the hemicellulose fractions. Taking advantage of the range of means available in genetic engineering and molecular biology, current laboratory studies are focused on improving the ability of the selected microorganisms to produce ethanol from these sugars.

Chapter 6 specifies the technical, financial and environmental aspects.

1.2.6.3 Alternatives: heat, electricity, fuels, chemistry

Remember that biomass, in particular that from agricultural and forest residues, as well as organic waste, can be converted into heat or electrical energy, fuel or raw material for chemistry.

In France, bioenergies represent about 4% of primary energy consumption. Wood supplies the largest share as heat energy, for about 9 million toe. The rest comes from incineration of organic waste, methanisation, biofuels.

Apart from municipal waste incineration which already provides thermal and electrical energy, bioelectricity production systems such as biogas electric motors or cogeneration with steam turbine, for example in the paper industry, are still relatively rare.

In order to reduce greenhouse gases and energy dependence, massive use of biomass will become essential. It offers major advantages in these areas: it recycles CO_2, it is renewable and it is produced locally. A recent study has shown that biomass is the renewable energy source which has the highest growth potential [21].

However, the resource is limited, firstly due to the land areas available, constraints regarding the collection and distribution of thermal energy in case of forest, yields, water resource, competing uses (food, wood and pulp and paper industries, etc.), soil organic matter requirements, etc., and secondly to the price of competing fossil energies which sets the mobilisation level of forest biomass in particular.

Under these conditions, how will biomass be allocated to the various uses? We must not forget that most plant resources are dedicated in priority to food purposes. Several factors will be involved, in particular the markets, prices, compared CO_2 gains, logistics and incentive schemes.

In the short term, two critical sectors have an impact on the reduction of greenhouse gases: residential and tertiary buildings on the one hand, transportation on the other. However, the greenhouse effect should not lead to conflicts over the use of renewable raw materials.

A. Thermal use

In the collective and tertiary housing sector, almost 80% of the final consumption is dedicated to low-temperature thermal use. Biomass combustion, which meets this requirement, is undoubtedly the most efficient way of obtaining energy from biomass. For example, combustion of 4 m^3 of wood produces the equivalent of one toe. In the private housing sector, the domestic market in fuel wood is globally stagnant.

The collective and tertiary housing sector is a high potential market; it is closely linked to the development of thermal networks in which large, very efficient systems can be shared by connecting several buildings to a single boiler. It is nevertheless dependent on investment grants.

While wood fuel is competitive compared with the other energies, wood boilers require greater initial investment. Consequently, thermal energy will remain a major outlet, firstly for residual biomass (by-products of the wood industry, forest and agriculture, straw, etc.) and possibly in the long term for dedicated lignocellulosic crops such as short rotation coppices or energy crops (miscanthus, etc.).

B. Bioelectricity

The potential of bioelectricity is limited due to the heat utilisation constraint and to external procurement costs, which increase rapidly for installations that are necessarily large, but also by the low efficiencies of current techniques (15 to 20%). Gasification techniques, which should double the electrical efficiency, represent an interesting direction for the potential development of bioelectricity production, while guaranteeing a good return on the plant resource.

These techniques are at preindustrial stage. A true market opening would trigger their industrialisation. It depends mainly on the electricity buy-back rate being sufficiently attractive, which is not always the case at present.

The bioelectricity market is *a priori* of interest for industries generating biomass resources (e.g. paper mills and agribusiness industries) or special situations (bagasse plants). Concerning the reduction of greenhouse gases, production of electricity from biomass would only prove interesting as backup during peak periods, to replace coal, fuel oil or gas.

C. Biofuels

While there are other alternatives to oil, gas and coal to produce energy, it remains difficult to find a substitute for liquid fuels. Over 95% of the transport sector relies on fossil hydrocarbons.

To reach the 2050 objective of reducing emissions from the developed countries by a factor of 4, CO_2 emissions from the transport sector would have to be reduced by a factor of 7.

Apart from reducing fuel consumption in cars and developing new engines (the hybrid vehicle is a good example), it is expected that biofuels will play an increasing role in reducing CO_2 emissions in transportation.

D. Plant chemistry

Biomass will also be required for the petrochemistry sector, since it is the only renewable raw material that can be substituted for oil. In view of the inevitable increase in the price of oil, the organic chemistry industry will need to turn towards plant based raw materials.

This is a strategic stake for both agriculture and the chemical industry, which would likely undergo radical technological changes within the next 20 to 30 years.

1.2.6.4 Conclusions

In conclusion, what are the perspectives for these various utilisations of biomass? Obviously, they are not interchangeable. We also know that biomass can only represent a partial solution to the energy problem since the requirements are disproportionate with respect to the land areas available. In France, the equivalent of almost all the utilised agricultural land would be nceessary to meet the total fuel consumption. Drive or eat: it's up to you!

Although heat is the most economical way of using biomass, with the lowest cost per tonne of CO_2 avoided, biomass will still be required for many years to come to produce liquid fuels and even possibly to power fuel cells. In the short term, these are the two major outlets for biomass. Production of bioelectricity, which relies on thermal conversion, is more to be seen as a backup option.

Lastly, biomass should eventually be used as a basis for numerous "bioproducts" such as lubricants, solvents, surfactants, biopolymers, fibrous materials, etc.

The development of bioenergies and plant chemistry depends largely on the price of hydrocarbons. So far, biofuels and bioproducts have been globally more expensive than those produced from oil.

In the meantime, financial incentive systems based on subsidies or remuneration of externalities (carbon emission permits, etc.) are required.

As a result of these considerations, projections for France are:

- a global biomass utilisation potential increasing from 11 Mtoe/year in 2000 to 26 Mtoe/year in 2050,
- a physical limit in the region of 30 Mtoe/year that will be difficult to exceed, taking into account the quantities of resources and land available,
- a significant change by 2020 in biomass utilisation towards real industrial chains focusing in particular on biofuels, syngas and higher added value chemistry.

The areas dedicated to this type of production can be expected to increase from 0.5 Mha in 2000 to 3.5 Mha in 2020 and 6 Mha in 2050.

1.3 THE FIRST STEP IN BIOFUEL PRODUCTION: PRODUCTION OF BIOMASS

1.3.1 Competition with food agricultural crops

Should agricultural land be cultivated to produce food resources or energy, including biofuels? Obviously, supplying the food market will always take priority.

Over the last few years, sometimes violent criticisms have been made on the decision of numerous countries to continue producing first generation biofuels such as ethanol and biodiesel from cereals, sugar plants and oilseeds, grown mainly for human and animal consumption.

However, such frontal competition may be mitigated by an optimization of land use by the agricultural sector. Several years ago, for example, a cultivatable fallow land scheme was set up in Europe to produce industrial and energy resources in order to best use the excess land due to surpluses in foodstuff production. This situation had been a key point in the launch of biofuel manufacture. In addition, this represented a good way for the United States and Brazil to better control and regulate their massive productions of corn and sugar.

We must not forget that the production of biofuels always goes hand in hand with the commercialisation of food materials, such as cakes or draff, and oils, for example in the processing of maize.

With the 2007-2008 economic crisis, the prices of plant materials, like other materials, rose sharply. Could this be linked to the production of biofuels which had reduced the land areas available for food crops? Some "experts" were quick to suggest this possibility.

Several other reasons obviously played a major role, for example:

- the increased consumption of meat food at the expense of vegetarian food, in the emerging or developing countries. This resulted to a sharp rise in the use of agricultural land. To produce a given quantity of proteins, it takes 6 to 7 times more area for white meat proteins and 8 to 9 times more for red meat proteins, compared with vegetable proteins,
- there is a clear trend worldwide to replace animal oils by vegetable oils in human food,
- bad regional harvests, as was the case in Australia, where wheat yields were reduced due to water shortages,
- non-negligible speculation in the field of raw materials often based on trivial information.

The economic crisis saw a dramatic increase in the price of cereals. While first generation biofuels could have an impact on the price surge for maize and sugar, both raw materials involved in the production of ethanol, it is hard to imagine that this could have also been the case for wheat, rarely used for this application, and even less so for rice.

After the crisis, the prices of plant raw materials dropped considerably, although there was a substantial increase in the quantity of first generation biofuels manufactured in 2008 and 2009.

In addition, arable land is still available in some regions, especially Africa.

In the medium term, the growth in the production of first generation biofuels is expected to stop, to be replaced by that of second generation biofuels, obtained by converting non-food lignocellulosic materials. Competition for land will still apply to these bioenergy feedstocks, but will be mitigated by their high energy yields per hectare and their potential to be grown on marginal or degraded land, unsuitable for food crops.

1.3.2 Biodiversity

Intensive crops on arable land, such as those of woody species, may impact biodiversity, which must be preserved and maintained. Remember that the UN declared 2010 as the international year of biodiversity. Currently being evaluated, biodiversity must then be included in a crop certification system, taking into account an environmental assessment which is as complete as possible, in a framework of validating a sustainable development activity, such as bioenergy, for example,

1.3.3 Biomass, biofuels and soil organic matter

Will the production of biomass for energy purposes deplete the soil organic matter and simultaneously generate emissions of CO_2 from the soil carbon stock? This type of scenario could counteract the advantages of replacing fossil fuels by biofuels in terms of greenhouse gas emissions, even leading to a negative balance, as recently suggested [22]. This complex issue, which is far from being settled, depends on numerous factors including the amounts of biomass exported from soils, as well as the type and history of the soils. One thing is clear: clearing forests or permanent pastures to grow annual crops is not the right option.

1.3.3.1 Biomass and CO_2 compatibility

The carbon contained in biomass is said to be renewable, biogenic or short cycle. It is released into the atmosphere as CO_2 after the biomass has been used, but since it has just been taken from the atmosphere by plant photosynthesis, it does not contribute to the accumulation of greenhouse gases. On the fields on where the energy plants are grown, not all the CO_2 fixed by photosynthesis is converted into energy. A variable part of the biomass produced remains on the plot (as roots, straw, dead leaves, etc.). Eventually, these residues will be partly stabilised in the soil organic matter, which means that the filed functions like a "sink" of atmospheric CO_2. In terms of CO_2 emissions mitigation, a compromise must therefore be made in management of the biomass produced between this "sink" function and the substitution of fossil energies. This substitution occurs at the expense of the carbon sink strenght on the field since the biomass used to generate energy is exported from the field. In particular, the biofuel balance in terms of greenhouse gas emissions must include this compromise and the change in soil carbon stocks (including the aerial biomass in the case of forests or short rotation coppices).

Table 1.4 lists values of carbon fixation and stocks in the biomass and the soil for several ecosystems worldwide, and provides an indication of the annual carbon fluxes and the potential losses related to turning forest land over to agricultural crops. This conversion corresponds to a land use change, i.e. a change in the type of vegetation which develops in the ecosystem and above all its possible management by man.

Table 1.4 Quantities of atmospheric CO_2 fixed annually worldwide and by different types of ecosystem, and carbon stocks in the soil and in the associated biomass [23] (1 Gt = 10^9 tonnes).

Carbon fluxes/stocks	Planet	Wet tropical forest (Amazon)	Temperate forest (around Paris [France])	Cereal crop (around Paris [France])
FLUXES				
Net primary production	10 Gt C/year	5 t C/ha/year	3 t C/ha/year	2.8 t C/ha/year
STOCKS				
Biomass	70 Gt C	200 t C/ha	80 t C/ha	6 t C/ha
Soil	2 000 Gt C	150 t C/ha	100 t C/ha	60 t C/ha

1.3.3.2 Evolution of soil carbon stock

Agricultural and forest soils contain a few percent by weight of organic matter from roots and debris or residues produced by the plants whose growth they support (cereal straw, leaves and small branches, etc.). Composed of about 60% carbon, this organic matter is decomposed by the microorganisms present naturally in the soil. Some of the decomposed carbon is released into the atmosphere as CO_2, by a respiration process. The evolution of the soil carbon stock is therefore the result of the balance between the inputs of organic matter (residues brought by the plants) and the outputs (related to organic matter decomposition). This evolution depends on farming practices (tillage, straw management, succession of crops, etc.) and land use changes.

Two types of factor are likely to be involved during the production of biomass for energy. Firstly, it overall increases the fraction of carbon taken up by ecosystems which is removed from these ecosystems, which reduces the soil carbon sink. Secondly, since additional land needs to be farmed to increase biomass production, bioenergy may lead to land use changes by various cascading compensation effects. These two effects are discussed in the following paragraphs.

1.3.3.3 Bioenergy feedstock and land use changes

The main land use changes concern the transition between permanent pastures, forests and crops for the managed ecosystems, and between natural (savannas, primary forests, peatland, etc.) and managed ecosystems.

Agricultural crops are generally annual, i.e. they are resown every year and the land ploughed, although simplified tillage techniques without deep ploughing are currently tending to spread in temperate zones. As a result, the soils associated with these crops (including those currently used to produce biofuels) contain less organic matter than pastures or forests. In France, average soil carbon losses due to conversion of pasture and forest into agricultural land are estimated at 25 and 30 tonnes of C per hectare respectively [23].

Given these dry losses, it takes between 20 and… 110 years, depending on the biofuels considered, to amortise the carbon debt contracted following this type of land use change. This is certainly too high as a price to pay to take over new arable land for biofuels. However, this only concerns annual crops, such as wheat and rapeseed. The situation is different for perennial crops, including those which produce lignocellulose such as the miscanthus and *switchgrass*, since they have a better capacity to increase carbon in the soils. This would minimise the carbon losses caused by the land use changes [24]. When converting low-productivity permanent pastures into energy crops, the soil carbon stock could even increase fairly rapidly following the land use change. It should nevertheless be checked that the organic matter accumulated during the life cycle of these crops (typically 20 years) is not depleted too rapidly when the field is ploughed. Taking into account the diversity of land use changes and the temporal dynamics of changes in soil carbon stocks, a recent study estimated a 15-year alfback time at best, with the objective of meeting global requirements for liquid fuels through the use of biomass in 2100 [25].

Finally, we must mention the case of fallow land since, until recently, energy crops were grown on this type of land. Fallow land can be fixed (i.e. frozen for several successive years) or rotating (the field is frozen one year, with traditional annual crops being seeded the year before and the year after). It is difficult to obtain precise statistics on the distribution of the two types of fallow land, but it is estimated that about 30% of fallow land areas cannot be cultivated under good conditions and are therefore fixed [26].

On the rotating fallow land, vegetable cover is grown in the autumn so that the soil is not left bare, which would allow the nitrates to migrate to groundwater. Since energy crops have a higher biomass productivity (aerial and underground) than temporary cover crops, we may consider that they form a small carbon sink of about 0.15 t C/ha/year. Greater carbon losses may be observed if fixed fallow land is ploughed to grow annual plants, but no studies have yet been conducted on this subject.

1.3.3.4 Effect of cultivation practices

For energy crops, the practices affecting the dynamics of soil organic matter are mainly those which determine the biomass yield (e.g. fertilisation or irrigation) and the biomass fraction left on the field. Combination of these 2 variables determines in fact the amount returned to soil and contributing to its stock of organic matter. Other practices are involved, such as tillage (which speeds up the decay of organic matter in the soil) and the duration for which the soil is covered. From this point of view, perennial crops offer a certain number of advantages, since they occupy the soil for longer than annual crops and are not tilled during their cultivation cycle. With energy crops, however, the entire plant is generally harvested, which means that very little biomass is returned to the soil. Overall, the interest for soil organic matter seems to prevail [24], but more experimental tests are required to confirm this trend.

Apart from dedicated energy crops, export of food plant harvest residues, such as wheat straw or corn stover, is considered to be a high potential resource worldwide. This potential must nevertheless be adjusted to preserve soil organic matter. The topic remains controversial in the United States, where erosion (and losses of organic matter) are sensitive issues [27]. In France, a recent study conducted in Picardy has shown that the level of removal acceptable from this point of view would be in the region of 25% of the total resource – or even a little more if fields are managed on a case-to-case basis.

Of the annual crops currently used to produce biofuels, rapeseed could be interesting for the soil organic matter since it is a straw supplier (which *a priori* cannot necessarily be used for energy purposes due to harvesting constraints).

In conclusion, biomass production may have a relatively high impact on the stocks of soil organic matter and ecosystems in general, which should in principle be included in the greenhouse gas balances of bioenergy chains. Unfortunately, numerous unknowns remain both concerning the effects related to direct (e.g. transition from annual crops or permanent pastures to perennial energy crops) and indirect land use changes. Acquisition of new experimental data and integrated modelling of the ecosystems producing biomass should help overcome these barriers and identify the most interesting scenarios to ablate greenhouse gas emissions and to improve the environmental and economic performances of biofuels.

1.4 VISION OF THE REFINING INDUSTRY

The vision of the refining industry with respect to any new technology or solution is generally highly influenced by the way it is perceived by companies in terms of technical relevance and economic viability, and therefore of the possibility of "banking" on these innovations in the long term. For biofuels, security of supply is also a concern.

In Europe, the oil industry was initially highly sceptical when biofuels appeared. The following main reasons were put forward:

- it is a problem of agricultural policy which has nothing to do with the petroleum industry, which does not need these products to manufacture fuels. In any case, the potential resources are insufficient to meet market requirements,
- due to the progress made in terms of automotive vehicle depollution, biofuels have very little to offer in the fight against automotive pollution,
- they are economically uncompetitive, and the situation is unlikely to change in view of the cost price difference excluding VAT between fuels of petroleum origin and biofuels,
- subsidies via tax exemptions, whether direct or indirect, are economically unhealthy: they distort the competition and create economically metastable situations,
- production of biofuels represents a side activity for agriculture, making it a precarious resource of uncertain availability.

Apart from the fact that the CO_2 balance of the European ethanol-fuel industry is slightly less favourable than that of the VOME industry, the reluctance of most oil companies with respect to ethanol-fuel was due to a certain number of specific disadvantages, in particular problems related to the incorporation of untreated ethanol in gasolines: if there is any humidity in the circuits, addition of ethanol forms a water/ethanol mixture which separates from the rest of the gasoline, potentially leading to vehicle malfunctions. This is accompanied by corrosion of the circuits, possibly resulting in blocking.

Moreover, addition of ethanol significantly increases the gasoline vapour pressure. Splash blending, which consists in adding ethanol at the last minute, when leaving the terminal before going to the service stations, increases the mixture vapour pressure. When mixed, the apparent vapour pressure or mixture vapour pressure of ethanol lies between those of butane and pentane. There are currently no satisfactory solutions to this problem.

The solution which would consist in "preventively" reducing the vapour pressure of gasolines produced by the refineries to compensate for the effect of subsequently adding ethanol on leaving the terminal, is unrealistic. The extent of this "preventive reduction" would have to be known in advance and, in practice, all gasolines would have to be formulated with the same margin on the vapour pressure.

In addition, reducing the vapour pressure of the basic gasoline, before adding ethanol, would be economically penalising, which would reduce the benefit obtained by producing the added ethanol.

Incorporating ethanol in gasolines at the refining stage would be faced with the same obstacles, with the need for an agreement across the profession to adopt homogeneous practices, to avoid distorting the competition and the risk of exceeding the vapour pressure specification during distribution. The mixing law between an "ethanolated" gasoline and a normal gasoline is in fact such that the vapour pressure of the mixture is higher than the average of the two gasolines.

The simplest solution, least penalising from the financial point of view, would consist in granting a waiver on the gasoline vapour pressure specification if ethanol is added, but this would be in total contradiction with the other objective of minimising emissions of volatile organic compounds.

Another reluctance concerns the cost and financial risks of using ethanol as ETBE: ETBE is a synthetic product, manufactured from ethanol and isobutene obtained by refining oil (see Chapter 2). It is used in most European countries, except Sweden.

Addition of ethanol as ETBE to gasolines does not raise any problems, which is not the case with pure ethanol. ETBE is an excellent component of gasolines. Compared with ethanol, it offers a favourable vapour pressure (below the specification) and is fully compatible with the other components.

This operation also offers the advantage of converting isobutene into gasoline, while eliminating its undesirable properties as a fuel: olefinic nature and high vapour pressure.

In other words, the gain obtained by converting ethanol into ETBE is based on the synergy between ethanol and isobutene.

However, manufacture of ETBE involves investment in dedicated units. Firstly, sufficient isobutene resources are necessary, which is not the case with all refineries, as well as guaranteed supplies of ethanol, to avoid penalising the profitability of the investment. French supplies of isobutene from oil refining (catalytic cracking and steam cracking) represent about 400 000 tonnes/year. If all the isobutene available in France was to be used to manufacture ETBE, this would correspond to a limited use of ethanol of about 400 000 t/year. The same limits exist throughout Europe.

Concerns on the reliability of biofuel supplies have diminished with time.

Lastly, the other point to be taken into account is the imbalance between fuel supply and demand: Europe has surplus gasoline and insufficient Diesel, exactly the opposite to the situation in the United States. As a result, it is more relevant to privilege VOME production in Europe, while *a priori* ethanol production is more adapted to the United States. This is also true in Brazil, where the CO_2 balance and the economic balance are better in case of ethanol produced from sugar cane.

1.4.1 A key point for the future: the impact of helping reduce CO_2 emissions

For the European petroleum industry, whose main challenges are related to the imbalance of the fuel market and the need to minimise its CO_2 emissions, while biofuels represent more an opportunity than a threat, they are unable to cope with the scale of the problems to be solved.

The interest of biofuels, and probably their future, seem increasingly related to their ability to help reduce CO_2 emissions. From this point of view, however, the interest of biofuels depends not only on each production pathway, but also on the situation of the fuel market in each region, as well as the pressure which will be applied at global level to control CO_2 emissions.

Incidentally, the appearance of an economic cost related to CO_2 emissions following the application of the European directive on carbon trading will, in the future, generate a bonus in favour of technologies reducing these emissions.

All things being equal otherwise, this will help reduce the price difference between biofuels and petroleum-based fuels, in proportion to the price allocated to CO_2 multiplied by the quantity of CO_2 avoided by the production of biofuel.

The quantity of CO_2 avoided per tonne of biofuel depends on the production method considered and must therefore be established on the basis of a relevant comparative balance. In particular, the CO_2 emissions to be assigned to the manufacture of conventional fuels may differ significantly depending on the local situation.

In the United States, for example, the emissions assigned to gasoline, whose market is in deficit, are higher than those assigned to Diesel, which is consistent with the fact that US refiners convert Diesel into gasoline. In Europe, however, we observe the opposite situation, such that the bonus in terms of CO_2 balance is less favourable than in the United States for ethanol and more favourable for VOME.

The price per tonne of CO_2 avoided will be determined by the market, according to the law of supply and demand, under the carbon trading scheme, and will depend on the emissions quotas allocated by the States. The lower the quotas, the higher the market price of CO_2 will be.

Note that the likely trend towards increasingly strict quotas should be of more benefit to the production of VOME and ethanol from sugar cane, compared with the production of ethanol from beet or cereals, whose CO_2 balance is less favourable.

1.4.2 Comparison of biofuels with the other non-conventional fuels

The interest of biofuels must be assessed in view of their merits, compared with the alternative solutions: other uses of agricultural raw materials (in the food chain, chemistry, etc.) and other substitutes for conventional fuels of petroleum origin, such as the alternative fuels produced from fossil energies.

The second section of this comparison, more directly concerned with the petroleum industry, is discussed below.

1.4.2.1 Comparison criteria

As a general rule, the comparison between the various fuels must be based on the following four key criteria:

- impact on CO_2 emissions,
- extent of the resource and reliability of the supply,

– implementation, especially compatibility with existing equipment and ease of use,
– cost price, compared with the use value [1].

In the future, the impact on emissions of traditional pollutants (organic compounds, sulphur and nitrogen oxides, particulates), which until now has been an important criterion when discriminating between fuels, will decrease significantly due to the progress made in the prevention of this type of pollution, as shown by the performance of the vehicles currently manufactured. In addition, since the fuel properties likely to affect this type of pollution are defined in specifications, the financial impact of these properties is already taken into account when estimating the use value of the product considered.

Apart from VOME and ethanol of agricultural origin, used as alcohol or ether, which correspond to the biofuels currently used, there are various types of alternative solutions to the conventional fuels, gasoline and Diesel produced from oil, which can be classified and evaluated as follows depending on the origin of the energy, the resources available and the impact on CO_2 emissions.

1.4.2.2 Alternative fuels obtained from fossil energies

A. Compressed or liquefied Vehicle Natural Gas (VNG)

Its use induces strict constraints regarding filling, storage and autonomy for the vehicle. Consequently, it is used mainly by captive fleets (buses, utility vehicles, taxis) in order to reduce urban pollution. It makes a modest contribution to reducing CO_2 emissions. In addition, to offer an advantage in terms of greenhouse effect, it is essential that there are no gas leaks.

With the progress made in terms of vehicle depollution, the advantage of driving on gas can be expected to decrease, especially in cases where domestic fuel oil is used for heating, while this product represents a potential source of Diesel given relatively limited processing.

The well-to-wheel balance for a private vehicle shows that the vehicle must be fitted with a perfectly optimised gas engine to perform better than a Diesel engine, whereas for heating, the balance is still more favourable when gas is used rather than domestic fuel oil.

In contrast, the contribution to reducing greenhouse gas emissions is much greater when biogas (methane obtained from anaerobic fermentation of organic matter) is synthesised and used. Its use as fuel has developed significantly over the last few years.

B. Synthetic fuels derived from natural gas – the GTL (Gas-to-Liquids) pathway

This pathway mainly concerns Diesel obtained by Fischer-Tropsch synthesis or DME (dimethylether), synthesised from methanol. The products manufactured can also be obtained from coal (CTL, Coal-to-Liquids) and in both cases, the pathway involves the manufacture of syngas, a mixture of hydrogen and carbon monoxide.

The quality of the products obtained is excellent; they contain no sulphur or aromatic compounds and have a very high cetane number. However, the CO_2 balance is unfavourable compared with conventional fuels, even when natural gas is used, since the energy

1. Use value: price of the reference fuel, corrected by the greater or lesser cost of producing the alternative fuel in view of its properties or its impact on CO_2 emissions.

efficiency of the manufacturing process is poor, in the region of 60% compared with about 90% for the conventional processes.

In addition, these products are characterised by high production cost, in case of GTL, or very high distribution costs, in case of DME, whose vapour pressure is equivalent to that of Liquefied Petroleum Gas (LPG).

C. Liquefied petroleum gas (LPG)

Use of LPG as fuel was first developed in order to reduce urban pollution. Then, as its advantage compared with conventional fuels declined, its advocates claimed an advantage in terms of CO_2 emissions. This is not at all the case. There is an advantage if we only consider the CO_2 emitted by the vehicle, at least compared with gasoline but, even in this case, the balance is at best break-even.

LPG is in fact a by-product of oil refining and processing of natural gas. It is a mixture of C3 and C4 hydrocarbons, containing more or less olefins. Since the substitutions between gasoline and LPG do not affect the energy efficiency of the vehicle or that of the petrochemical facility, overall, the energy balance and the CO_2 balance are unchanged.

In view of the precautions required during its use and the resulting additional costs in the logistics chain and the vehicles, fuel-LPG can be seen less and less as a competitive solution. Reduced to the distance travelled, its actual cost price for the community, i.e. excluding taxes and subsidies, is about the same, or even greater, than that of biofuels, and time is on their side in view of the room for progress with the biofuels pathway, their favourable impact on the CO_2 balance and the steadily rising cost of oil.

In conclusion, none of the alternative fuels of fossil origin represents a satisfactory solution to replace the conventional fuels, in view of all the comparison criteria to be taken into account.

1.4.2.3 BTL *(Biomass-to-Liquids)*

This is the counterpart of GTL, the only difference being that the primary energy consists of biomass instead of natural gas (see Chapter 5).

This process offers numerous advantages and could potentially even be the ideal solution. It combines the advantages of each of the other alternative processes without their disadvantages:

- favourable CO_2 balance, like the other biofuel processes, and unlike the fossil origin fuel processes,
- production of a “Fischer-Tropsch” Diesel type fuel matching the requirements of the European market, like the Fischer-Tropsch processes based on gas or coal,
- high production potential, like the processes based on gas or coal, unlike the VOME process,
- lastly, a sustainable production potential like the other renewable energies, unlike the processes based on fossil energies.

The main handicaps of the BTL process are due to its low overall energy efficiency, significantly less than that of the GTL process, and an economic balance penalised by the cost of collecting and storing the raw material. If this process is to be developed, its performance in the above areas will probably have to be improved.

1.5 THE VISION OF THE CAR MANUFACTURERS

All manufacturers are concerned about the "sustainable development" of the car. Their reflection on the concept of "sustainable mobility" – mobility is an essential part of economic and social development –, which is not limited to simply taking into account the progress made in technical fields, aims at defining new stakes and commitments designed in particular to improve the environmental performance of automotive transport.

This improvement should lead to control of polluting emissions and reduction of greenhouse gas emissions, including CO_2.

1.5.1 Control of polluting emissions

The main pollutants emitted by road transport, qualified as primary pollutants, are carbon monoxide CO, nitric oxide NO, Volatile Organic Compounds VOC (mainly unburnt hydrocarbons present in exhaust gases, also known as HC), sulphur dioxide SO_2 and particulate matter PM. These emissions are also generated by sources other than transport, for example, industry, residential activities, agriculture and forestry.

Some of these compounds react in the environment to produce other "secondary" pollutants. This is the case with NO for example, which will change into nitrogen dioxide (NO_2) and the VOCs and NO_x which, depending on the photochemical activity and weather conditions, will cause the formation or destruction of tropospheric ozone O_3.

1.5.1.1 European regulations

The European Commission has defined depollution regulations (or standards) related to road transport, applied in France since1970.

These standards define values in grams of pollutant generated per kilometre travelled for the various private vehicles and in g/kWh for heavy goods vehicles. They have been designated as "Euro" since 1992 when Euro 1 came into effect, followed by Euro 2 in 1997, Euro 3 in 2000 and Euro 4 in 2005. Euro 5 became effective for all new vehicles in January 2011. Euro 6 should be specified in 2015.

The effect of these regulations is shown on Figure 1.8 as regards HC + NO_x and on Figure 1.9 for Diesel particulates. Application of these regulations must reduce HC + NO_x emissions by about 97% between 1971 and 2005 and particulate emissions by at least 90% during the same period, for all new vehicles sold from 2005.

With Euro 5, in 2011, HC + NO_x emissions will be reduced to 0.16 g/km for gasoline vehicles and to 0.23 g/km for Diesel vehicles. The figure for particulates emitted by Diesel vehicles will be 0.005 g/km. NO_x emissions will be reduced even further with Euro 6 in 2015, down to 80 mg/km.

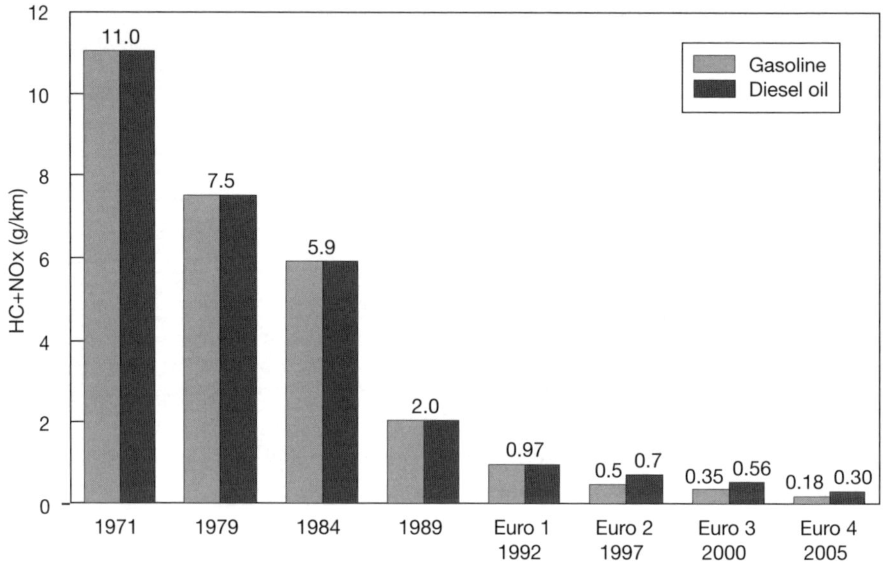

Figure 1.8

Emission regulations for private vehicles in Europe (HC + NO_x).
Source: CCFA (French Automobile Manufacturers Committee)

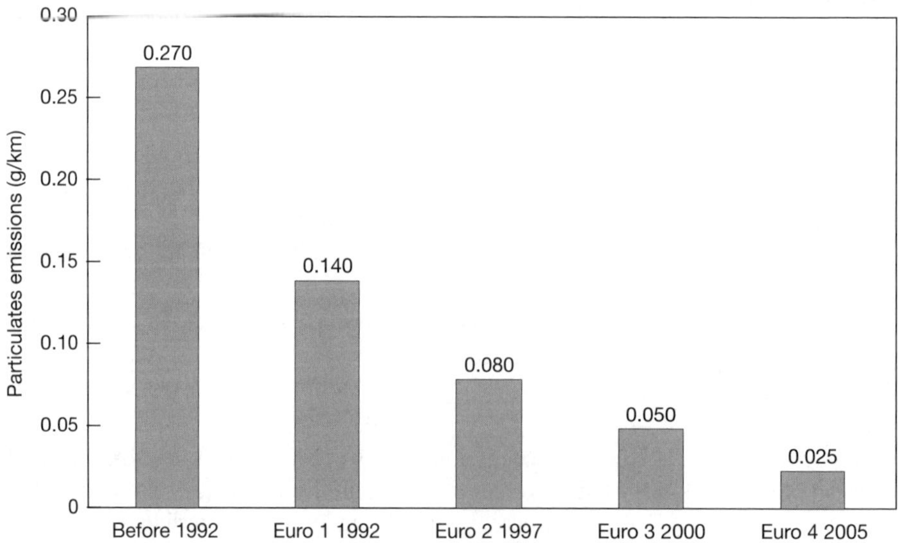

Figure 1.9

Emission regulations for Diesel vehicles (particulates).
Source: CCFA

1.5.1.2 Progress made

A study conducted by the CITEPA (Interprofessional Technical Centre for Studies on Atmospheric Pollution) on behalf of the MEDD (Ministry of Ecology and Sustainable Development) shows that between 1990 and 2008, despite higher traffic, pollutant emissions due to road transport dropped.

Examples of reductions measured:

- 29% for PM10 particulates (diameter $\geq 10\ \mu m$) and 30% for PM2.5 particulates (diameter $\geq 2.5\ \mu m$),
- 86% for CO,
- 97% for SO_2,
- 86% for VOC,
- 41% for NO_X

These results reflect the efficiency of the new technologies implemented by car manufacturers in order to meet the imposed limits. The low rate at which the vehicle fleet is renewed is the main obstacle to reducing several types of pollution.

1.5.1.3 Expected progress

Progress is expected, especially regarding the reduction of NO_X and particulates.

With the reduction in sulphur contents in the new fuel formulations (50 ppm from 2005 then 10 ppm in 2009), the efficiency and lifetime of all post-treatments will be improved. Euro 5 imposes a requirement that pollution control devices can last for a distance of 160 000 km.

European manufacturers continue developing new particulate filters to limit emissions of particulates of all sizes, including the smallest, which are less than 1 µm.

These reductions in polluting emissions, in particular on Diesel vehicles, will nevertheless lead to higher energy consumption and consequently CO_2 emissions, which must be taken into account in the environmental balances.

With the application of Euro 5 in 2011, emissions expressed in tenths of a gram/km for gaseous pollutants and in hundredths of a gram/km for particulates will have little or no negative impact on air quality which is improving to comply with public health requirements.

As a long-term perspective, the CITEPA proposed a scenario describing the trend in pollutant emissions by road transport in France, until 2020, year 1990 being chosen as reference [28].

Figure 1.10 summarises the trends. It shows that by applying Euro 4 then Euro 5 standards progressively over time to all vehicles, emissions by road transport continue to decrease, resulting in 97% to 74% reductions depending on the pollutant, despite a traffic increase estimated at about 2% per year between 1990 and 2020.

Comparable results are obtained in a study conducted for all OECD countries [29].

Assuming that all private vehicles respect Euro 5 at the start of 2011, emissions should be reduced by 53% for NO_X, 85% for particulates, 77% for CO and 81% for VOC compared with 2007.

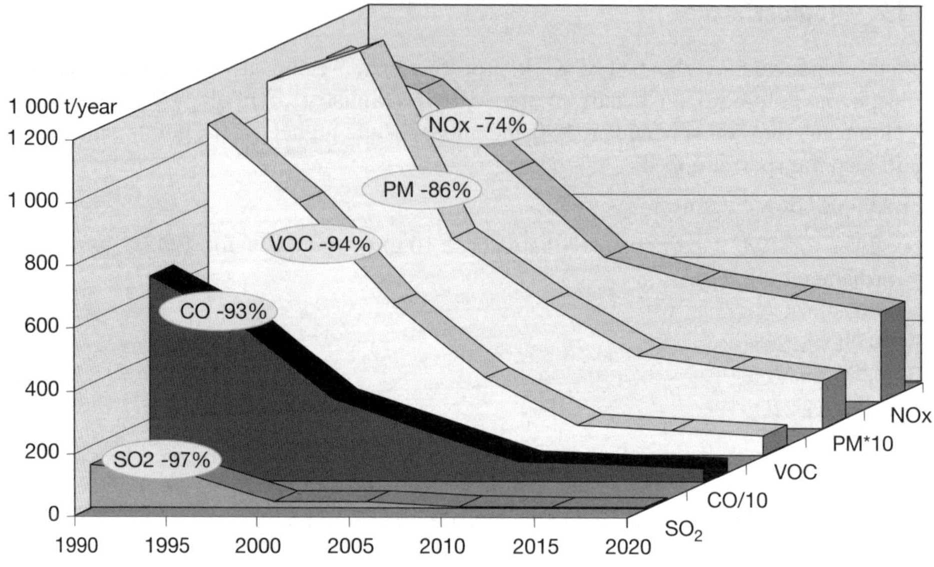

Figure 1.10

Road transport emissions. History and 2020 perspective.
Source: CCFA (ref. CITEPA)

1.5.1.4 Interest of biofuels

A. Gasoline engine

Since ethanol contains oxygen, adding this alcohol to gasoline has a positive but low impact on HC and CO emissions.

Ethanol is moderately volatile with a boiling point of 78°C, higher than the initial distillation point of the gasoline cut. In contrast, the ability of ethanol to form azeotropes with some light hydrocarbons (Table 1.5) has a major impact on the vapour pressure. This results in greater hydrocarbon emissions by evaporation, leading to an increase in the formation of

Table 1.5 Ethanol-light hydrocarbon azeotropes.

Hydrocarbon	Boiling point (°C)	Azeotrope composition (% weight of ethanol)	Azeotrope boiling point (°C)
n-pentane	36	5	34
n-hexane	69	21	59
Benzene	80	32	68
Cyclohexane	81	29.2	65
Toluene	111	68	77
n-octane	126	88	77

Source: CRC Handbook of Chemistry and Physics, 74th Edition

ozone. If 5% ethanol is mixed with gasoline, the quantities of compounds evaporated may increase by up to 50%. Being less volatile, if ETBE is added instead, this main disadvantage is avoided.

B. Diesel engine

Absence of sulphur and aromatic compounds in biodiesel has a positive substitution effect. Adding this biofuel to Diesel also has a positive impact on the local environment by reducing HC and particulate emissions, but a negative impact on NO_x. However, European specification EN 590: 2009, which imposes a maximum limit of 7% VOME added to Diesel for general distribution, has little effect on these pollutants.

1.5.2 Reduction of CO_2 emissions

The other major concern of car manufacturers committed to a "sustainable development" approach is related to the reduction of greenhouse gas emissions. Reduction of CO_2 emissions represents the major priority for the years to come.

The transport sector (road, air, sea, including agricultural tractors and worksite plant) is currently responsible for about one third of greenhouse gas emissions, a figure which is tending to rise. This percentage amounts to 34% for EU27. In France, since a large proportion of the electricity is produced by nuclear power stations and, to a lesser extent, hydroelectricity, with no CO_2 discharge, this values reaches 39%.

1.5.2.1 Progress made

CO_2 emissions depend firstly on the technologies used and the progress made by car manufacturers in terms of engines, and secondly on how the vehicles are used (driving style, speed, etc.).

A study conducted by the European Automobile Manufacturers Association (source: CCFA) shows the reduction of CO_2 emissions for the period between 1994 and 2009 (Figure 1.11), calculated in grams of CO_2 produced per kilometre travelled. This trend is in line with European regulation 443/2009 which sets the average CO_2 emissions for new passenger cars at 130 g/km by 2015.

In France, while CO_2 emissions from gasoline engines dropped from 188 g/km in 1995 to 157 g/km in 2008, i.e. a gain of 16.5%, the average figure for all types of engine dropped from 186 g/km to 154 g/km over the same period, i.e. a gain of 17.5%. This slight increase in the gain can be attributed to the more widespread use of Diesel engines, which discharge less CO_2, at the expense of gasoline engines. Since 2005, however, the differences in CO_2 emissions between Diesel and gasoline vehicles have dropped considerably.

In 2003, Diesel cars represented 67% of new registrations and the average CO_2 emission of new cars sold in France by PSA and Renault was 148.5 g/km. Deployment of the Diesel offer, with vehicles emitting less than 120 g CO_2/km, will help reduce CO_2 emissions even further. However, dieselisation of the automobile fleet could be held back by rising vehicle costs, due to the application of new environmental regulations. These new regulations could also have a negative affect on Diesel consumptions.

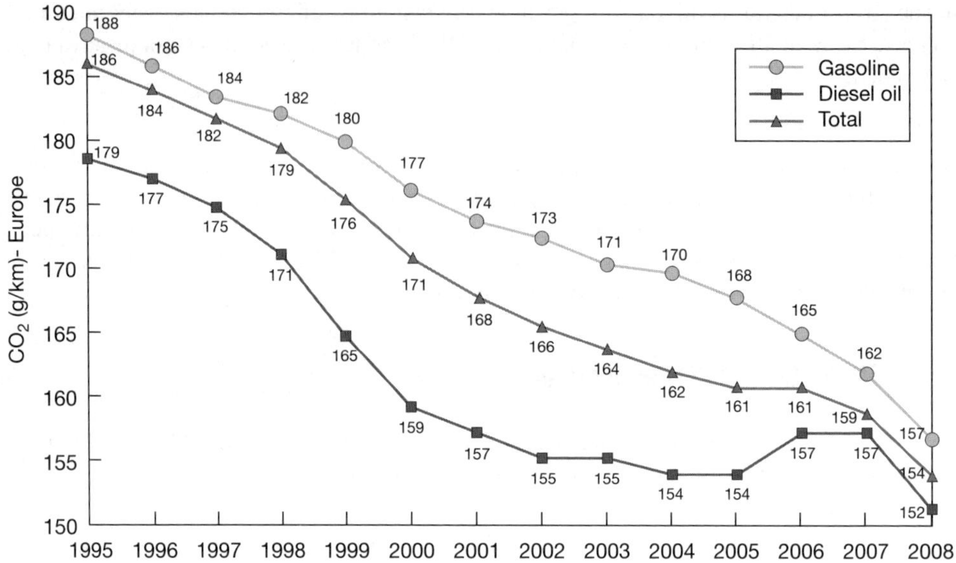

Figure 1.11

Reduction of CO_2 emissions (g/km travelled).

In France, nearly one vehicle out of two sold in 2009 emits less than 120 g CO_2/km (CCFA data).

1.5.2.2 Expected progress

The ACEA announced that in Europe, average CO_2 emissions from new vehicles registered in 2008 were less than 140 g/km. With the aim of reaching the CO_2 target set by the above-mentioned European regulation, European car manufacturers are developing new engine technologies offering, amongst other things, better energy efficiency, and consequently lower CO_2 emissions. These approaches adopted to improve the greenhouse gas balance must not penalise, quite the contrary, the simultaneous breakthroughs made on pollutant emissions.

Several options are considered:

- downsizing: reducing the swept volume of gasoline engines and friction losses would reduce CO_2 emissions by 20%. Combined with suitable supercharging, this decrease could reach 25% which, in terms of energy consumption and CO_2 emissions, would bring the performance of gasoline engines in line with that of Diesel engines. Note, however, that Diesel engines can also benefit from downsizing,
- stop and start vehicles: the engine stops automatically when the vehicle is stationary (traffic lights, traffic jam, etc.), restarting when the accelerator is pressed to drive on. Fuel savings are estimated at 10% in urban cycle and 5% in mixed cycle (road/urban),

– hybridisation of gasoline and Diesel-electric engines: in case of gasoline-electric hybridisation, for example, from starting up to a speed of 50 km/h the vehicle is driven by the electric motor alone. Above this speed, the thermal engine takes over. Consumption may drop down to 3.9 litres of gasoline per 100 km (standardised European cycle), corresponding to CO_2 emissions of 89 g/km. With the Diesel-electric engine hybridisation, CO_2 emissions would also be slightly less than 90 g/km,
– new Diesel HCCI (Homogeneous Charge Compression Ignition) processes.

Figure 1.12 shows the estimated CO_2 gains produced by implementing each of these new technologies.

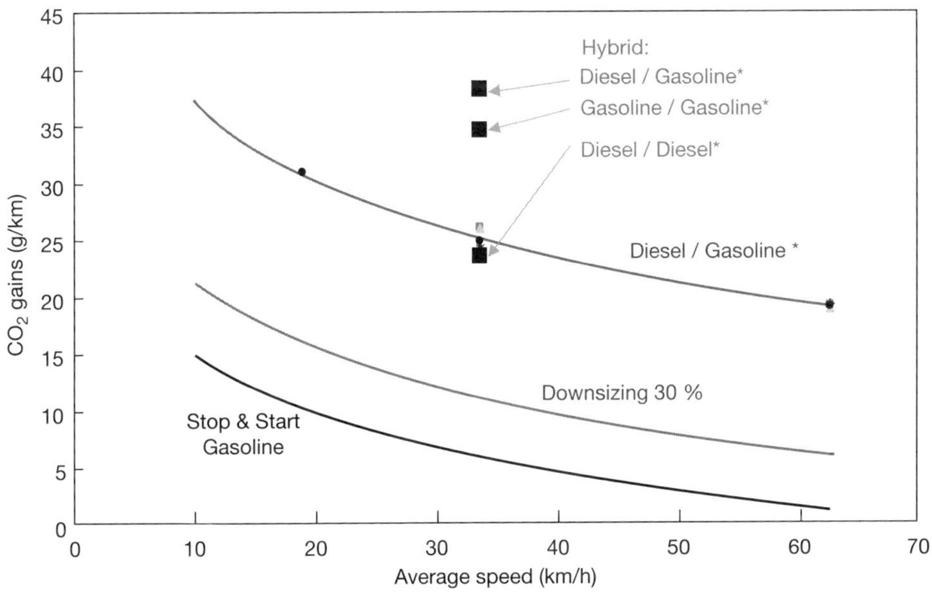

Figure 1.12

CO_2 gains: Comparison of technologies.
* Reference vehicle, same power.
Source: CCFA

1.5.2.3 Interest of biofuels

Use of other fuels, such as natural gas, hydrogen and the biomass fuels, represents another way for the transport sector to improve the greenhouse gas emission balance.

Even though the use of natural gas offers good performance in terms of CO_2 emissions, roughly equivalent to that obtained with Diesel engines, it seems highly likely that the transport sector will continue to privilege liquid fuels. The difficulty of storing gas on board vehicles increases in the order: LPG, VNG, hydrogen.

Hydrogen could replace some of the fossil origin fuels, but not before 2030, a date agreed by most experts.

A mixture of 80% natural gas and 20% hydrogen is nevertheless being evaluated on captive fleets in the United States and in France as well as in the northern European countries. Its use leads to lower pollutant and CO_2 emissions.

Biofuels already represent a promising alternative which, although suffering from limited availability, can be blended with fossil origin fuels without the need for specially adapted engines or specific distribution networks.

Their qualities, faults and performance are detailed in Chapters 2 and 3.

Vegetable oil esters blend perfectly with Diesel, like the hydrocarbons which will be obtained via the BTL process (Biomass-to-Liquids, see Chapter 5), which is currently developing.

Although the maximum legal rate of incorporation of esters in Diesel is currently set to 7%, validations on captive vehicle fleets have demonstrated that values of up to 30% could be reached.

CO_2 emissions for vegetable oil methyl esters are about 70% less than those produced with Diesel (Chapter 3). With the BTL process, gains of about 90% have been estimated (Chapter 5). A 90% gain corresponds to a tenfold drop in emissions.

Ethanol can be used in engines, before or after conversion into ETBE, blended in various proportions in gasoline, with no major difficulties to adapt the existing systems. This has been demonstrated with the large-scale implementations of these biofuels in Brazil and the United States, and on a smaller scale in Europe. If ethanol is added in large proportions, however, it is important to check that the engines and injection systems are correctly adjusted and that the plastic parts and seals are compatible with the new fuel formulation. This has also been demonstrated by the explosion of the FFV (Fuel Flexible Vehicles) market in Brazil which is the only one which currently exists on this country's market.

Gains in CO_2 emissions, calculated with respect to those produced by gasolines, are in the region of 50% for ethanol manufactured from wheat or beet, while reaching a maximum of about 25% for ETBE. For ethanol produced from sugar cane, the gains are in the region of 90% (Chapter 2). With the process for production of ethanol from lignocellulosic biomass, currently under development, gains of at least 75% would be obtained (Chapter 6).

In 2008, use of biofuels in France can be credited with about 6.4 Mt of CO_2 avoided out of a total of 126 Mt emitted by transportation. With the application of the new European directives laying down a biofuel incorporation rate of 5.75%, between 7 and 8 Mt of CO_2 will be avoided annually.

1.5.3 Future trends in fuels

1.5.3.1 Characteristics of conventional fuels

It is a well-known fact that fuel quality has a direct impact on the quantities and types of pollutant emitted by engines. Fuels must also comply with a range of specifications designed to guarantee minimum quality to users and participate efficiently in correct engine operation.

Before the 1990s, the specified fuel criteria were designed to guarantee safe distribution, good storage stability and engine operation with no major combustion and clogging problems. Then, during the 1990s, the Euro 1 and Euro 2 standards laid down first specifications, but the key step in the changes made to these specifications, which is worth mentioning, concerns the European Auto-Oil program which brought together manufacturers, refiners and legislators to set up the Euro 3 standards, defining specifications allowing optimum operation of the associated engine technologies.

1.5.3.2 Gasolines

With the Euro 1 anti-pollution law, spark ignition engines were obliged to fit 3-way catalysts, which were totally incompatible with some of the compounds contained in gasolines, lead being one of the most well-known.

In addition, sulphur reduces catalyst activity and increases light-off temperatures (temperature at which the catalyst reaches 50% of its activity). In this context, the lead content specification was reduced to a very low value near zero (0.013 g/l) and the sulphur content was limited to 500 mg/kg in 1992 to allow optimum operation of the post-treatment systems installed. This is a major turning point since the fuel specifications now introduce limits affecting pollutant emissions via the post-treatment system. Reduction in sulphur contents also has a direct impact on pollution by decreasing SO_2 emissions. The Euro 2 regulation, enforced in 1996, had no major consequences on the specifications.

As pointed out in the preamble, the objective of the European Auto-Oil programme was to determine Euro 3 anti-pollution standards and implement fuel specifications ensuring optimum behaviour of the technologies required to comply with these standards. In addition, and this is once again a new approach, the specifications were also drawn up to have a direct impact on emissions, both from the exhaust and as regards emission of vapours, to limit and better control the discharges of toxic compounds and gases affecting the formation of tropospheric ozone. In 2005, the EN228 specifications introduced:

- maximum sulphur content of 10 mg/kg,
- benzene content reduced to 1% by volume,
- maximum olefinic compound content of 18% by volume,
- maximum aromatic compound content of 35% by volume,
- almost complete elimination of lead, with content below 0.005 g/kg.

In addition, new constraints on the vapour pressures were introduced to obtain better control of discharges by evaporation. This point is extremely important, especially when incorporating certain alcohols.

Table 1.6 shows the respective impacts of the various characteristics specified.

What could the next regulations be? Proposing another reduction on the olefinic compounds could be a possibility, further reducing the aromatics could appear to be too restricting to maintain the specific octane numbers on the entire gasoline pool.

Table 1.6 Direct or indirect impact of the various characteristics specified on polluting emissions.

Characteristics specified	Impact on		
	emissions at the exhaust	emissions by evaporation	operation of the post-treatment system
Sulphur content			+++
Benzene content	++	++	
Aromatic content	+++		+
Olefin content	++	+++	
Lead content			+++
Vapour pressure		+++	

1.5.3.3 Diesels

Diesel specifications have undergone continuous changes since the end of the 1980s with modification of the cold characteristics, re-evaluation of the cetane number from 46 to 49, and a first limitation of the sulphur content to 2 000 mg/kg in 1994. A new trend has been associated with the Euro 2 regulation, applied in 1996. Manufacturers responded to this law by massively introducing oxidation catalysis on Diesel vehicles. As already mentioned, catalyst activity is highly sensitive to sulphur. In addition, and this is a very important phenomenon, the possible conversion of SO_2 emitted during combustion into SO_3 on the oxidation catalyst from temperatures of 300-350°C generates sulphuric acid which forms sulphates on the particulates. In this case, the sulphur in Diesel considerably increases the weights of the particulates emitted. To control this phenomenon, in 1996 the sulphur content in Diesel fuels was therefore reduced to 500 mg/kg.

During the 1990s, France stood out from the rest of Europe by opting for the development of VOME via widespread incorporation in Diesel at contents less than 5% by volume. In 1997, the French public authorities introduced this possibility in specification EN590.

The fallouts of the European Auto-Oil program for diesels, via the Euro 3 anti-pollution regulation, led to new limits on polluting discharges, in particular particulate emissions. Fuel quality must not only ensure optimum operation of the post-treatment system but also help reduce emissions directly. In 2009 therefore, EN590 specifications laid down:

- sulphur content reduced to 10 mg/kg,
- minimum cetane number of 51 (previously 49). The higher cetane number reduces particulate emissions,
- upper limit of the density 845 kg/m^3 at 15°C,
- new definition of the distillation curve designed to limit the heavy fractions,
- new regulation on the content in polyaromatic compounds, limiting them to less than 11% by weight.

1.5.3.4 Trend in fuel characteristics

The trend in fuel characteristics is closely related to the trend in technologies developed to comply with anti-pollution standards. In this context, the main manufacturer associations: ACEA, JAMA, ALLIANCE and EMA drew up the "World Wide Fuel Charter" which defines the minimum qualities of the main fuel categories to ensure optimum operation of the technologies developed to meet the main statutory constraints. This charter identifies 4 fuel classes, the fourth (Cat 4 WWFC) corresponding to the requirements of the most advanced vehicles. Tables 1.7 and 1.8 shows the trend in the specifications compared with those taken into account by this fourth fuel category. We observe a rapid increase in the

Table 1.7 Trend in the main gasoline specifications and request by the manufacturer associations.

	Year 2000	Year 2005	Cat 4 WWFC
Sulphur content (mg/kg)	150	50 to reach 10 in 2009	5-10
Aromatic content (% vol.)	< 42	< 35	< 35
Olefin content (% vol.)	< 18	< 18	< 10
Benzene content (% vol.)	< 1	< 1	< 1
Lead content (mg/l)	5	Id	ND
Vapour pressure (kPa)	45-90	45-90	
RON/MON [a b]	95/85	95/85	95/85
Density (kg/m^3)	720-775		715-770

[a] RON: Research Octane Number, characterises the resistance to auto-ignition of gasolines, is measured on CFR engines under conditions simulating low engine speed and low load.
[b] MON: Motor Octane Number, measured at high load.
These two octane numbers are evaluated by comparison with reference fuels: value 100 with isooctane, value 0 with n-heptane.

Table 1.8 Trend in the main Diesel specifications and request by the manufacturer associations.

	Year 2000	Year 2005	Cat 4 WWFC
Sulphur content (mg/kg)	350	50 to reach 10 in 2009	5-10
Cetane number measured [a]	> 51	> 51	> 55
Density (kg/m^3)	820-845	820-845	820-840
Cetane number calculated [b]	> 46	> 46	> 52
Polyaromatic hydrocarbon content (% by weight)	> 11	< 11	< 2
Total aromatic content (% weight)	–	–	< 15

[a] Cetane number measured: on standardised CFR engine, with prechamber; does not take into account fuel additivation; evaluated by comparison with two reference hydrocarbons: value 100 with n-hexadecane, value 0 with α-methylnaphthalene.
[b] Cetane number calculated: evaluated by correlation and based on a formula including the fuel distillation characteristics and density.

severity of the characteristics having a direct impact on polluting emissions. In these proposals, the olefin content in gasolines should not exceed 10% in order to control the reactivity of the discharges with respect to the formation of ozone, while for diesels, several measures were suggested to reduce particulate emissions, e.g. higher cetane number, total aromatic content kept below 15% by weight, polyaromatic hydrocarbon concentration less than 2% by weight. To comply with this last constraint, refiners could reduce the final distillation point of the Diesel cut, which would make it easier to desulphurise. This operation would nevertheless result in lower Diesel production, probably leading to a shortage, in a context where Diesel demand is tending to rise, and tension on prices.

1.6 BIOFUELS IN THE WORLD

1.6.1 North America

For several years, the United States has been the leading world producer of biofuels, mainly ethanol and, to a lesser extent, biodiesel.

The United States is divided into 5 PAD (Petroleum Administration for Defence) districts. More than 90% of ethanol added to American gasolines is produced by the "Midwest" district, PAD II.

Splash blending ethanol into gasoline makes the fuel more volatile, resulting in higher emissions of volatile organic compounds from the vehicle (see Chapter 1.5).

This effect is noticeable from 0 to 5% incorporation, then stabilises at higher rates: for 3% blends, the increase in volatility expressed as vapour pressure is about 60 to 80 mbar. In the United States, this effect is taken into account since gasolines containing up to 10% ethanol are granted a waiver allowing a vapour pressure of more than 1 psi (68 mbar).

Without this authorisation, a gasoline base of volatility reduced by the same order of magnitude would have to be formulated, which would mean removing the lighter components from the gasoline pool, especially cuts C_4 (butanes) and C_5 (pentanes). All things being equal otherwise, one of the consequences would be a reduction in gasoline availability. With the ban on MTBE in California, for example, replacing it by ethanol and removing the light cuts C_4 and C_5 would result in a fuel shortfall of 102 000 bbl/day.

The possibility of incorporating 85% ethanol (E85) in gasolines is strongly considered, especially since a law was voted by both chambers in October 2004, extending the Volumetric Ethanol Excise Tax Credit (VEETC) to blend E85.

Ethanol is almost exclusively manufactured from corn, using two types of process (see Chapter 2.2): wet milling and dry milling, the first steadily being replaced by the second.

Calculated over the month of June 2004, ethanol consumption amounted to 222 000 barrels/day (35 300 m^3/day), compared with that of gasoline which totalled 9.4 M barrels/day (about 1.5 Mm^3/day, 540 Mm^3/year), i.e. a substitution percentage of 2.4% by volume.

In 2005, 4 billion gallons of ethanol were produced from 13% of the American corn harvest, i.e. 2.7% of the national gasoline consumption. Apart from ethanol, 11 Mt of cattle feed and 250 000 tonnes of oil were also produced from corn. According to the USDA, the United States does not intend to devote more than 30% to 35% of the national corn to produce ethanol. Production of first generation fuels (ethanol and biodiesel) is estimated to peak at 15 billion gallons per year. The energy independence and security act of 2007 promotes the development of second generation biofuels, which are designed to cut greenhouse gas emissions by at least 50%.

The United States aims to reduce gasoline consumption by 20% by 2020 and 30% by 2030, through the use of first and second generation biofuels and by reducing vehicle fuel consumption.

In Canada, production and use of ethanol-fuel are developing, but most imports still come from the United States.

1.6.2 Latin and Central America

Brazil is the world's second largest producer of biofuels, mainly ethanol.

In addition to its forest wealth, which could represent a considerable source of bioenergy for the future, Brazil has extensive areas available for cultivation, with 90 Mha of agricultural land.

Amongst the producers of agricultural raw materials, Brazil is the first world producer of sugar (16%), the second of soya (26%) and the third of corn (7%).

Until the early 1990s, sugar cane was grown mainly in the north/north east regions. Since then, the cultivated areas have shifted towards to the centre-south regions around Sao Paulo, which currently supplies most of the sugar cane harvests.

The following data illustrate the continuing competitiveness of this crop. The annual production of sugar cane has increased steadily over the past 20 years, obviously due to the land area devoted to the crop but also to the higher average yields, which have increased from 5.5 to 8 tonnes of sugar per hectare. Realistic objectives of increasing the quantities of sugar cane harvested per hectare indicate an average of 88 t/ha for 2010, compared with 85 t/ha in 2000, with maximum values of 115 t/ha.

The sugar contents of the cane should remain in the region of 14.5%, although higher values of up to 15.8% have already been obtained. These values are measures on the raw materials entering the factory, which only account for two thirds of the plant production per hectare, most of the remaining third being left on the field, where it is burnt.

The figures shown in the following insert provide a better picture of the energy potential for this crop.

The main reasons for the development of ethanol-fuel in Brazil have already been discussed. Maintaining employment in rural areas is another.

The manufacture of ethanol has increased steadily since the mid-1970s with the launch of the Proalcool programme. Although at this time only 15% of the sugar was converted into

Energy potential of sugar cane biomass in Brazil

One tonne of sugar cane can be used to produce:

– 150 kg sugar

or	85 l (67.3 kg) ethanol	i.e. 0.042 toe
	+280 kg bagasse with 50% dry matter	i.e. 0.056 toe
	+280 kg lignocellulosic residues	
with 50% dry matter		i.e. 0.056 toe

Overall, one tonne of sugar cane (0.154 toe) has approximately the same energy equivalent as a barrel of oil (159 l).

The harvest obtained from one hectare of sugar cane is theoretically equivalent to 85 barrels of oil, provided that all parts of the plant can be upgraded, which is not currently the case. Although bagasse is already used to supply heat and electricity, both for sugar refineries and distilleries and for external uses, the same is not true of other lignocellulosic residues which are mostly burnt on the field. 50 to 60% of these lignocellulosic residues could be recovered and upgraded.

The theoretical energy potential of cane cultivation could represent half the Brazilian oil consumption.

ethanol, ten years later the figure had risen to 70% while, over the same period, the global production of sugar rose regularly.

This percentage remained stable until the mid-1990s, before declining to a current figure of about 50%.

In the 1980s, cars ran on hydrated ethanol (92%). This was largely due to state grants which subsidised the purchase of vehicles running on pure alcohol.

In the 1990s, it became more profitable to export sugar than convert it into ethanol and, as a result, the subsidies provided under the Proalcool programme were gradually withdrawn. The Brazilian government nevertheless maintained the obligation to incorporate between 20 to 25% of 99.7% pure anhydrous ethanol in gasoline.

No cars run on petrol only. New tax incentives have led to the development of the new Fuel Flexible Vehicles (FFV) which can run on ethanol blended with gasoline in any proportions. This new type currently represents most of the automotive market.

Apart from the fact that it offers a means of regulating the price of sugar (Brazil is the world's leading sugar producer and exporter), the development of alcohol consumption, as replacement for petroleum products, has generated currency savings.

This industry plays a significant role in maintaining employment in rural areas.

Lastly, the environmental impact is far from being negligible if we consider the rate of substitution of biofuels in gasolines and Diesel in the near future. These 100% renewable biofuels improve the greenhouse gas emission balance. In addition, this balance can only improve when, especially in the case of sugar cane, all parts of the plant are recovered for energy purposes.

Oilseeds are grown in the five Brazil's main agricultural regions, at least as regards soya. The other plants are rapeseed, sunflower, cotton, castor and babassu, the latter two being grown mainly in the arid north eastern region.

Encouraging the production of these two plants is one way of promoting the economic development of this region and maintaining its rural population.

The Brazilian harvest of soya seeds has risen dramatically over the last few years.

Biodiesel production is still in its infancy. Most synthesised esters are ethyl esters. As we have seen above, this is largely due to the availability of ethyl alcohol at competitive costs, whereas methanol is mainly imported.

In addition, compared with methyl esters, ethyl esters offer the advantage of being 100% vegetable origin and renewable.

A Probiodiesel programme was launched in 2002. It groups the main players of the sector, including the producers of the agricultural resource, research organisations, the national oil company Petrobras, car manufacturers and the Ministry of Science and Technology.

The law on mandatory incorporation of 2% biofuel in Diesel came into effect in January 2008, the content being raised to 3% later that year. The 2013 forecasts target 5% addition of biodiesel.

Many other countries in this region, mainly Argentina which implements a large-scale development programme on the production of soya and its conversion into biodiesel, are involved in the field of biofuels, primarily ethanol.

1.6.3 Europe

Energy and transportation are the foremost concerns of the European countries, which intend to propose measures guaranteeing clean energy, in sufficient quantities and at a reasonable cost, for their populations. Special attention is paid to road transport, a sector still expanding strongly, in a global document on the mobility of persons and goods [30].

In 2001, the Commission proposed an action plan to promote biofuels and substitution fuels in road transport.

In 2003, the European Council and Parliament validated a proposal of directives made in 2001 as part of the action plan to promote substitution fuels [31]. Directive 2003/30/CE lays down increasing objectives on the consumption of biofuels in transportation. Consumptions of biofuels should represent at least 2% in 2005 and 5.75% on 2010 (percentages calculated on the basis of energy) of global consumptions of gasoline and Diesel used in transportation. These objectives are given as guidelines and are not mandatory. The Member States must nevertheless inform the Commission of the measures taken to reach them.

In 2003, directive 2003/96/CE on taxation was also validated, in which Member States may apply total or partial exemptions on biofuel excise duty.

These two texts complete the earlier directive 98/70/CE on fuel quality, which already legalised the incorporation of up to 5% ethanol or 15% ETBE in gasoline and up to 5% VOME in Diesel, for regular sale at the pump. Higher rates are perfectly compatible with current engines, but in this case, information must be provided at the pump.

The European Union laid down objectives in terms of reducing greenhouse gas emissions via the renewable energy Directive, voted in December 2008, which defines three objectives for 2020:

- increase energy efficiency by 20%,
- reduce greenhouse gas emissions by 20%,
- use 20% renewable energies in the total energy consumption.

For transportation, this Directive stipulates 10% consumption renewable energies, including biofuels, in 2020. Concerning biofuels, the Directive set incorporation rates of 5.75% in energy content by 2010 and 10% by 2020. These objectives are not mandatory. Each country can set its own objectives but must provide evidence of the progress made in this field. In 2007, the average incorporate rate, in EU27, was 2.6%, with large disparities between the countries, the most advanced being Germany, France and Sweden (*source*: *EurObserv'ER Biofuels Barometer* 2008).

Reaching this objective by 2020 depends on several variables, including the commercialisation of second and third generation biofuels and the use of renewable energy in electric vehicles. Sustainability criteria on the biofuels will also be taken into account. Compared with fossil origin fuels, the reduction of greenhouse gas emissions for biofuels is set to at least 35% by 2010, 50% by 2017 and 60% for new facilities after 2017. Social criteria as well as those related to the consequences of land use changes will then have to be taken into account.

The European Directive on fuel quality, published in 2008, allows a blend of 10% ethanol in gasoline and 7% biodiesel in Diesel from April 2009.

Enforcement of fallow land, since it was being used to produce energy crops, was cancelled in 2009. The subsidies granted for these crops stopped in 2010.

1.6.4 The other biofuel producing regions

In Asia, China is the world's second largest energy consumer and the first automobile producer, with a vehicle fleet that has grown by a factor of 6 in 10 years. Biofuels represent a promising solution to reduce the country's energy dependence, expected to reach 80% by 2020, as well as to reduce greenhouse gas emissions and offer an economic opportunity for rural populations. The 2007 renewable energy development plan sets forth ethanol production objectives of at least 10 Mt (mainly from lignocellulosic substrates) and 2 Mt of biodiesel, by 2020.

India announces high biofuel incorporation rates, close to 20% in 2017. A plan has been set up to support the rural economy, with the development of energy crops in relatively arid regions, especially oilseeds, that do not compete directly with food uses.

Japan is constantly trying to increase the incorporation rate of ethanol or ETBE in gasolines, like Thailand which replaces MTBE by ethanol produced from manioc, corn and molasses.

Malaysia and Indonesia have considerably developed palm tree plantation, to produce oil for food, chemical and energy (biofuel) purposes.

In Australia and New Zealand, it is estimated that the ethanol incorporation rate in gasolines will reach several percent in 2015.

In Africa, the production of biofuels would certainly help develop local economies by creating jobs and bringing energy resources, generally scarce, which would also help develop agricultural activity. This must be carried out while making sure that priority is always given to food crops.

Ethanol production is developing slowly but steadily. Considerable hope is being placed on the cultivation of the most drought and disease resistant crops, such as Jatropha curcas.

REFERENCES

[1] Baudry de Saunier L (1902) L'illustration, n° 3093, pp 103, n° 3094, pp 430-431, n° 3095, pp 151.

[2] Guibet JC (1997) Carburants et moteurs – Technologies – Énergies – Environnement – Editions Technip, Paris – Tome 2, pp 624-627.

[3] Stern R, Guibet JC et Graille J (1983) Les huiles végétales et leurs dérivés. Carburants de substitution (analyse critique). Revue de l'Institut Français du Pétrole, **38**, 1, pp 121-136.

[4] Energy Information Administration/US Department of Energy (2003) International Energy Outlook 2003, DOE/EIA, Washington.

[5] International Energy Agency (2004) World Energy Outlook (2004) OECD/IEA, Paris.

[6] European Commission, Directorate-General for Research Energy (2003) World Energy, Technology and Climate Policy Outlook 2030 – WETO. European Commission, Luxembourg.

[7] World Energy Outlook 2009. www.iea-org/country/

[8] Groupe d'Experts Intergouvernemental sur l'Évolution du Climat – GIEC – (2001) Changement climatique 2001: les bases scientifiques, OMM/PNUE, Shanghai.

[9] Protocole à la convention-cadre sur les changements climatiques adopté le 11 décembre 1997. UNFCCC, Kyoto.

[10] Rivière C and Marlair G (2010) The Use of Multiple Correspondence Analysis and Hierarchical Clustering to Identify Incident Typologies Pertaining to the Biofuel Industry. Biofuels, Bioproducts and Biorefining, **4**, 1, pp 53-65.

[11] Brandes E and Frobese DH (2009) Ethanol-containing Automotive Fuels-a Safety Concept for Petrol Stations in Germany, **73**, 1, pp 25-32.

[12] Marlair G *et al.* (2007) Biofuels for Transport in the 21st century: Why Fire Safety is a Real Issue. Proceedings of the 10th International Conference on Fire and Materials, San-Francisco, CA, États-Unis, 29-31 January 2007.

[13] Report entitled: Response to Ethanol-based Incidents – From Performance Test Results. www.ethanolresponse.com

[14] Acheson B *et al.* (2011). Proceedings of the Sixth Int. Seminar on Fire and Explosion Hazards, Leeds, 11-16 April 2010, Ed. D. Bradley et al., Research Publishing, 596-608, 2011.

[15] Koseki *et al.* (2001) Evaluation of the Burning Characteristics of Vegetable Oils in Comparison with Fuels and Lubricating Oils. Journal of Fire Sciences, **19**, 1, pp 31-44.

[16] Anonyme (2009) Ethanol: Shipping it Through Pipelines. FYI, Energy API.

[17] Marlair G *et al.* (2010) On the Fire and Explosion Risk Profile of Alternative Fuels to Kerosene in Civil Transportation Aircrafts. Congrès Interflam, 5-7 July, Birmingham, UK.

[18] www.caafi.org

[19] Concawe, Eucar, European Commission Joint Research Center (2004) Well-to-Wheels Report. European Commission, Bruxelles.

[20] JRC, Eucar, Concawe (2007) Well-to-Wheel Analysis of Future Automotive Fuels and Powertrains in the European Context.

[21] Prévot H (2003) Energie et changement climatique : diviser par trois en trente ans nos émissions de gaz carbonique. Annales des Mines, Août 2003, pp 21-31.

[22] Searchinger T, Heimlich R, Houghton RA, Dong F, Elobeid A, Fabiosa J, Tokgoz S, Hayes D and Yu T-H (2008) Use of US Cropland for Biofuels Increases Greenhouse Gases Through Emissions from Land-Use Change. Science, 319, pp 1238-1240.

[23] Arrouays D, Balesdent J, Germon J, Joyet P, Soussana J et Stengel P (Ed.) (2002) Stocker du carbone dans les sols agricoles de France ? INRA Expertise Scientifique Collective.

[24] Anderson-Texeira KJ, Davis S.C, Masters MD and Delucia EH (2009) Changes in Soil Organic Carbon under Biofuel Crops. GCB Bioenergy, 1, pp 75-96.

[25] Melillo *et al.* (2009) Indirect Emissions from Biofuels: How Important? Science, 326, pp 1397-1399.

[26] Sourie JC et Tréguer D (2005) Le développement des biocarburants en France : une analyse de la compétition entre colza ester et colza alimentaire à l'horizon 2010. Oléagineux, Corps Gras, Lipides, 12, pp 94-97.

[27] Wilhelm WW, Johnson Jane MF, Karlen DL and Lightle DT (2007) Corn Stover to Sustain Soil Organic Carbon Further Constrains Biomass Supply. Agron J, 99, pp 1665-1667.

[28] CITEPA (2002) Estimation des émissions de polluants du trafic routier en 2010 et 2020. Étude réalisée à la demande du Ministère de l'Équipement, des Transports et du Logement. Mai 2002.

[29] World Business Council for Sustainable Development-Sustainable Mobility (2004) www.wbscd.org

[30] Commission européenne (2001) Livre blanc, La politique européenne des transports à l'horizon 2010 : l'heure des choix. Office des Publications Officielles des Communautés Européennes, Luxembourg.

[31] Commission européenne/Direction Générale de l'Énergie et des Transports (2001) Promotion des biocarburants et autres carburants de substitutions dans les transports routiers: plan d'actions, présentation succincte des nouvelles propositions.

2

First generation biofuels for spark ignition engines: ethanol and ETBE

D. Ballerini, D. Casanave, A. Forestière,
S. Lacombe, X. Montagne, K. Surla

This chapter provides a complete description of first generation biofuels for spark ignition engines, i.e. those that can be substituted for or added to gasoline. It describes the agricultural raw materials required to produce biofuels, the conversion processes, the characteristics of the products and their applications in engines. The environmental and economic aspects are then developed.

2.1 RAW MATERIALS OF VEGETABLE ORIGIN

While ethanol is still produced on industrial scale from ethylene of petroleum origin, mainly in Saudi Arabia, most ethanol is manufactured by fermenting raw materials of vegetable origin.

The raw materials for alcohol production can be divided into three broad categories:

- sugar plants, such as beet, cane and to a lesser extent *sweet sorghum*, which contain sugars that can be directly fermented into ethanol,
- cereal plants such as corn, wheat and barley, and tubers such as potato and manioc, in which sugars are present as a polymer, starch, hence their other name: amylaceous plants. The starch must be hydrolised into sugar monomers before being converted into ethanol. The hydrolysis step, conducted for many years by chemical pathway in the presence of acid, has been conducted by enzyme pathway for at least two decades,
- lignocellulosic plants such as agricultural residues, including cereal straw, or forest residues, dedicated crops such as Short Rotation Coppices (SRC), including poplar, willow and eucalyptus, etc., and miscanthus. Triticale, a type of wheat, may be included in the latter category since its is cultivated for the lignocellulosic part as opposed to the grain. Two of the three main components of these plants are sugar polymers, cellulose and hemicelluloses, which must be hydrolised into monomers before they can be fermented into ethanol. The third, lignin, can be used as a source of energy. The qualitative aspects and availability of this category of raw material will be discussed in Chapter 4, while Chapters 5 and 6 deal with all aspects related to its conversion into liquid biofuels.

One other plant used for alcohol production is worth mentioning: the Jerusalem artichoke, whose tubers contain fructans, fructose and glucose polymers (fructose/glucose ratio = 3). These polymers are also known as inulin. The total sugar content of the Jerusalem artichoke is about 14%. The part of the plant above ground could represent an abundant source of lignocellulosic biomass.

2.1.1 Alcohol-producing potential of sugar plants

Of the three sugar plants taken as examples, sugar cane offers the best alcohol-producing potential. In addition, like sweet sorghum, it generates lignocellulosic biomass, bagasse, which can be used to produce energy (cogeneration) or syngas by thermochemical pathway (Chapter 5) or also ethanol after biological conversion (see Chapter 6).

The sugars obtained by extraction (diffusion process) from sugar beet “cossettes” or by grinding and pressing sugar canes are directly fermented into ethanol.

The main results, listed in Table 2.1, show that between 4 and 8 m^3 of ethanol can be generated by growing these sugar plants on one hectare of land, i.e. between 2 and 4 toe/ha.

Table 2.1 Sugar plant yields and alcohol-producing potentials.

Sugar plant	Agricultural yield t/ha	Plant sugar content (wt%)	Ethanol yield l/t plant	Ethanol yield m^3/ha	Energy yield toe/ha
Sugar beet	70-75	15	92	6.5-7.0	3.3-3.5
Sugar cane [a]	80-90	14-15	85	7.0-8.0	3.5-4
Sweet sorghum[a]	50-60	16	80	4.0-5.0	2-2.5

[a] *Sugar cane and sweet sorghum also produce bagasses which can be used as a source of energy.*

2.1.2 Alcohol-producing potential of amylaceous plants

Of the three cereals mentioned in Table 2.2, corn has the highest starch content and therefore the highest ethanol production potential. Apart from starch, barley contains relatively large quantities of pentosans (about 10%) which, in the current state of knowledge, cannot be fermented into ethanol.

In addition to the cereals mentioned above, we must mention the potato, whose yield is about 4 m^3 of ethanol/ha, and manioc, with 66% starch content, which can be used to produce 370 l of ethanol/t of manioc.

Starch granules consist of a mixture of amylose and amylopectin. Amylose, which represents 20 to 30% of the starch, is a polymer made up of D-glucose units, connected together by α1-4 bonds, whose degree of polymerisation varies between 1 000 and 5 000, and which is mainly composed of linear chains with a few α1-6 branch points.

Table 2.2 Cereal crop yields and alcohol-producing potentials.

Cereal	Agricultural yield t/ha	Starch content of the grain (wt%)	Ethanol yield l/t grain	Ethanol yield m^3/ha	Energy yield toe/ha
Wheat	7.2-8.3	62-65	370	2.7-3.1	1.4-1.6
Corn	7.2-8.5	72	400	2.9-3.4	1.5-1.7
Barley	5.0-7.0	56-59	320	1.6-2.2	0.8-1.2

Amylopectin, the majority compound of starch, has a branched structure with high molecular weights and degrees of polymerisation between 10^4 and 10^6.

These polymers must be hydrolised into glucose, the substrate for ethanol fermentation. Although numerous yeasts exhibit amylolytic activities, unfortunately they are not very good fermenters for the conversion of glucose into ethanol

For many years depolymerisation of starch was carried out by acid hydrolysis at pH = 2 in the presence of hydrochloric acid, at a temperature of between 140°C and 150°C.

This simple and inexpensive reaction nevertheless presents several disadvantages, starting with low glucose yields. Reversion reactions occur almost inevitably in the presence of acid, in fact, leading to the synthesis of oligosides which are difficult or impossible to convert into ethanol. The other disadvantage is financial, with high investment costs for reactors built from corrosion-resistant materials.

With the development of efficient and inexpensive enzymes, enzymatic hydrolysis has been progressively replacing acid hydrolysis for over twenty years and has now become the only starch depolymerisation technique.

It takes place in two successive steps. The first consists in liquefying the starch suspension in aqueous phase heated to 105°C in the presence of a thermostable enzyme, α-amylase, at a pH of between 6 and 7. About 5 mg/l of Ca^{++} ions must be present in the medium to increase the enzyme stability. The liquefaction times are about 1 hour. The starch suspension contains between 20 and 40% of dry matter. α-amylase is produced on industrial scale by bacteria of the *Bacillus* (*B. licheniformis*) species.

After cooling to 60°C, a second enzyme, amyloglucosidase or glucoamylase, is added to hydrolyse the liquefied starch into glucose. The pH is adjusted around 5. Saccharification takes at least 4 hours, under optimum conditions. The amyloglucosidase is produced by fermentation on industrial scale by fungi, such as *Aspergillus niger*.

To prevent the appearance of non-fermentable reversion products, due to the transglycosylation activity of glucoamylase when this enzyme is in the presence of high glucose concentrations, this sugar must be converted into ethanol as rapidly as possible. Ethanol fermentation can therefore be implemented rapidly while cooling from 60°C to 30-35°C (optimum fermentation temperature) or during the second step of enzymatic hydrolysis, carried out in this case not at 60°C but 35°C, allowing the pH to drop from 5-5.5 to 4-4.5, to prevent glucose from building up (simultaneous saccharification and fermentation).

The activities and production of these enzymes have been considerably improved over the last few years. The overall mean consumption of enzymes is between 250 and 300 g per m^3 of ethanol manufactured from corn.

2.2 ETHANOL PRODUCTION PROCESSES

2.2.1 Ethanol fermentation

2.2.1.1 Principle

Ethanol is produced by biological pathway. The sugars are converted into ethanol through the fermentative action of microorganisms, yeasts and bacteria. Chemical synthesis by hydration of ethylene is used less and less frequently.

The fermentative processes involved in the production of alcoholic beverages, for example wine and beer, used by man since the sixth millennium BC, were still considered as "spontaneous" in 1815, when Gay-Lussac determined the stoichiometric equation for the conversion of glucose into ethanol:

$$C_6H_{12}O_6 \longrightarrow 2\,C_2H_5OH + 2\,CO_2$$

The theoretical yield calculated from this equation, also known as the Gay-Lussac yield, indicates that 51.1 kg of ethanol can be manufactured from 100 kg of glucose.

Before 1850, other scientists, including Cagniard-Latour, relate the production of ethanol to the development of microorganisms. In 1857, Pasteur specified and quantified the anaerobic fermentative processes. He demonstrated that co-products are always associated with the synthesis of ethanol and carbon dioxide. About 3% of the sugars are converted into glycerol, 0.5% into succinic acid, 0.5-0.8% into fusel oils (a mixture of amyl and propyl alcohols and their isomers) and 0.8-1% are consumed to allow the fermentative microorganisms to develop.

Taking these sugar losses into account, the maximum yield, or Pasteur yield, corresponds to the production of 48.4 g of ethanol from 100 g of glucose, i.e. 94.7% of the theoretical yield.

On industrial scale, the ethanol production yield is generally between 90% and 92% of the theoretical yield.

2.2.1.2 Microorganisms

While numerous species of microorganisms are known for their ability to ferment sugars into ethanol, yeasts belonging to genera *Saccharomyces* (such as *S. cerevisiae*, *S. carlsbergensis*, *S. bayanus*, *S. uvarum*), *Schizosaccharomyces* (the most active definitely being *S. pombe* which can ferment lactose directly, unlike *S. cerevisiae*), and *Kluyveromyces* (such as *K. fragilis*), appear to have the best alcohol-producing potential.

Saccharomyces strains offer several advantages:

– they are tolerant to high ethanol concentrations, up to 10-12% by volume,

- they are not inhibited by high sugar contents,
- they can withstand high osmotic pressures,
- they have demonstrated their stability at industrial scale,
- they can develop and ferment sugars at acid pH values of 3-4, conditions which repress the appearance and development of bacterial contaminants, the main ones belonging to the genus *Lactobacillus*,
- due to their size, they are easy to separate from the culture medium, by centrifugation or filtration, before being recycled in fermentation.

Some strains of amylolytic yeasts naturally contain starch hydrolysis enzymes, but they produce little or no ethanol. With the development of techniques and knowledge in genetic engineering and molecular biology, more and more studies are dedicated to the construction of amylolytic strains of *S. cerevisiae*.

Other research work focuses on obtaining strains of thermophilic yeasts (strong interest in case of amylaceous plants, to conduct the saccharification and fermentation steps simultaneously at temperatures between 50 and 60°C), strains with higher productivity and more tolerant with respect to ethanol.

Amongst the bacterial species, strain *Zymomonas mobilis* stands out with good ethanol production yields. Some of its characteristics, however, are likely to penalise it at industrial stage, for example:

- its susceptibility to contamination by other bacteria due to its inability to develop at $pH < 5$,
- ethanol yields not exceeding 88% of the theoretical yield, when it is cultivated on industrial substrates such as sugar cane molasses or juices, with the synthesis of co-products such as levans and sorbitol,
- its tolerance to ethanol, slightly less than that offered by yeasts,
- the difficulties encountered during separation from the fermentation medium, due to its very small size compared with yeasts. Flocculating strains exist, however, but unfortunately their aggregation properties vary depending on the fermentation conditions.

In view of all these points, it is still too risky to consider extrapolation to industrial stage of ethanol fermentation with *Z. mobilis*.

2.2.1.3 Ethanol fermentation parameters

The parameters which have a significant influence on ethanol fermentation are the reaction product, the ethanol itself, temperature, oxygen, mineral elements and pH.

A. Ethanol

In view of its potential action on the growth and fermentative capacity of the microorganism, ethanol is undoubtedly the main factor limiting ethanol fermentation.

In the presence of high ethanol contents, about 8 to 10% by volume, the so-called ethanol-resistant yeasts, such as *S. cerevisiae*, can adapt to this new environment by modifying their membrane lipid composition. This membrane modification tends to increase the cell permeability to ethanol, favouring the diffusion of alcohol to the outside of the yeasts and reducing its intracellular concentration, the main factor inhibiting the microorganism's metabolic processes.

B. Temperature

The effect of temperature, on the viability and growth of the yeast as well as on its ethanol productivity, increases as the alcohol contents in the fermentation medium rise.

The viability of the microorganism cells decreases as the temperature rises. The optimum growth temperature decreases as the ethanol concentrations in the fermenter rise, while the specific ethanol productivity, at equal alcohol concentrations, increases with temperature.

C. Oxygen

Micro-aeration, which favours development of the microorganisms, makes the strains more resistant to ethanol and increases the sugar-to-ethanol conversion yield.

D. Mineral elements

The mineral elements required by the microorganisms for growth, maintenance and renewal are firstly nitrogen, supplied as ammonia or amino acids, followed by phosphorus, potassium and magnesium.

2.2.1.4 Fermentation step

Fermentation, the central step in the conversion of sugar-containing substances into ethanol, was conducted discontinuously or semi-continuously for many years. This simple implementation was justified by the small size of the industrial facilities. In addition, they only operated intermittently, primarily with molasses, an inevitable by-product in the manufacture of sugar from beet, then with concentrated sugar syrups. The sugar refineries operated only 3 to 4 months per year. It also represented a way of controlling problems of bacterial contamination more efficiently.

Fermentation times were long, however, and productivities low, due to problems of growth inhibition by high sugar concentrations at the start of the fermentative process and inhibition by ethanol at the end of the conversion process.

As production capacities increased, together with the need to make equally higher energy savings while being able to include all the ethanol production steps: preparation of the substrate – fermentation – distillation – treatment of effluents and co-products, with the constraint of operating all year round with one or more substrates, continuous processes have progressively been taking over for at least the last twenty years.

This type of operation avoids intermediate storage, sources of potential contamination, and stops between two cycles. It means that the operating schedule can be constant over time and optimised. As a result, this is also true for the dimensioning of the utility supply units (electricity – heat – process water). Moreover, a more concentrated sugar supply can be used.

The shift to continuous operation also increased the specific productivity (g ethanol $\times$ g^{-1} yeast $\times$ h^{-1}) and productivity per unit volume (g. ethanol $\times$ l^{-1} $\times$ h^{-1}) as well as the sugar-to-ethanol conversion yield, while minimizing the problems due to inhibition of the fermentative metabolism of the yeasts by ethanol.

The continuous cascade fermentation technology progressively took over, especially for the construction of large units whose production capacities could reach several million hectolitres. Figure 2.1 shows an example of this type of process, where the fermenters are directly connected to the starch liquefaction and saccharification reactors (production of ethanol from cereals). It is also illustrated by the photograph of an industrial unit (Photo 2.1).

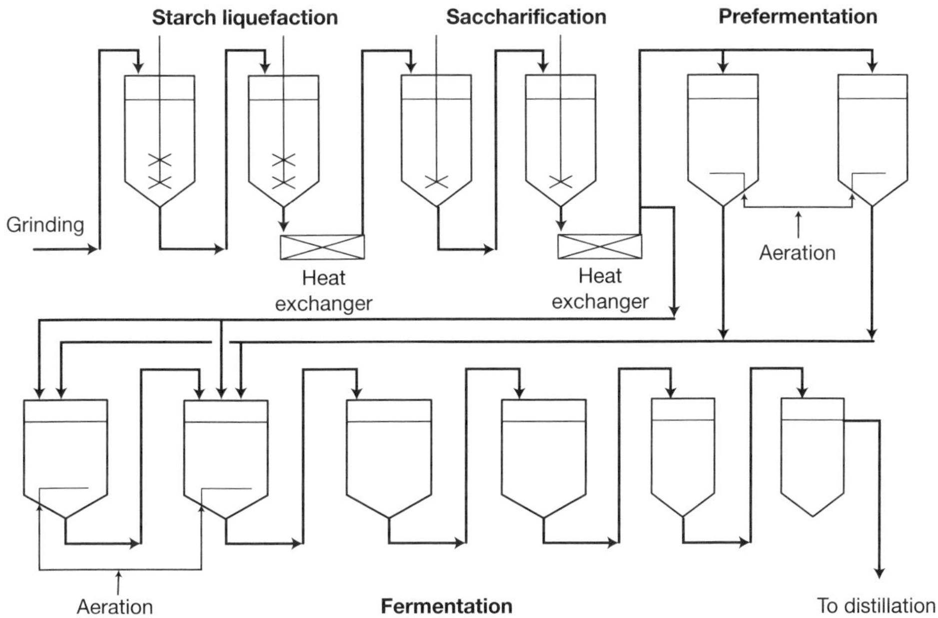

Figure 2.1

Diagram of continuous fermentation process (*dry milling*).
Cascade fermentation.

A small proportion of the hydrolysate supplies the prefermenter(s) for the production of fresh microorganism cells, most of it being fed into the first two reactors of the cascade (about 2/3 of the feedstock in the first and 1/3 in the second). Sugar conversion is not complete in these two reactors. This is a means of increasing the alcohol productivity, since the alcohol content at equilibrium is not too high. Obtaining a concentration between 4 and 5% v/v in the first reactor represents a good compromise between productivity increase and ethanol inhibition. The reaction is completed in the two or three other fermenters set up in series.

To improve the physiological state of the cells, a small amount of aeration is provided in the first two reactors. To make the mixture more homogeneous and improve the fermentation rate, a stirring system can be implemented, in the first two fermenters only.

Photo 2.1

Plant producing ethanol from cane molasses, located in Thailand. The fermenters.
Source: Maguin Interis

The carbon dioxide released by ethanol fermentation is recycled in the last reactors of the cascade to homogenise the medium and in particular to prevent the yeasts from depositing.

When the fermentation medium contains only very small quantities of solid elements in suspension, apart from yeasts, the yeasts can be separated from the fermented wort by centrifugation or filtration before at least 50% is recycled to the top of the cascade. To prevent development of bacterial contamination, the yeast suspension is washed in the presence of sulphuric acid. Recycling yeasts is an attractive way of increasing the productivity per unit volume and the ethanol conversion yield since the quantity of sugars required for the production of yeasts is reduced, and that required for the production of ethanol is increased by the same amount.

While the ethanol content in the first reactor, as mentioned previously, rarely exceeds 4-5% v/v, the content leaving the cascade is at least 10% v/v.

The sugar-to-ethanol conversion yields are approximately 92% of the theoretical yield.

Ethanol fermentation is exothermic. 1.2 MJ of heat are released for each kilogram of ethanol produced. While not a problem with current procedures, evacuation of the heat produced could certainly cause difficulties in the future if productivities are doubled or trebled. Another potential problem is entrainment of ethanol in the gaseous effluents, with a simultaneous increase in the production and the rates of CO_2 released per unit area.

With cereals as source of sugar, the saccharification and fermentation steps could be carried out simultaneously, after liquefying the starch. Under these conditions the glucose is converted into ethanol as soon as it is produced. Since it does not build up in the medium, glucose does not inhibit saccharification and does not lead to the appearance of non-fermentable reversion products. In addition, problems of contamination by other microorganisms are reduced, since the sugar residence time is very low.

With saccharification-fermentation taking place below the optimum saccharification reaction temperature, fermentation times are longer. This means that more reactors (5 to 7) are required in cascade than with processes where these two steps take place in turn.

2.2.2 Ethanol extraction

2.2.2.1 Distillation

Anhydrous ethanol is obtained in two steps (Figure 2.2):

- firstly, traditional distillation producing ethanol at a content similar to that of the azeotrope. Fusel oils are tapped off in the middle of the column,
- secondly, azeotropic distillation, leading to anhydrous ethanol (99.8% v/v), generally using entrainers such as cyclohexane, diethylether and *n*-pentane. Cyclohexane is generally chosen. This type of process is progressively being abandoned in units currently operating, to be replaced by dehydration on molecular sieves.

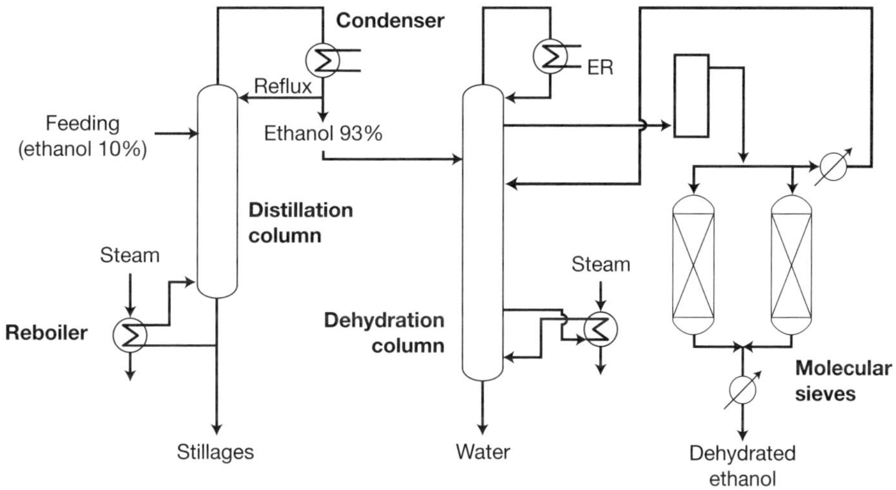

Figure 2.2

Ethanol distillation.

The energy consumption of traditional facilities, with a feedstock of 10% v/v ethanol, is 9 MJ/kg of anhydrous ethanol including 3.8 for azeotropic distillation. This consumption is equivalent to one third of the calorific value of ethanol.

Energy expenditure has been considerably reduced by adapting the various systems such as multiple effect vacuum distillation, thermo-compression or mechanical recompression of vapours on a case by case basis. With the latter technology and a diagram where the vapour at top of column is recompressed then condensed in the reboiler, the energy expenditure has been reduced to 6 MJ/kg of anhydrous ethanol.

Further energy savings have been made for the entire process, by including ethanol recovery and processing of effluents (concentration of stillages and drying of co-products).

2.2.2.2 Other technologies

Concerning the ethanol dehydration step, azeotropic distillation has now been overtaken by dehydration on molecular sieves, using Pressure Swing Adsorption (PSA), in a cycle where gas is adsorbed under high pressure and the sieve regenerated at low pressure. This energy-saving technology consumes only 0.1-0.2 MJ/kg of anhydrous ethanol.

Membrane separation processes based on vacuum pervaporation have been developed. The liquid phase is in contact with one side of the membrane, the vacuum is applied on the other. The energy consumed, including the prior traditional distillation, is estimated at 5 MJ/kg of ethanol.

2.2.3 Effluents and co-products

The manufacture of ethanol from sugar beet generates dry pulp with 90% dry matter, used as animal feed. The pulps are the solid residue recovered after extracting sugar from the beet during the diffusion step. 70 kg of ethanol and 54.5 kg of dry pulp are produced per tonne of beet, i.e. 0.78 t pulp/t ethanol.

The stillages (also known as spent wash), salt-rich liquid effluents still containing organic matter that cannot be fermented into ethanol, are recovered at the bottom of the conventional distillation column. When authorised, stillages can be spread on agricultural land as soil conditioner, up to a maximum of 3 tonnes per hectare. They can also be concentrated for use in animal feed formulations, or used in methanisation to improve the distillery energy balance.

When sugar cane is the raw material, apart form stillages, the co-product is bagasse, a solid residue of the sugar cane grinding/pressing steps. Bagasse is currently used as an energy source to produce electricity and heat in the factory, but also for other users.

Two processes are available to produce ethanol from cereal plants (corn, wheat): wet milling and dry milling (see Chapter 2.2.4.2).

With wet milling, the oily part of the corn germ is extracted before recovering the proteic part, the corn gluten meal, which has a protein content of about 60%. This proteic meal is turned into animal feed.

The residual raw material is converted into ethanol. The non-fermentable part, recovered after distillation, is dried to produce another type of cattle feed, corn gluten feed, containing less proteins (20%).

With dry milling, the entire grain is processed. The effluents collected after distillation, containing 10% solid matter, are centrifuged. Some of the clarified liquid phase (the stillages) is recycled during the liquefaction step and the rest concentrated. This concentrate is then remixed with the solid phase recovered after centrifuging for drying.

The product obtained, draff (also known as Distiller's Dried Grain Solubles – DDGS), is used as animal feed. DDGS produced from corn contain 30 to 32% proteins and 8 to 9% fat. Those produced from wheat ethanol fermentation contain between 35 and 40% proteins, with a lysine content higher than that of corn DDGS, and 4 to 5% lipids. Due to their composition, wheat DDGS can be sold for 10% more than corn DDGS.

370 l of ethanol and 370 kg of DDGS can be produced from 1 tonne of wheat, i.e. 100 kg of DDGS per hl of ethanol. 400 l of ethanol and 320 kg of DDGS can be produced from 1 tonne of corn, i.e. 80 kg of DDGS per hl of ethanol.

As mentioned earlier, distillation and concentration/drying of effluents and co-products are integrated to optimise energy expenditure.

2.2.4 Process diagrams

The main European manufacturers of ethanol production plants are Abengoa, Maguin-Interis, Technip and Vogelbusch.

2.2.4.1 Production of ethanol from sugar plants

Figure 2.3 shows the various steps involved in the production of ethanol from sugar beet or cane.

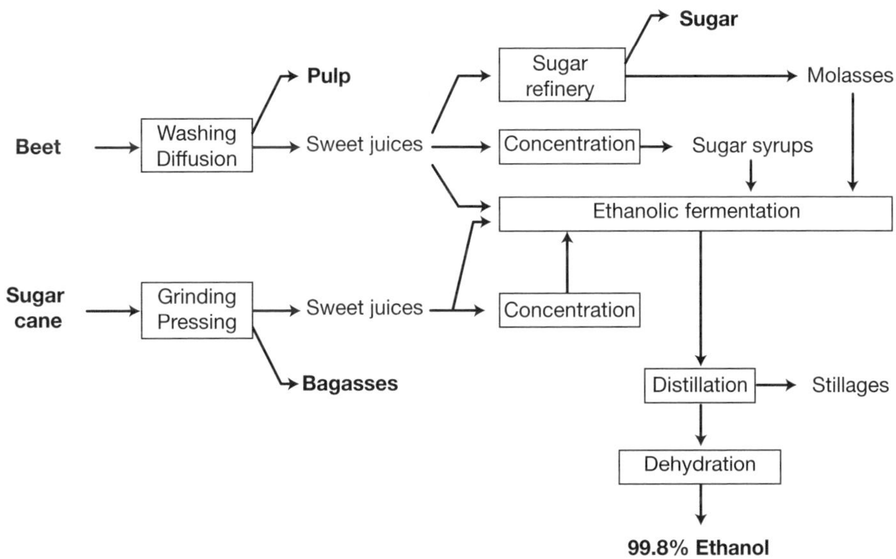

Figure 2.3

Production of ethanol from sugar plants.

After washing the beet to remove all the earth clinging to them, they are chopped into cossettes. The weight of earth, which could represent up to 40% of the weight of the beet, has been reduced in two ways: harvesting means and modification of the beet morphology, to make it much more regular. Washing techniques in the factory have been improved. The sugars are then extracted by hot water diffusion. Ethanol fermentation is carried out using these juices, concentrated or not as syrup. Concentration protects the juices against bacterial contaminations and increases their storage time. Fermentation can also be carried out from the inevitable sugar refinery by-products, the molasses. Molasses contain about 50% by weight of sugars, as well as mineral elements and non-fermentable organic products.

The fermentation juices are then distilled to produce ethanol. Some of the distillation effluents, stillages, are recycled and the rest concentrated or not before being evacuated.

The sugar canes are ground and pressed to obtain the fermented sweet juices and a solid residue, bagasse, a source of energy, electricity and heat.

2.2.4.2 Production of ethanol from amylaceous plants

Two types of process are still currently implemented. However, with the increasing quantities of ethanol produced from corn, mainly in the United States, wet milling (Figure 2.4) is progressively being replaced by dry milling (Figure 2.5).

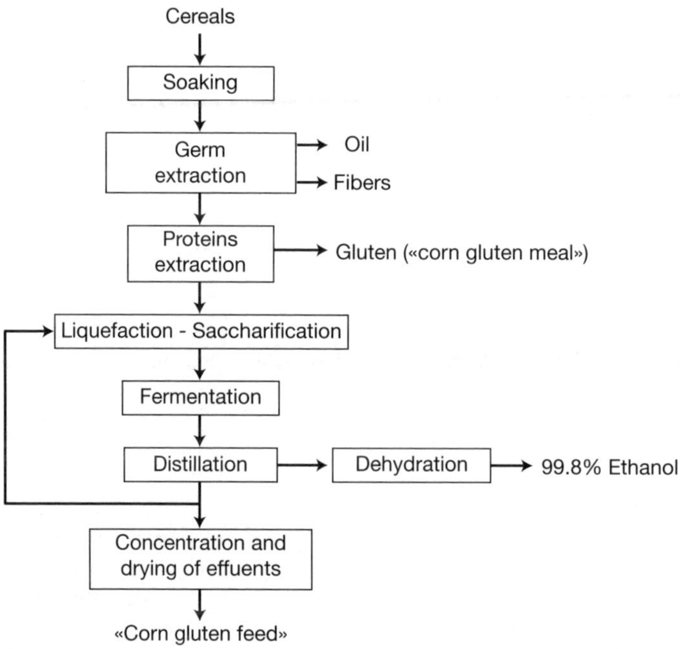

Figure 2.4

Production of ethanol from cereals (wet milling).

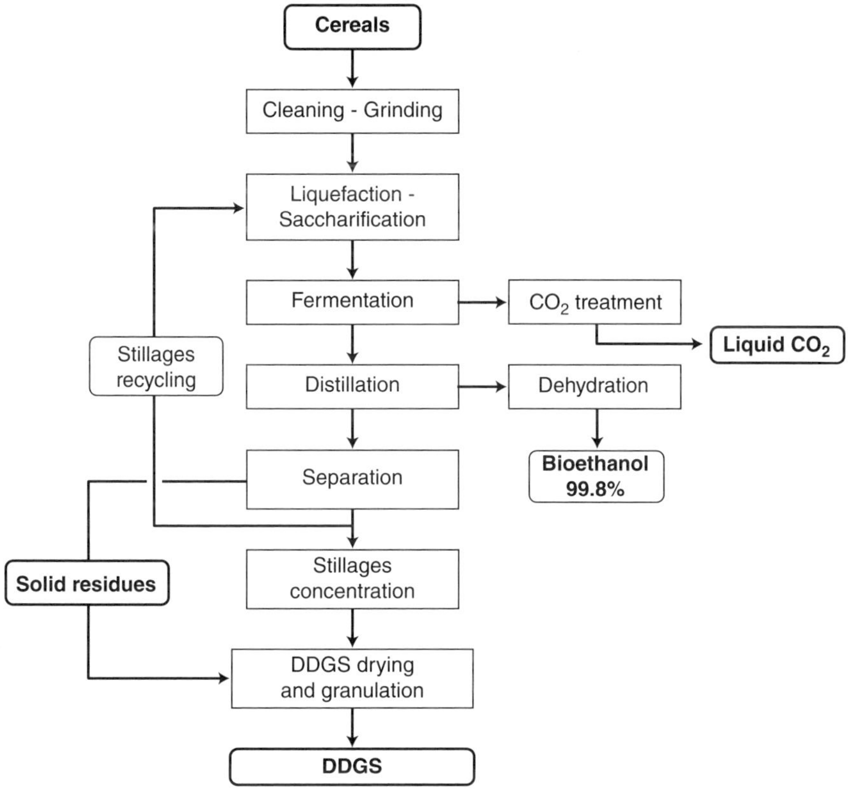

Figure 2.5

Production of ethanol from cereals (dry milling).

With wet milling, ethanol production is associated with the valorisation of other products and co-products, some being extracted at the start of the process: oil extracted from the cereal germ, fibres and proteins or gluten used in human or animal food.

Some of the residual amylaceous fraction is converted into ethanol and into a by-product called corn gluten feed, sold as animal feed. Ethanol fermentation is dependent on other uses of starch: manufacture of fructose, whose demand fluctuates with the season, especially as regards the sale of soft drinks, and numerous other industrial applications of starch, generally based on the higher quality fractions, A- and B-starches.

Ethanol fermentation therefore represents the outlet for lower quality amylaceous fractions (C-starch and some of the B-starch) or amylaceous surpluses when demand drops for the other applications.

Consequently, operation of an ethanol production unit implementing the wet milling process must take into account variations in availability, quality and cost of the fermentable raw material.

With the dry milling process, which does not depend on other uses of the raw material, the cereals are only converted into ethanol and DDGS.

2.2.5 Energy balance

2.2.5.1 Production of ethanol from sugar beet

Energy balances [1] are calculated for an industrial plant producing ethanol from sugar beet juice and EP2 (B molasses, low grade run-off resulting from the second sugar crystallisation step, after centrifugation).

The data are shown in Table 2.3.

Table 2.3 Energy consumptions per hl of ethanol produced from sugar beet.

	Sugar beet juice		EP2	
	Steam (kg)	**Electricity (kWh)**	**Steam (kg)**	**Electricity (kWh)**
Diffusion	25	11		
Fermentation		6		6
Distillation [a]	80	3	80	3
Ethanol dehydration [b]	55	1	55	1
Concentration [c]			75	9
Total	160	21	210	19

[a] *Double-effect thermocompression system.*
[b] *Molecular sieve or azeotropic distillation associated with a thermocompression system.*
[c] *Concentration with a multiple-effect thermocompression system (6 effects).*

From sugar beet juice, the primary energy consumption related to the production of ethanol [2] represents 30 to 33% of the alcohol energy content. This ratio increases to 39-40% when EP2 is used.

Energy consumptions can be improved by methanisation of stillages. For example, 1.1 m^3 of stillages [3] are generated per hl of ethanol produced. Taking as calculation assumptions an 85% COD abatement rate and a production rate of 0.3 m^3 methane per kg of COD

1. The data shown in this subchapter were provided by UNGDA (the French National Federation of Alcohol Distillers Associations).
2. The primary energy consumption is calculated using the following equivalences: 1 kWh = 8 MJ, 1 kg HP steam = 3.1 MJ.
3. The average characteristics of the stillages are as follows: COD (Chemical Oxygen Demand): 30 $g \times l^{-1}$, dry matter content: 26.5 $g \times l^{-1}$, suspended matter content: 6.8 $g \times l^{-1}$.

eliminated, the potential energy production by methanisation is about 12 MJ/kg COD or 336 MJ/hl of pure ethanol. In this case, the primary energy consumption drops down to about 23% of the energy content of the alcohol produced.

2.2.5.2 Production of ethanol from amylaceous plants

The data [1] correspond to the operation of an industrial unit producing ethanol from wheat, by dry milling.

This factory, which operates continuously, can produce about 3 000 hl of anhydrous ethanol per day, or 1 Mhl/year operating round the clock 333 days per year.

The daily matter balance is as follows:

- consumption of 798 t of wheat,
- production of 238 t of ethanol (+ 7 hl of fusel oils) and 284 t of DDGS with 90% dry matter.

Optionally, the CO_2 released from fermentation can be recovered and liquefied for use in the food industry.

The energy consumptions per hectolitre of anhydrous ethanol manufactured are detailed for each step of the process in Table 2.4.

Table 2.4 Energy consumptions per hl of anhydrous ethanol produced from wheat.

Process steps	Steam (kg) [a]	Electricity (kWh)
Grinding		5.7
Liquefaction-saccharification	40	3.0
Fermentation		5.6
Distillation	Coupling with stillage concentration	0.6
Dehydration	34	1.9
DDGS separation		8.3
Stillage concentration	210	4.9
DDGS drying and granulation		9.8
Total	284	39.8

[a] *Steam 2.5-4 bar.*

The process water consumption is about 0.4 m^3/hl of anhydrous ethanol produced.

About 13 m^3 of natural gas are consumed during the DDGS drying step for a quantity corresponding to the production of 1 hl of anhydrous ethanol.

1. All data shown in this subchapter were provided by Maguin Interis.

Condensate water produced during the stillage concentration step is recycled in the cooling circuits to compensate for about 75% of the losses. The net water consumption is therefore considerably reduced and represents about 0.1 m^3/hl of anhydrous ethanol produced.

The data can be extrapolated linearly up to production capacity levels of 5 000 hl/d.

2.2.6 Potential ethanol production improvements

Several research pathways – preparation of raw materials, fermentation and separation of alcohol from the aqueous phase – are being investigated.

2.2.6.1 Pretreatment of amylaceous products

- Search for new enzymes (amyloglucosidases) capable of working at high temperature (above 60°C), to reduce the risks of contaminating the sugar solution,
- increase the dry matter content of the ground cereal or meal suspension while trying to reduce viscosity. If successful, the energy expenditure associated with the enzyme hydrolysis steps and the quantities of cooling water will be reduced.

2.2.6.2 Ethanol fermentation

- Select yeasts producing less or no glycerol, to improve the ethanol conversion yield,
- increase yeast ethanol productivity, while remaining within reasonable limits, however, to avoid complicating the technologies required to cool the fermenters (low temperature heat evacuation) and recover the ethanol vapours entrained in the fermentation gaseous effluents. A good way must be found to reduce the size of the facilities and therefore the investments,
- search for ethanol-tolerant yeasts to increase the final ethanol content in the fermentation wort. Provided that this research pathway does not result in lower yield and productivity, which is absolutely essential, it should lead to energy savings during the distillation steps,
- search for thermo-tolerant yeasts capable of producing ethanol, with no loss in yield, at temperatures greater than 40°C. This study finds an application, especially when ethanol is manufactured from amylaceous materials. This would considerably enhance the simultaneous saccharification-fermentation option since the risks of contaminating the sugar solutions would be reduced and the heat exchanges improved due to the greater temperature difference between the fermentation medium and the cooling water.

2.2.6.3 Improved energy performance

- By optimising the integration of the various process steps, including the co-product conditioning and drying steps,
- by reducing the external supply of fossil origin energy through optimising the energy conversion of residues originating from the cultivation of alcohol-producing plants (cereal straw, corn stalks, sugar cane bagasses) and of the effluents generated by the process (e.g. methanol fermentation of stillages).

2.3 ETHER FUELS

Very few ethers are currently used in the composition of fuels.

The list can be reasonably reduced to five, all being used in the formulation of the gasoline pool: MTBE, ETBE, TAME (TertioAmylMethylEther) and ETAE (EthylTertioAmylEther) to which we must add DIPE (DiIsoPropylEther).

Introduction of heavy ethers such as DPE (DiPentylEther) in the Diesel pool has not yet been commercialised, although its cetane number is greater than 100 in mixtures.

MTBE is by far the most widely used ether. It is in fact the petrochemical product which has witnessed the strongest growth over the last 25 years. From laboratory scale in 1975, production reached 20 Mt/year in 1998 [1] and [2]. The world capacity of TAME is still below 2 Mt/year.

MTBE and TAME are derivatives obtained respectively by adding methanol to isobutene, or 2-methylpropene, and to isoamylenes, in fact a mixture of 2-methylbut-1-ene and 2-methylbut-2-ene.

If ethanol is added to the same hydrocarbons instead of methanol, we obtain respectively ETBE and ETAE. In 2010, world production of ETBE reach 2.5 to 3 Mt/year, versus 0.6 Mt/year in 2005, while ETAE will only continue to be used commercially on very specific situations.

DIPE (diisopropyl ether), obtained by dehydration of isopropanol, itself derived from hydration of propylene, has not been developed at significant scale. Only two industrial units are currently operational worldwide [3]. Its economic use as fuel is penalised by the current price of propylene. The reactions used to synthesise the various ether fuels are illustrated on Figure 2.6.

Ethers produced by addition of methanol to the unsaturated hydrocarbons of FCC (Fluid Catalytic Cracking) gasoline C_6 and C_7 cuts [4a] should probably be added to this list. They are not pure compounds but complex mixtures, always obtained in mixture in hydrocarbon cuts. Some industrial units nevertheless produce TAME from wider FCC gasoline cuts and co-produce branched methoxyalkanes and therefore ether fuels.

In terms of total production for the ether fuels, the situation did not changed significantly between 2005 and 2009 [4b]. Although MTBE production has dropped in North America, it has been largely compensated by the massive arrival of new players such as Russia, with about 30 production units, and China, which has built about 40 units. Chinese production exceeded 3 Mt/year in 2009. What has happened in Europe, however, is that apart from Great Britain, Italy and Greece, most MTBE units have been converted to ETBE. TOTAL, for example, has converted all its MTBE units to ETBE [4c]. In recent years, Spain is the only country that has actually built a new ETBE unit, at Huelva (Andalousia). Japan is now considering switching over to ETBE.

TAME has not achieved the expected success. Note that at Porvoo (Finland) company Neste Oy converted its TAME production unit into ETAE. TOTAL, with a licence to operate the Neste Oy process, has started an ETAE production unit of 200 000 t/year in Germany. This unit processes a hydrocarbon cut larger than the traditional C5 cut and therefore produces heavier ethyl ethers.

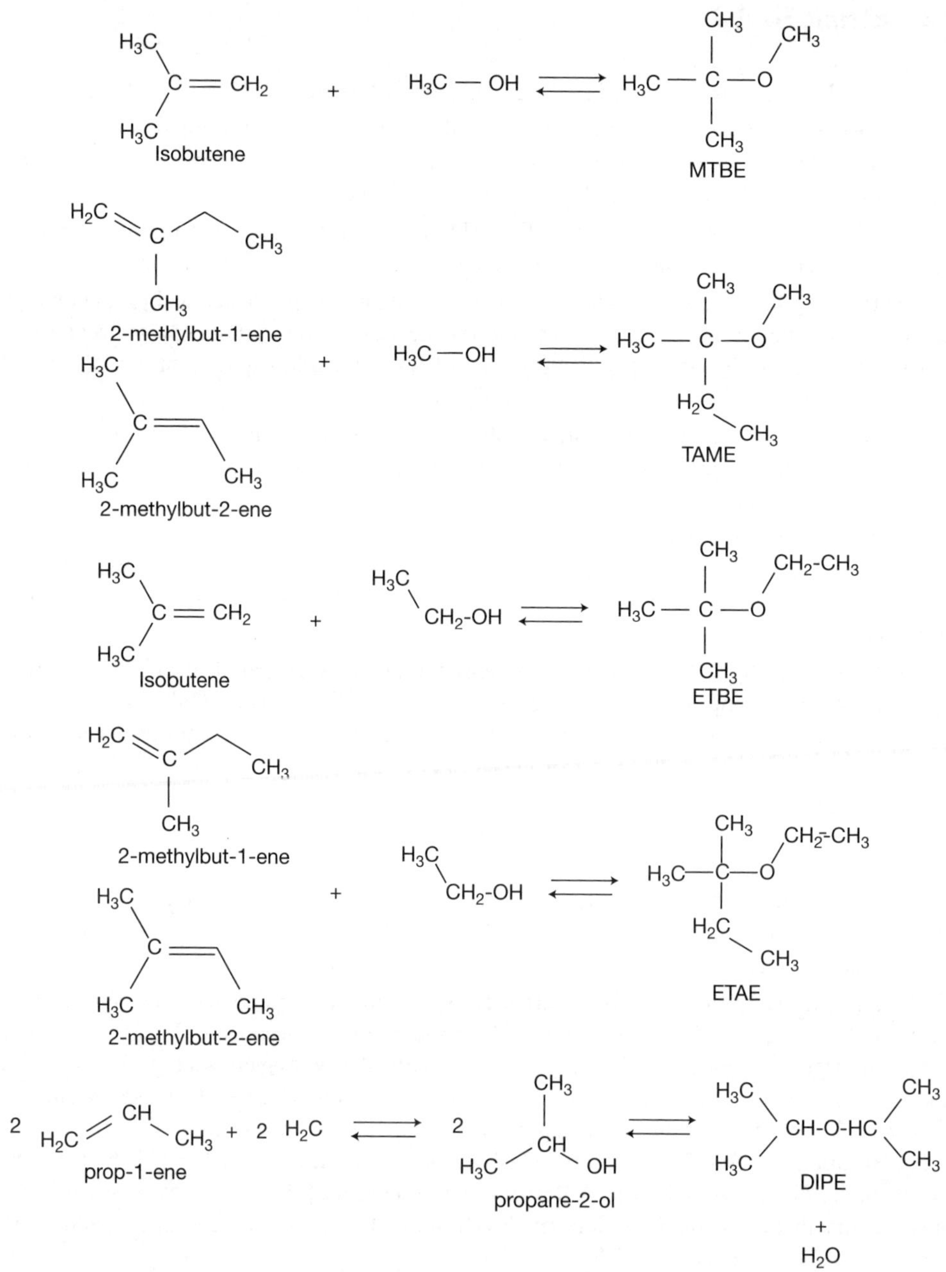

Figure 2.6

Synthesis of ether fuels.

Industrial production of MTBE first started in Italy at the Ravenne Refinery in 1975, but North America has been for many years the largest consumer of this compound. When the Clean Air Act came into effect, MTBE was widely used in the reformulated gasoline pools. From 1995, use of this ether started to cause problems, firstly in California which at the time was the leading MTBE consumer and importer (see Chapter 1).

Due to its structure, biological degradation of MTBE is much slower than that of hydrocarbons and, in addition, it can be detected in trace quantities (a few ppb) in water.

Although no studies have provided formal evidence of toxicity at this level of pollution, MTBE was banned first in California, then in a large number of US States. The ban has been extended to TAME.

Some southern states such as Texas and Louisiana, which have large MTBE production units and which are connected by barges to the midwest states, large producers of corn ethanol, could also switch to ETBE, definitely turning this ether into a commodity product which would certainly be offered at a very attractive price, in some European countries in particular.

Export could represent the main outlet for ETBE since ethyl ethers, although less persistent in the environment [5a] than MTBE, have also been banned as fuels in the United States.

This possibility for massive export does not seem to be taking shape, however. In contrast, Brazil, which retains a decisive advantage in terms of ethanol production costs, would export large quantities of ETBE to Europe, via Spain, Portugal and the Netherlands.

In Europe and possibly throughout the world, apart from the United States which has excluded the use of ethers, the real question concerns the simultaneous use of ETBE and ethanol. The existing ETBE units should be able to survive, but it is unlikely that new facilities will be built.

Finally, the shortage of isobutene will not be responsible for stopping the development of new ETBE capacities. For example, Company Lyondell has still not converted its giant units in France and the Netherlands producing mainly propylene oxide, to ETBE mode. However, the newcomers on the MTBE market, such as Russia and China, can be expected to switch over completely or at least partially to the production of ETBE.

Company Methanex, the leading global methanol producer, has seen that the share of methanol intended for MTBE production continues to decline, although without collapsing [5b].

2.3.1 Sources of raw materials for the production of ether fuels

Methanol can be considered as a commodity. It is produced in giant units, the largest being able to produce up to 5 Mt/year in a single reaction train. Today, very few remain. These units are preferably located near gas production fields. Methanol is in fact produced from natural gas which is first converted into syngas before conversion into methanol.

The methanol is then distilled, and grade A is almost exclusively used to manufacture ether fuels.

The ethanol used to manufacture ether fuels is exclusively of agricultural origin. Unlike methanol, therefore, the ethanol incorporated in fuels is obtained from a large number of different sites. En France, for example, only ethanol produced from sugar beet or wheat is currently authorised [6] for incorporation in the fuel pool.

Ether fuels are produced by mixing an alcohol and an unsaturated hydrocarbon. The hydrocarbons generally used are contained in the C_4 and C_5 cuts obtained firstly by catalytic cracking or steam cracking of heavier hydrocarbons and secondly through specific operations. These specific operations (Figure 2.7) include the propylene oxide production pathway, which involves coproduction of tertiobutanol (less than 5 units worldwide), which is dehydrated into isobutene, and dehydrogenation of isobutane leading directly to isobutene (about 10 units worldwide).

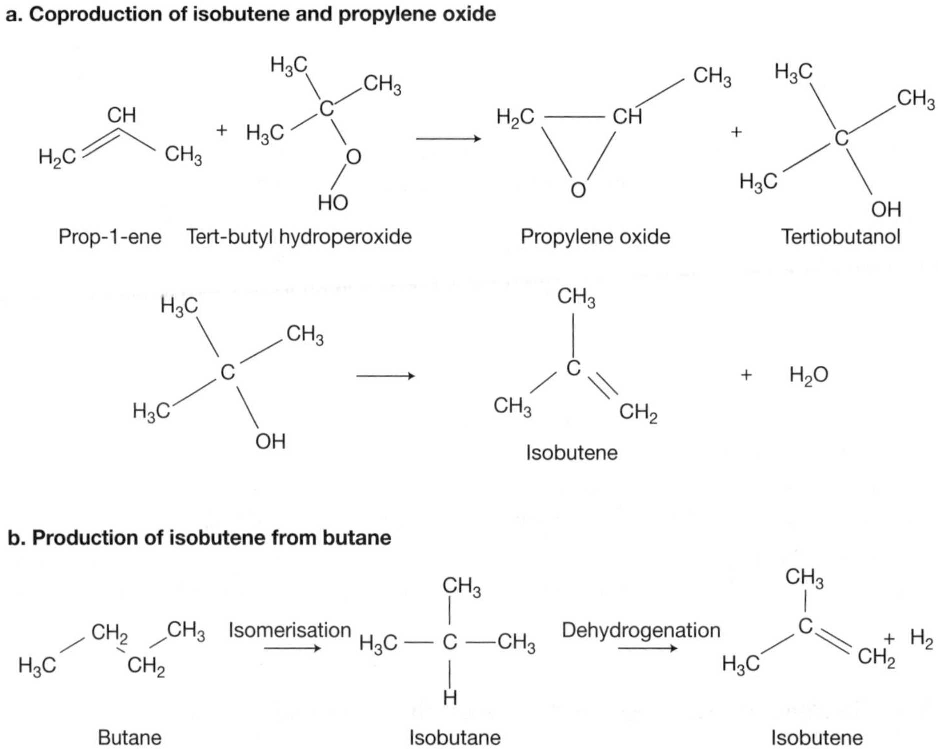

Figure 2.7

Various isobutene synthesis pathways.

The capacities of the industrial MTBE production units using the first of these pathways can reach up to 500 000 t/year. Similarly, the isobutane dehydrogenation pathway is in fact included in a chain of processes which, starting from butane contained in natural gas, includes isomerisation into isobutane, and generally ends with an MTBE production unit. All the links of this chain are part of giant units which can supply up to 700 000 t/year of MTBE.

The ether production units, operating on C_4 and C_5 cuts from catalytic cracking, are either located in refineries or on the steam cracker petrochemical site, generally when the steam cracker is next to a refinery. While these units are generally smaller, rarely producing more than 150 000 t/year of ethers, they are also more numerous, with about 200 worldwide. The refinery etherification units incorporate their production directly in their gasoline pools whereas the giant units are forced to export their production. As a result, MTBE is a commodity product with its own market, whose price is set at Rotterdam.

2.3.2 Etherification reaction

The chemical reactions leading to the production of ether fuels by addition of alcohols (mainly methanol and ethanol, but also other oxygenated compounds such as propanol, isopropanol, the *n*-butanols, 2-methylpropanol and even glycerol) on the double bond of branched olefins are balanced and catalysed by acids.

Le Châtelier's law stipulates that to favour the formation of ethers, the reaction must be conducted with excess alcohol, especially since the olefin is generally diluted in an unreactive matrix and the olefin and/or alcohol have increasing molecular weights.

On industrial scale, the reaction is conducted with molar excesses of alcohol (up to 10%), such that a mixture of ethers and hydrocarbons which have not yet reacted with the excess alcohol is recovered at the outlet of the etherification reaction section, which must be separated and obviously recycled.

Addition of alcohol on a branched olefin is an exothermic process. Methanol and ethanol are added reversibly on the isobutene with the production of about 40 kJ of heat per mole of ether formed.

Although ethanol is more nucleophilic than methanol, the rates of formation of the corresponding ethers are substantially the same. These rates are probably controlled by the rate of diffusion of isobutene to the catalytic acid sites partially solvated by the alcohol molecules. The rate of formation of the carbocation therefore controls the entire kinetics of the etherification reaction, illustrated by the following equation:

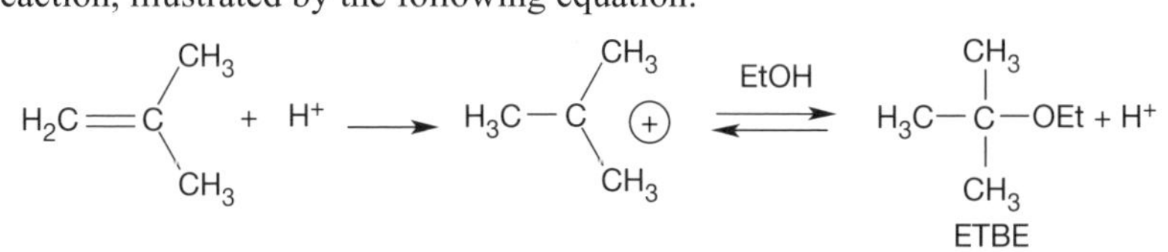

At equilibrium, the formation of ETBE is unfavoured compared with that of MTBE, as shown on Table 2.5 which lists the values of equilibrium constants as a function of

Table 2.5 Equilibrium constants (K) for the formation of MTBE and ETBE.

Temperature °C	40	50	60	70	80	90	100
KMTBE l/mole	117.5	71.6	45	29	19.2	13	9
KETBE l/mole	98.6	58.7	36	22.7	14.7	9.8	6.6

temperature [7]. In a given production unit, all things being equal otherwise, since the chemical equilibrium is more favourable for the formation of MTBE than for that of ETBE, this means that the expected conversion of isobutene in the latter case is significantly less. The synthesis reactors are in fact operated so that the composition of their effluents is as close as possible to that corresponding to thermodynamic equilibrium.

To improve the conversion rate, it is always possible to work with a larger excess of ethanol with respect to isobutene. This means that more alcohol will have to be recycled with the disadvantage that, if this alcohol is extracted from the raffinate by water (see 2.3.2.4), the water will also have to be recycled. Pure methanol can be obtained at the top of the alcohol recycling column while ethanol is always accompanied by at least 4 to 5% by weight of water. The water recycled to the resins is a moderator of the ether formation reaction, like the alcohols in fact. The expression of the direct rate constant of the etherification reaction clearly shows the negative effects of water and excess alcohol [7].

$$k = \frac{k_0}{1 + \alpha[Alcohol]^a + \beta[H_2O]^b}$$

α and β are related to the respective affinity of water and alcohol for the resins, or to the resistance to transfer of isobutene through the film formed by these inhibitors around the active site. Consequently, the ETBE formation rates will always be lower than those obtained for MTBE (Table 2.6).

Table 2.6 Conversion of isobutene into MTBE and ETBE.

	Conversion of isobutene (wt%)	
	Main reactor	**Reactor + catalytic column**
MTBE	97.5	99
ETBE	89	95

The loss of isobutene conversion is not detrimental if sulphuric alkylation is carried out downstream from the ETBE unit. In case of hydrofluoric alkylation, the octane number of the alkylate obtained is lowered by at least one point.

If the ethanol excess is too large, in an attempt to improve the isobutene conversion rate, the rate of reaction drops and the effect is opposite to that expected since the thermodynamic equilibrium is no longer reached at reactor outlet.

2.3.2.1 Secondary reactions

Methanol is a very pure product obtained from syngas, itself generally obtained by steam reforming of methane. The only noticeable impurity – although in principle it is eliminated by distillation – is methyl formate. The water content is about a hundred ppm. The only problem could be the dissolved oxygen content. Some rare cases of seawater pollution may occur during sea transport, however, with a catastrophic effect for the catalyst activity.

Agricultural ethanol is conveyed to the refinery by rail or by sea and may therefore also contain dissolved oxygen. It comes from various origins, however, each with its own characteristics. In France, the alcohol may be obtained from sugar beet or wheat. The content in impurities, such as esters, acetaldehyde and acetic acid, is variable.

The esters go through the ETBE production unit without being converted and end up in the ether formed. The same applies for acetic acid. The esters may hydrolise slowly during storage and form acetic acid, which is likely to cause corrosion. The content of these two pollutants is strictly controlled and it is recommended to limit their concentrations to values less than 150 ppm by weight and 15 ppm by weight, respectively.

Acetaldehyde is another disturbing impurity. Its content is limited to 150 ppm in ethanol. This aldehyde is converted into acetal during the etherification reaction and ends up in ETBE. During storage, this acetal may release acetaldehyde which will be oxidised into acetic acid (Figure 2.8).

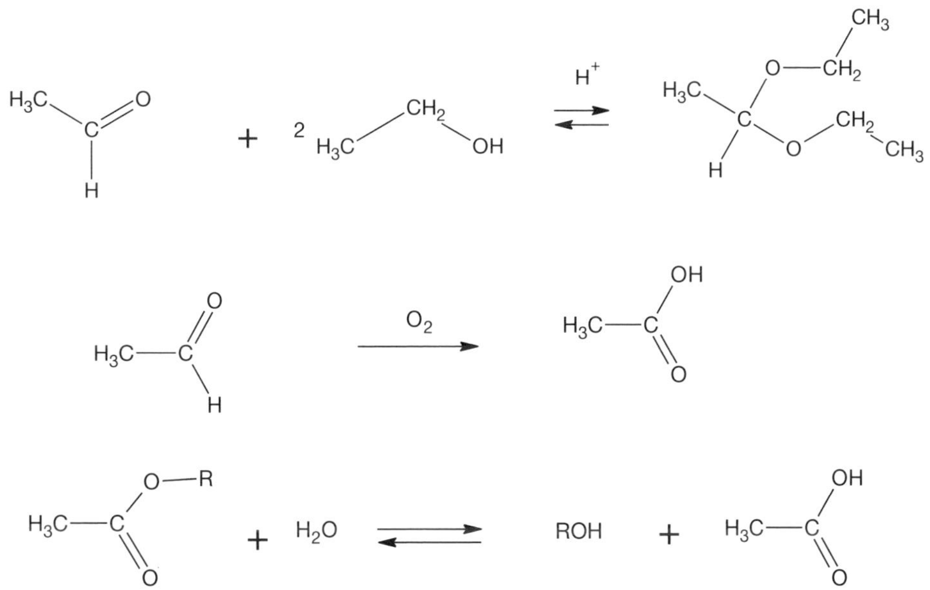

Figure 2.8

Acetaldehyde conversion reactions.

The ethanol water content is also of paramount importance. We have seen the effect of this contaminant on the ETBE formation kinetics and, consequently, on the isobutene conversion. The reaction of water on isobutene is also balanced, forming, as described in the following equation, tertiobutylic alcohol (TBA), a solid alcohol at ambient temperature:

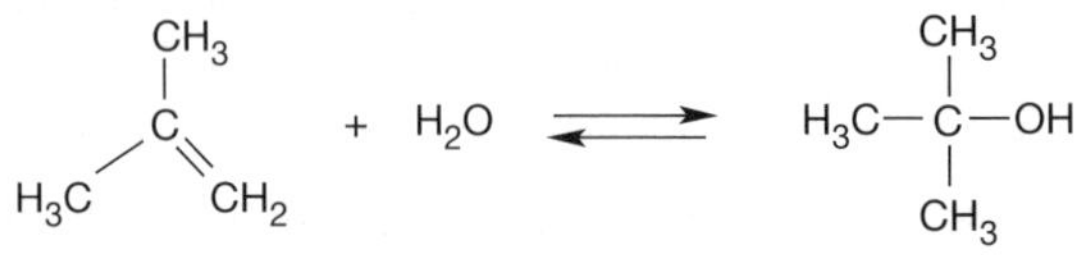

While TBA concentration is limited to 0.5% by weight in MTBE, in practice, it is close to 1.5% weight in ETBE. It is therefore essential to limit the water content in the alcohol delivered to the refinery to 3 000 ppm. Under these conditions, as much as 50% of the water entering to the ETBE unit comes from the fresh alcohol supply, 25% from recycling the excess ethanol and 25% due to the fact that the hydrocarbon feedstock is saturated with water.

Dehydration of alcohols on the resins is observed when synthesising ether fuels. Due to its low boiling point (– 25°C), the dimethylether (DME) produced is found in the raffinate leaving the MTBE unit. The raffinate generally contains between 250 and 450 ppm by weight of DME, depending on the age of the resins.

Under the same operating conditions, ethanol is far less likely to form diethylether. The traces of diethylether (boiling point 35°C) formed, are found at the bottom of the debutaniser (see 2.3.3) mixed with ETBE. These traces have no impact on the quality of the ether fuel.

The other secondary reactions are the same, both qualitatively and quantitatively, whether the etherification unit operates in MTBE or ETBE mode. More or less equal quantities of di-isobutene and oligomers of the same family, secondary alcohols and secondary alkyl ethers are synthesised. The latter compounds are formed by addition of water and alcohol on linear olefins (but-1-ene or but-2-enes) (Figure 2.9).

2.3.2.2 Etherification catalysts

All organic or mineral acids can favour the addition of alcohols on hindered olefins to form ethers. The literature describes the use of heteropolynuclear acids and zeolites. Industrially, only the ion exchange resins, in particular cross-linked polystyrene sulphonic acids, are used (Figure 2.10).

These resins are commercial products available as beads from 0.2 to 1 mm diameter, mainly used in demineralisation of water. In the field of catalysis for the manufacture of ether fuels, the main suppliers are Bayer, Dow Chemical, Purolyte and Röhm & Haas.

The etherification reaction is therefore carried out on a supported catalyst, whose swelling rate, adjusted by cross-linking, and mechanical resistance are essential properties.

These catalysts exhibit relatively poor temperature stability, however, since, already below 100°C, the decomposition reaction resulting in loss of sulphur trioxide (Figure 2.11), even if very slow, cannot be neglected. Deactivation of industrial catalysts can be partly attributed to this sulphur loss reaction.

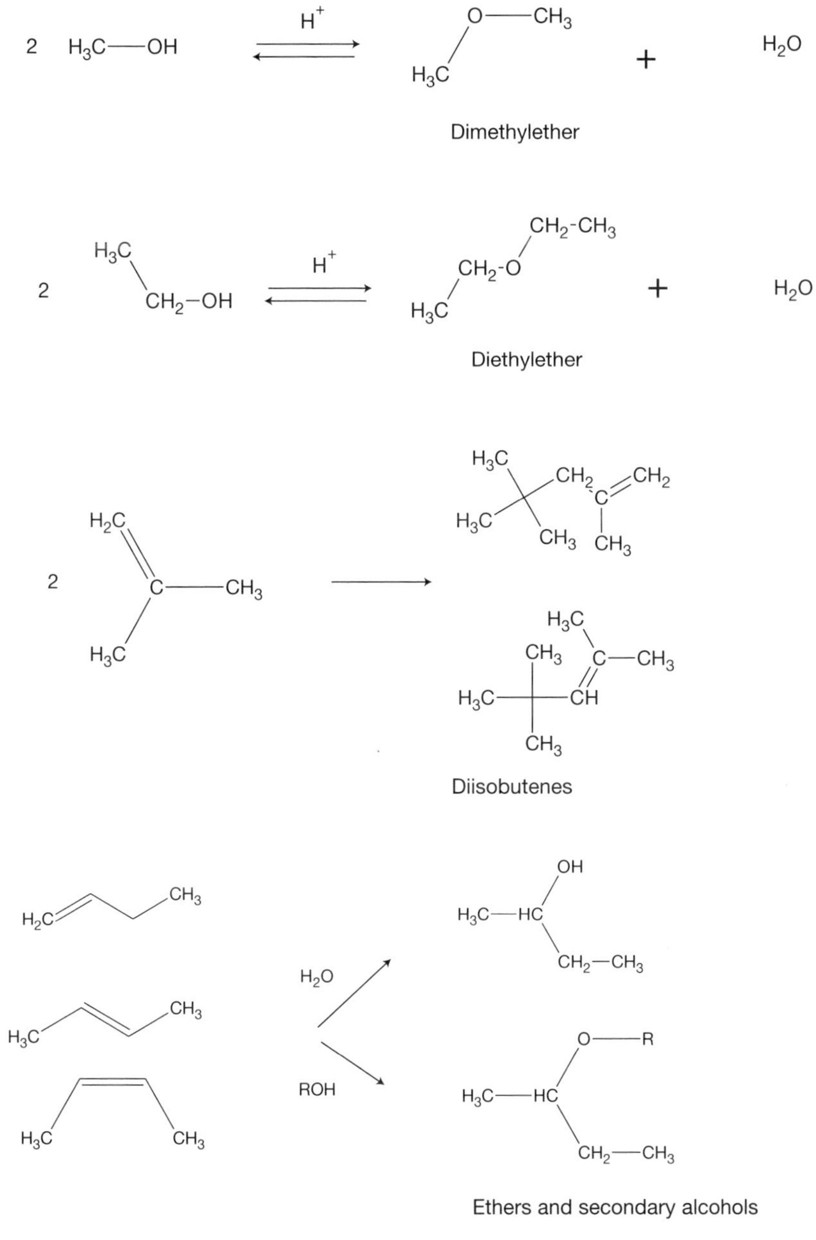

Figure 2.9

Other secondary reactions.

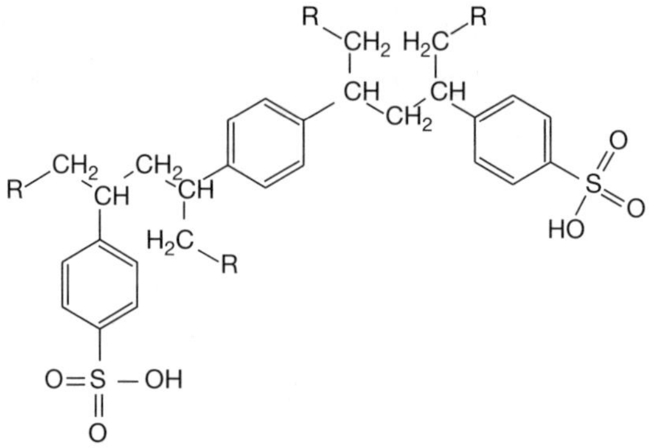

Figure 2.10

Structure of cross-linked polystyrene sulphonic acid resins.

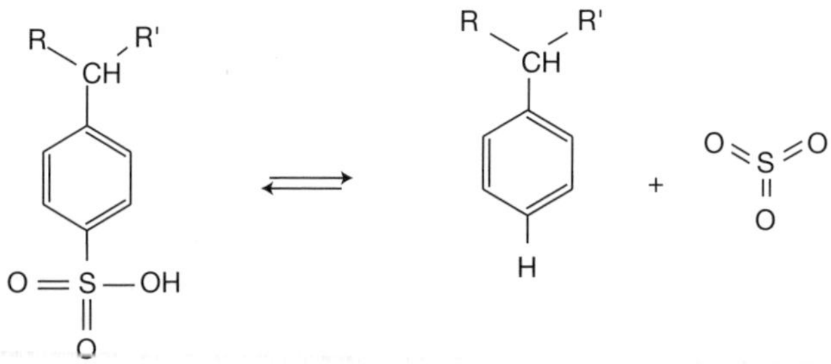

Figure 2.11

Resin deactivation by sulphur loss.

These resins are extremely sensitive to basic derivatives and metals, which globally neutralise the acid function through the formation of salts. In addition, nitriles found as trace compounds in the refinery feedstocks to be etherified are also slowly hydrolised into organic acids and ammonia, which poisons resins [8] and [9]. Moreover, in the presence of iron, the oxygen found either in the cut to be etherified or more generally in the alcohol favours rapid link splitting of the resins, resulting in destruction of the structures. Lastly, the presence of these strong acid functions also accounts for polymerisation of olefins, forming products which precipitate in the medium and build up on the catalyst grains. These deposits prevent the reactants from accessing the catalytic sites.

To avoid these catalyst deactivation reactions, the temperatures inside the reactor must not be too high and the temperature gradient limited.

In normal operation, the temperature at reactor outlet must remain below 80°C to avoid sulphur loss and deposit of polymers. The oxygen is eliminated, especially in the alcohol, by nitrogen stripping in the reflux drum of the water-alcohol separation column. The fresh alcohol therefore transits through this reflux drum. The nitriles and basic residues or mineral salts are eliminated by washing the hydrocarbon feedstock, generally by condensates.

If all necessary precautions are taken, the resins have a lifetime of about 3 years.

2.3.3 Etherification processes

Most units currently producing ETBE started life as MTBE production facilities. For example, in France, the Feyzin MTBE unit has been producing ETBE since April 1994 [10].

The conversion is relatively easy to implement, to such an extent that overall, no additional investments are required. Only the operating parameters need to be adjusted [11] and [12]. Since 1997, two units designed specifically to produce ETBE, according to a process developed by IFP Energies nouvelles [10], have started production in France.

We observe the same trend in Spain where four MTBE units, in other words all of them, have been converted to produce ETBE. It is only very recently that an additional unit, devoted to the production of ETBE, has been built.

New ether production units will have to be flexible enough to allow easy reconversion to di-isobutene production units, to cater for the situation where the government should decide to ban use of ether fuels, as is currently the case in the United States [13], [14]. Company UOP proposes this type of reconversion with the InAlk process [15] and has even implemented it in a Japanese refinery [16]. However, it is highly unlikely that the giant MTBE production units, built on natural gas fields near equally giant methanol production units, can be converted into ETBE plants. If necessary, these units will be converted into di-isobutene production units, which has already happened in Alberta (Canada) and in Venezuela [17]. In both cases the units were licensed by Snamprogetti, which also holds a licence for the production of di-isobutenes.

Generally speaking, the ether fuel production unit, whether MTBE or ETBE, consists of two main sections, after a stage where the hydrocarbon feedstock is washed with water to remove any basic impurities and the nitrile residues, the main poison of acid catalysts. Alcohol and hydrocarbon feedstock containing isobutene (or isoamylenes) are then mixed and treated in the reaction section which contains an ion exchange resin. The operation is carried out at a temperature of below 100°C. The etherification process must be carried out in liquid phase, which means that a pressure of about 1.5 MPa is generally required. The main licence holders propose various reactor technologies. Axens/IFP Energies nouvelles [18], for example, propose fixed bed or expanded bed reactors, while UOP proposes fixed bed reactors. For both licence holders, the cooled effluents are recycled. The Snamprogetti technology consists of water-cooled multitubular reactors while CdTech uses bubbling beds operating at lower pressure, the reaction heat being in this case eliminated by partial vaporisation of the mixture. In all cases, the same strategy is used. The reaction temperature is controlled to avoid runaway resulting in uncontrolled polymerisation of the olefins in the feedstock.

This reaction section is followed by the distillation section (debutaniser or depentaniser) to recover ether of the required purity at the bottom of the column and hydrocarbons containing unreacted alcohol at the top. The light hydrocarbon cuts, especially olefins, form azeotropes with the alcohols. As a result, most of the alcohols are recovered at the top of the column, even though they have the highest boiling point of the mixture [7].

This property has in fact found a major application. From the mid 1980s, CdTech marketed a catalytic column technology (CdColumn) which increases the conversion of reactive olefins by placing the catalyst in the top part of the debutaniser and takes advantage of the fact that there is no more ether in this area to generate the production of additional ether by shifting the chemical equilibrium. The same catalyst as that in the main reaction section is placed in bags located in the column in an area where both distillation and reaction take place. Other licence holders [19] also propose similar technologies. UOP with Katamax [15] places structured packings containing resins. Axens/IFP Energies nouvelles proposes its Catacol technology [20-23] with, in the top section of the column, a sequence of catalyst beds. Condensed liquid from the top part of the column flows upwards through the beds. The beds are separated by distillation trays. This method, shown on Figure 2.12, in which the catalyst is used in bulk, represents by far the cheapest solution.

Industrially, ethers are obtained practically pure, whether or not by catalytic distillation. With this method, however, for alcohols higher than methanol, it is sometimes impossible to avoid the presence of larger quantities of alcohol in the mixture of ethers produced. To comply with the vapour pressure specifications in the fuel pool, it may be important to limit this quantity of alcohol at the bottom of this column. With MTBE, the methanol content in ether

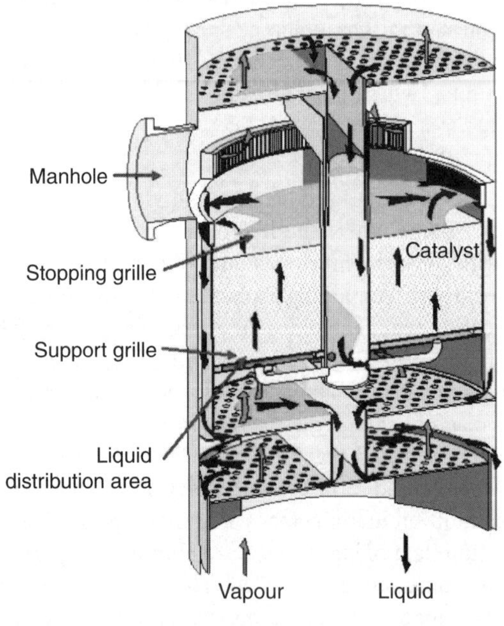

Figure 2.12

Internal components of a Catacol column (Axens/IFP Energies nouvelles catalytic distillation technology).

is limited to 1 000 ppm by weight. With ETBE, it is impossible to achieve less than 1% by weight of ethanol in ether.

Note that the higher the molecular weight of hydrocarbons and the higher the alcohol content at the top of the ether separation distillation column, the easier it will be to purify the corresponding ethers.

The raffinate from the etherification unit, which is the distillate from the distillation column, generally contains 1 to 5% by weight of alcohols.

In this case, two strategies can be considered to reduce the alcohol content. The method used most frequently in existing industrial units is to extract the alcohols by washing with water, which reduces the alcohol content in the purified raffinate by at least 50 ppm. This is extremely important since, very often, this raffinate will act as feedstock for an alkylation unit which is relatively intolerant to oxygenated compounds.

The aqueous fraction, which may contain up to 20% by weight of alcohol, is then distilled to produce virtually pure water at the bottom of the column (which is recycled to the extraction column) and the alcohol fraction at the top of the column. With methanol this fraction contains less than 50 ppm by weight of water but for all the other alcohols an azeotropic mixture is obtained with almost 10% by weight of water (Figure 2.13). This excess alcohol is recycled, with the disadvantage of recycling large quantities of water unless an alcohol dehydration system is used. Water inhibits the etherification reaction and therefore reduces the ether formation rate. Moreover, it competes with the alcohol in an addition reaction on the olefins (mechanism similar to that found in the formation of ethers), to supply tertiary alcohols contained in ethers and whose content in the gasoline pool has been limited by the government.

The other strategy (Figure 2.14) consists in adding an intermediate solvent in the hydrocarbon feedstock of the etherification unit in order to benefit from azeotropic distillation of the effluent from the etherification unit, to obtain at top of column the mixture of this solvent with the alcohol, in side stream the raffinate sent to any downstream alkylation units, and at bottom of column the ethers.

We can see that, in this case, it seems difficult to keep a low alcohol content in this raffinate, apart from using a second separation column to obtain the ethers at the bottom of the first column, the raffinate at the bottom of the second and the intermediate solvent containing the alcohol to be recycled at the top of the second. Obviously, if there is more excess alcohol, greater quantities of intermediate solvent will be required as well as a larger etherification unit, since addition of solvent amounts to diluting the content in reactive olefins in the feedstock. Neste Oy, which developed and commercialised the latter variant (NExETHERS technology) uses a C_3 cut as intermediate solvent. In this case, however, the azeotrope formed with the alcohol is limited to about 1% by weight of alcohol, which means that large quantities of this intermediate solvent must be used and redistilled.

In the end, with both technologies, the operator must limit the stoichiometric excess of alcohols with respect to the reactive olefins, and therefore limit the level of conversion of the reactive olefins.

Eliminating the alcohols from the etherification unit effluent, which would allow recycling without water or intermediate solvent, would solve the problem completely. In a

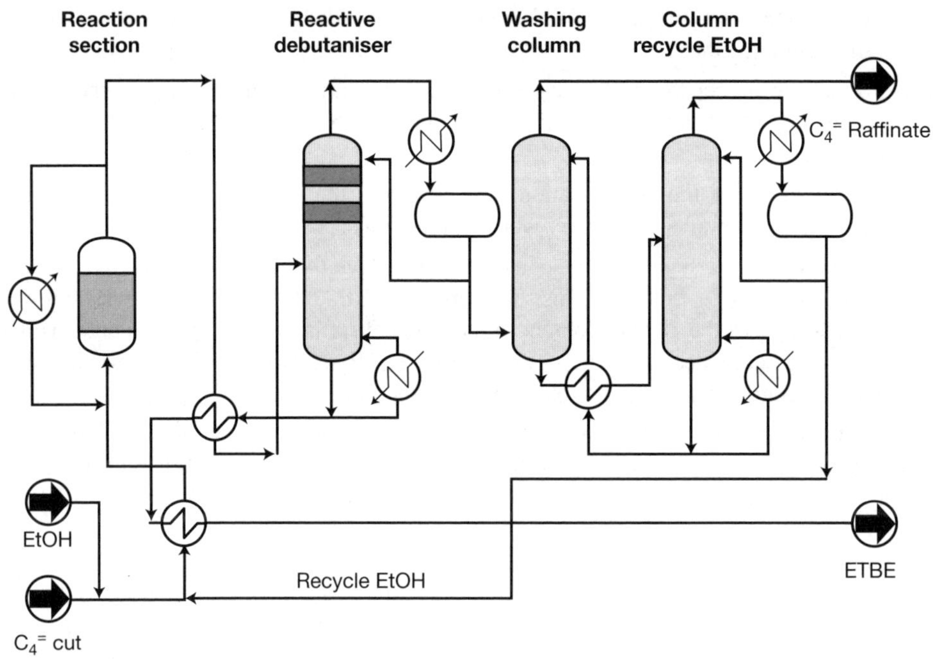

Figure 2.13

ETBE production unit with water extraction of the alcohol contained in the raffinate.

variant of its Ethermax process, UOP suggests the use of molecular sieves to do this. The alcohol is absorbed on numerous sieves which must be regenerated about ten times an hour, making the process difficult to implement due to its inherent operating constraints.

A more attractive technology, which still remains to be developed however, would also allow inexpensive conversion of a TAME unit into an ETAE unit (replacing methanol by ethanol). In this case, due to the low reactivity of ethanol on isoamylenes, large excesses of ethanol are required, which prove catastrophic in terms of recycling of water and/or auxiliary solvent.

This would also open the way for production of ethers from isopropanol (isopropanol is produced by hydration of propylene or hydrogenation of acetone, an excess product on the market due to its coproduction with phenol). We know that IPTBE (isopropylterbutylether) is by very far the best ether fuel [24].

This would leave the road open for the use of other heavier alcohols, alone or mixed, for example for the etherification of glycerol, co-product with VOME, whose use is becoming a matter of concern.

The other distinctions between the production of MTBE and ETBE are due to the different simple or crossed physical properties of these alcohols with their ether.

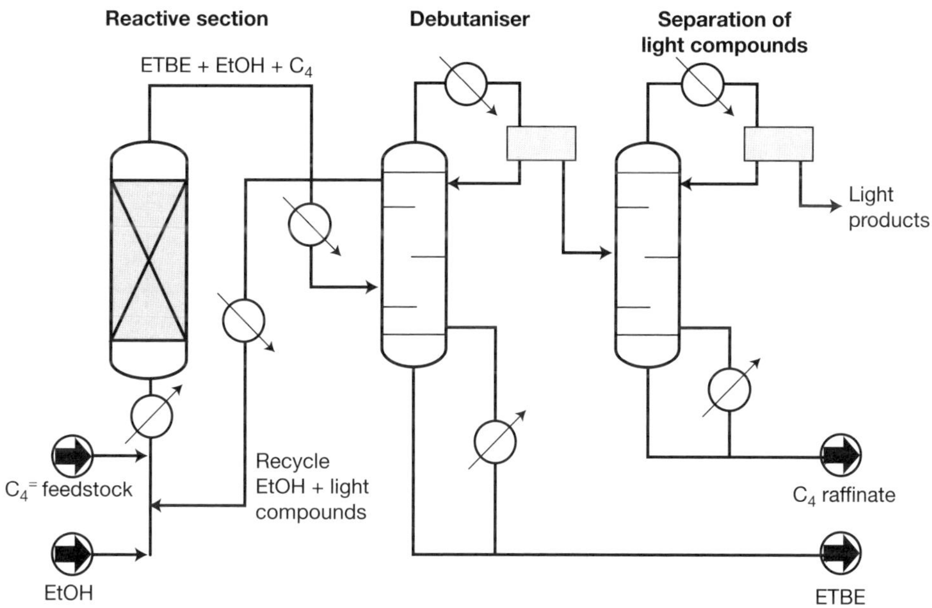

Figure 2.14

ETBE production unit: distillation pathway with elimination of alcohol contained in the raffinate.

Ethanol proves more difficult than methanol to extract from the raffinate at the top of the debutaniser. After water extraction, and all other things being equal, no more than 50 ppm by weight of methanol are left in the raffinate, generally with 5 ppm of MTBE (and virtually all the DME), whereas the alcohol content in case of ETBE may reach 200 to 300 ppm by weight and less than 5 ppm by weight of ETBE, but no diethylether. This ethanol content is a problem for the alkylation unit downstream from the ether production unit. The problem of the remaining alcohol, in particular ethanol, can be removed by installing an additional distillation column operating at low pressure (5 bar) with total reflux of hydrocarbons. The water saturating the raffinate (150 to 400 ppm by weight) is recovered in a boot of the reflux drum. This water contains practically all the alcohol which escaped water extraction upstream. In the presence of DME, this product is only partially condensed in the reflux drum and can be purged in gaseous form, resulting inevitably in loss of hydrocarbons. While solutions recommending the use of molecular sieves are also available, they make the operation considerably more complicated, especially since the saturating water must be eliminated before hoping to trap alcohols and ethers.

The debutaniser also behaves differently depending on whether MTBE or ETBE is being processed. With methanol, the alcohol/hydrocarbon azeotropes contain much more alcohol [25]. This is also true for the alcohol/ether azeotrope. The result of these physical properties is summarised in Figure 2.15 which, using a digital simulation, calculates the alcohol content in the liquid phase at liquid/vapour equilibrium at various levels in the debutanisation

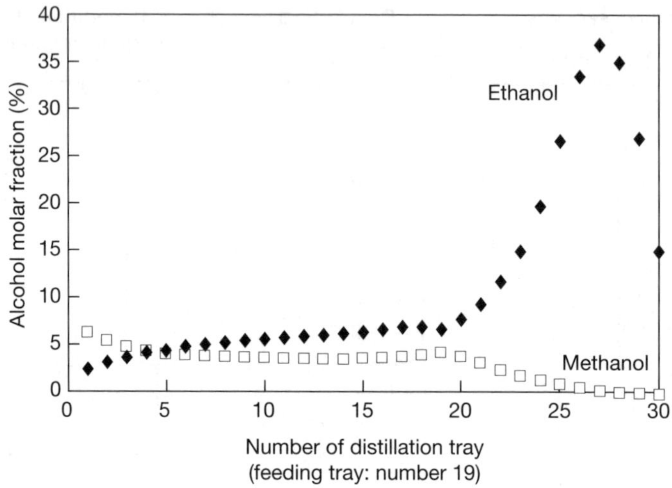

Figure 2.15

Alcohol concentration profiles in the debutaniser.

column of the etherification unit. Typically, at the top of the debutaniser we obtain between 2.5 and 4% by weight of ethanol whereas, at best, we obtain 2% by weight of ethanol in the equivalent raffinate. At the bottom of the debutaniser, we can easily obtain less than 1 000 ppm of methanol in the MTBE, whereas there will be at least 1.5% by weight of ethanol in the ETBE. In practice, about 5% by weight of ethanol is obtained on industrial scale.

This situation proved worrying for the refiners. They were concerned that once gasoline is formulated with ETBE, ethanol would mix with the water often present at the bottom of the storage tank which would mean that, when being drawn off, the water would require special treatment or that, inversely, too much water would accumulate in the fuel, causing the phases to separate in the tanks during cold weather. Experience has shown that these concerns were completely unfounded since ETBE acts as water compatibiliser. As soon as this became apparent, no-one insisted that the ethanol content in ETBE produced industrially should be kept below 5%. If necessary, ethanol and ETBE can be separated by distillation [25], organic membrane [26], or by the pervaporation-distillation combination [27a].

Lastly, as with all oxygenated fuels, mixing them with gasolines lowers the gasoline sulphur content, provided however that the hydrocarbon cuts used to produce the ethers are correctly desulphurised. An "extractive Merox" can be used to obtain ETBE or an MTBE containing less than 2 ppm by weight of sulphur. The residual sulphur derivatives are disulphides (dimethyl disulphide or methyl ethyl disulphide), which go through the etherification unit and end up in the ether, and sulphides (methyl and ethyl tert-butyl sulphides) obtained by addition of methyl and ethyl mercaptans on isobutene, a balanced reaction under the ether synthesis conditions [7]. These sulphides also act as inhibitors for the ether formation reaction.

2.4 MAIN INDUSTRIAL PLAYERS

The main European manufacturers of ethanol production plants include Abengoa, Vogelbusch, Technip and Maguin-Interis.

The major European producers [27b] are: Abengoa Bioenergia (Spain), Tereos (France) and CropEnergies AG (Germany). Each of these three companies has a European production capacity of 770 000 m^3 ethanol/year. They also have sites in the United States and Latin America. Abengoa for example has a capacity of 750 000 m^3 in the United States and 130 000 m^3 in Brazil. In 2008, Guarani, the Brazilian subsidiary of Tereos, produced 0.5 Mm^3 in Brazil.

Still in Europe, Cristanol (France), Agrano Group (Austria), Verbio AG (Germany), Agroetanol (Sweden) and IMA (Italy) come next.

The most important company in the United States is clearly ADM (Archer Daniels Midland), followed by POET and VeraSun Energy. Large companies also operate in India and China. There are also numerous other smaller companies throughout the world.

2.5 WORLD ETHANOL PRODUCTION

The world production of ethanol has developed dramatically, with an average annual increase of 20% between 2005 and 2008. This trend slowed down in 2009.

The United States has progressively become the leading global producer of ethanol fuel, ahead of Brazil. It is also the largest consumer.

In 2007, world production of ethanol was 60 Mm^3, distributed as follows: 42% in the United States, 33% in Brazil, 11% in Asia and 7% in the European Union (*source*: Wood Mackenzie). In 2008, production reached 77 Mm^3, out of which 77% was produced by the United States and Brazil. In 2007, 23% of the corn harvested in North America was converted into ethanol, a figure of about 80 Mt. The USDA predicts that a maximum of 35% of the corn harvested would be dedicated to ethanol production in 2018.

In 2008, consumption of ethanol fuel in the United States amounted to about 9 billion gallons (approximately 33 Mm^3). The production capacity was 45 Mm^3, with over 100 industrial units.

Through the Renewable Fuels Standard (RFS), the American energy policy plans increasing production objectives over the next few years, up to 15 billion gallons of first generation fuels, i.e. almost 60 Mm^3, mostly ethanol.

In Asia, India and China have major growth objectives in this field. In China, for example, ethanol production, which was about 2 to 2.5 Mm^3 in 2010, is expected to reach 12 Mm^3 in 2020.

Concerning Brazil and according to the UNICA (Brazilian Sugarcane Industry Association), ethanol exports dropped significantly during the period from April to October 2009, compared with the same period in 2008: about 20% to Europe, i.e. 665 000 m^3 versus

828 000 m^3. Exports to the United States dropped by 1.2 Mm^3 to 152 000 m^3 during the same period. The decline in exports can be largely attributed to the drop in the cost of oil and the application, requested by importers, of environmental certification schemes characterising this type of manufacture in Brazil. Note that ethanol exports from Brazil to Europe amounted to 1.5 Mm^3 in 2008, 1 Mm^3 in 2007 and 580 000 m^3 in 2006.

2.6 ETHANOL AND ETBE, FUELS

2.6.1 Ethanol fuel

It was obviously not by chance that ethanol was identified as a biofuel. Ethanol exhibits highly interesting physical and chemical properties for spark ignition engines, even through certain technical difficulties remain.

A European ethanol standard is currently still being produced (prEN 15376 final draft) (Table 2.7). This standard will apply to the ethanol used as fuel in 5% v/v mixture with regular gasoline.

Table 2.7 European standard prEN 15376/Bioethanol.

Properties	Units	Min.	Max.	Test method
Ethanol	% m/m	98.70	–	EC/2870/2000 method I
Saturated volatile monoalcohols (C3-C5)	% m/m	–	2.0	EC/2870/2000 method III
Methanol	% m/m	–	1.0	EC/2870/2000 method III
Water	% m/m	–	3.0	EN 15489
Inorganic chlorine	mg/l	–	20.0	EN 15484
Copper	mg/kg	–	0.1	EN 15488
Total acidity (as acetic acid)	% m/m	–	0.007	EN 15491
Appearance	–	Limpid and clear		Visual inspection
Phosphorus	mg/l	–	0.5	EN 15487
Non volatile compounds	mg/100 ml	–	10.0	EC/2870/2000 method II
Sulphur	mg/kg	–	10.0	EN 15485, EN 15486

Concerning E85, the situation is as follows:

- in Sweden, the most advanced country in the field of E85, a specific standard was introduced (SS 15 54 80) in 2006,
- in the rest of the EU, a draft standard (CEN Workshop Agreement CWA 15293) developed and adopted in 2005 should shortly lead to an appropriate standard which would include a large part of the Swedish standard,
- E85 can be transported under number UN 1993 corresponding to ethanol.

In the United States, ethanol fuel must comply with ASTM standard D 4806 (Table 2.8) which applies to mixtures with gasoline for use in spark ignition engines. The ethanol must be denatured. The maximum permitted content is 10% by volume, as specified in ASTM standard D 4814-88 for gasolines, with no technical modifications to vehicles.

At the same time, Flexfuel technology is developing more and more in the United States. In 2004, it was estimated that 188 service stations in 28 states were selling E85.

A specific standard, ASTM D 5798 (Table 2.9), has been drawn up for applications in spark ignition engines dedicated to high percentages of ethanol (E75-E85).

Table 2.8 ASTM standard D 4806 for denatured ethanol fuel.

Properties	Specifications	ASTM methods
Ethanol, % vol.	> 92.1	D 5501
Methanol, % vol.	< 0.5	
Gums, mg/100 ml	< 5.0	D 381
Water, % vol.	< 1.0	E 203
Denaturing agent content (% vol.)	1.96 – 4.76	
Inorganic chlorine, ppm	< 40	D 512
Copper, ppm	< 0.1	D 1688
Total acidity (as acetic acid), % by weight	< 0.007	D 1613
pH	6.5 – 9.0	D 6423
Appearance	Limpid and clear	

Table 2.9 ASTM standard D 5798 for ethanol fuel E75-E85.

Properties	Specifications	ASTM methods
Ethanol, % vol.	70 – 79	D 5501
Hydrocarbons, % vol.	17 – 30	D 4815
Vapour pressure, kPa	38 – 83	D 4953/5190/5191
Lead, mg/l	2.6 – 3.9	D 5059
Phosphorus, mg/l	0.2 – 0.4	D 3231
Sulphur, ppm		D 1266/2622/3120/ 5453
Methanol, % vol.	0.5	D 4815
Alcohols C_3-C_8, % vol.	2	D 4815
Acidity, % by weight	0.005	D 1613
Soluble gums, mg/100 ml	5	D 381
pH	6.5 – 9.0	D 6423
Insoluble gums, mg/100 ml	20	D 381
Chlorides, ppm	2	D 4929 B
Inorganic chlorine, ppm	1	D 512/2988
Copper, mg/l	0.07	D 1688 modified
Water, % by weight	1.0	E 203/E 1064
Appearance	Limpid and clear	

In Brazil, more than 20% of the vehicles are designed to run on 100% ethanol. The other vehicles are designed to run on 22% ethanol mixed with gasoline. The National Agency for Petroleum (ANP) defined the specifications for hydrated and anhydrous ethanol (Table 2.10). The analysis methods are developed by the Brazilian Association of Technical Standards (NBR) and the ASTM.

Table 2.10 Brazilian ethanol specifications.

Properties	Min/ Max	Anhy-drous ethanol	Hydrated ethanol	NBR method	ASTM method
Appearance		Limpid and clear		Visual	Visual
Colour		Colourless or yellow		Visual	Visual
Total acidity, mg/l	max	30	30	9 866	D 1613
Electrical conductivity, µS/m	max	500	500	10 547	D 1125
Relative density at 20°C, kg/m^3	max	791.5	807.6 – 811.0	5 992	D 4052
Alcohol content °INPM	min	99.3	92.6 – 93.8	5 992	–
pH		–	6 – 8	10 891	–
Evaporation residue, mg/100 ml	max	–	5	8 644	–
Hydrocarbon content, % vol.	max		3.0	13 993	
Chlorides, ppm	max	–	1	10 894/10 895	D 512
Ethanol, % vol.	min	99.3	92.6	–	D 5501
Sulphates, ppm	max	–	4	10 894/12 120	–
Fe, ppm	max	–	5	11 331	–
Na, ppm	max	–	2	10 422	–
Copper, ppm	max	0.07	–	10 893	–

2.6.1.1 Physical properties

The main physical properties of ethanol are summarised in Table 2.11.

A number of useful properties for spark ignition engines appear immediately:

- very high octane number. This characteristic offers significant freedom to optimise engine operation in terms of compression rate and ignition advance, for example, to reach high efficiencies and reduce fuel consumption,
- density close to that of gasolines,
- miscible in all proportions with gasoline,
- presence of oxygen, which produces a more homogeneous fuel/air mixture and therefore reduces the emissions of unburnt hydrocarbons and partial oxidation products (HC, CO),
- high latent heat of vaporisation which leads to what is commonly called the "cooling effect" which improves the combustion chamber filling efficiency.

Table 2.11 Physical properties of ethanol.

	Ethanol	Gasoline (standard)
Molecular weight (g/mol)	46.07	102.5
C (wt%)	52.2	86.5
H (wt%)	13.1	13.5
O (wt%)	34.7	0
Relative density (kg/m^3)	794	735-760
Latent heat of vaporisation (kJ/kg)	854	289
Boiling point (°C)	78.4	30-190
Net calorific value per unit weight (kJ/kg)	26 805	42 690
Net calorific value per unit volume (kJ/l)	21 285	32 020
Stoichiometric ratio [a]	8.95	14.4
RON	111	95
MON	92	85

[a] *Air/ethanol or air/hydrocarbon weight ratio of gasolines, calculated using the ethanol or gasoline combustion stoichiometric equation.*

In contrast, a number of disadvantages must be considered:

- the presence of oxygen in the molecule (30% by weight) results in a lower energy content than that of gasolines and therefore higher fuel consumption,
- the high latent heat of vaporisation may cause poor running when cold, especially when starting,
- ethanol forms azeotropes with the light hydrocarbons, which may cause volatility problems,
- ethanol is miscible with water, which may create demixing problems in mixtures with hydrocarbons,
- the high oxygen content of ethanol and its ability to form acetic acid by oxidation are responsible for incompatibilities with some metallic materials or polymers,
- aldehydes, which have an adverse impact on health, are emitted on combustion of ethanol in engines.

2.6.1.2 Octane numbers

Ethanol offers advantages due to its high octane numbers. Incorporated at contents of 5 to 10% by volume in gasolines, the octane numbers of the ethanol mixture are about 120-125 and 95-100 respectively for the RON and the MON. At 5% content, gains of about 0.8 to 1.3 can be obtained on the RON and the MON. This is a very important advantage giving ethanol true added value. Lastly, note that at high content, the high octane numbers remain a major advantage. The compression ratio can therefore be increased, which improves the thermal efficiency.

2.6.1.3 Volatility

Ethanol is moderately volatile with a relatively high boiling point (78°C), compared with the initial distillation point of gasolines. In addition, the vapour pressure, calculated according to Antoine's law, is low for ethanol compared with the light fractions of fuels (Figure 2.16).

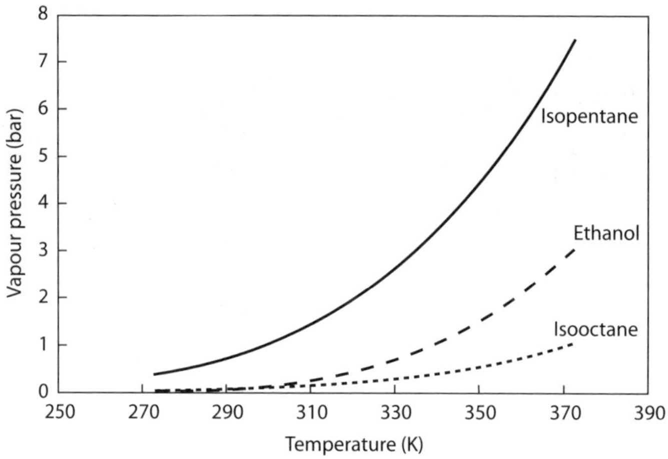

Figure 2.16

Vapour pressures of ethanol and components of a standard gasoline.

A physical property of ethanol must nevertheless be considered here: ethanol can interact strongly with some light hydrocarbons to form azeotropes. The main feature of these azeotropes is that their boiling points, and consequently their vapour pressures at a given temperature, are quite different from those of each of its components. With ethanol, the boiling points of the azeotropes are lower than the two initial boiling points.

For example, the boiling point of an azeotrope formed from ethanol and *n*-pentane is 34°C, which is less than those of ethanol (78°C) and *n*-pentane (36°C) (see Table 1.5).

The vapour pressure of gasoline bases increases regularly and in proportion to the quantity of ethanol added up to an alcohol content of about 5% by volume. For example, when 5% ethanol is incorporated in a standard gasoline, the vapour pressure of the mixture may increase by 80 to 90 mbar. This increase stabilises for mixtures containing 6% ethanol. For mixtures with a higher alcohol content, the vapour pressure then decreases (see Figure 2.17).

Table 2.12 shows that incorporating 10% ethanol by volume in various more or less light gasolines increases the vapour pressure of the mixture by 80 to 100 mbar [28].

In a context where pollutant emissions are controlled, emissions by evaporation are now regulated and the new fuel specifications which came into effect in 2000 have significantly reduced the vapour pressures of gasolines in order to limit these losses. Incorporation of ethanol goes against this trend.

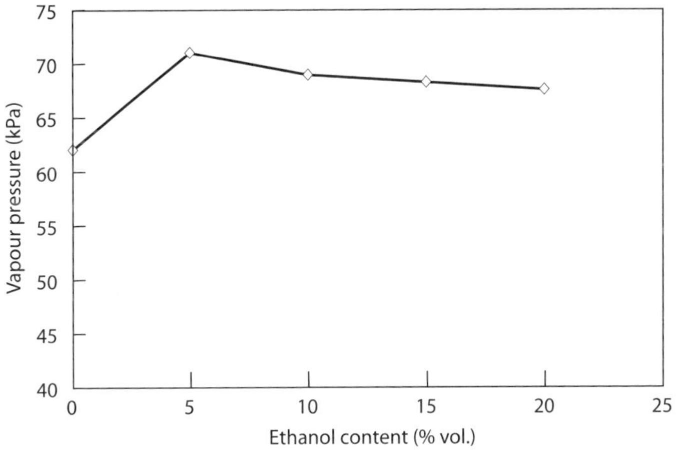

Figure 2.17

Graph of vapour pressure against ethanol content of the gasoline-ethanol mixture.

Table 2.12 Change in vapour pressure of mixtures of various gasolines and ethanol incorporated at 10% by volume.

Vapour pressure of the gasoline base (mbar)	Vapour pressure of the 10% by volume ethanol mixture (mbar)	Increase in vapour pressure (mbar)
373	469	96
524	607	83
697	780	83

To overcome this phenomenon, ethanol would have to be incorporated in a low-volatility gasoline or in a gasoline whose light fractions responsible for causing these azeotropes have been removed. This means that splash blending of ethanol is extremely difficult to implement.

Obviously, this phenomenon does not exist when pure ethanol is used. As already pointed out, however, the volatility of pure ethanol is very low, which may induce cold starting problems. The simplest way of avoiding this problem is to replace pure ethanol by mixtures containing 85% ethanol and 15% light hydrocarbons, known as E85.

2.6.1.4 Stability in the presence of water

Ethanol and water are miscible in all proportions. With standard gasolines, the presence of water in fuel is not a serious problem. Up to a concentration of 50 ppm at ambient temperature, water remains completely soluble, in fact. Above this value, water separates out naturally and the aqueous layer can be removed if necessary.

In the presence of ethanol, however, traces of water may generate complete demixing of the ethanol, resulting in highly visible clouding, especially at low temperature; a two-phase mixture then forms, with a first water-ethanol phase and another phase containing the hydrocarbons with a reduced ethanol content. This phenomenon is penalising in two respects. Firstly, there is a risk of increasing the volume of the aqueous layer at the bottom of the tank in service stations or vehicle tanks and, secondly, there is a risk of reducing the octane numbers of the mixtures concerned.

This phenomenon can be avoided by incorporating co-solvents like TBA, as in the United States in the 1980s, or by closely monitoring the storage conditions which must protect against humidity.

Lastly, this phenomenon disappears if ethanol at very high content is used. In this case, since water is totally miscible in alcohol, there is no risk of demixing.

2.6.1.5 Cold starting

We have already mentioned that the volatility of pure ethanol is very low. Its high boiling point (78°C) combined with high latent heat of vaporisation is penalising during cold starting [29, 30].

There are several ways of minimising these drawbacks, such as using E85 (85% ethanol) instead of pure ethanol. Adding 15% light hydrocarbons significantly increases the volatility, which makes cold starting easier. This modification must be carried out with caution, however, to avoid losing some of the advantages of ethanol (octane number, reduced HC and CO emissions).

Other technical solutions have been studied, especially in Brazil. The methods used most frequently include heating the injection system, in particular the injection rail, and using a second fuel tank containing a highly volatile gasoline to start the engine for a few seconds before switching over to ethanol.

2.6.1.6 Compatibility of materials

Ethanol causes swelling and reduces the mechanical performance of rubber components, since rubber absorbs fuel [30], possibly leading to operating incidents and leaks. This situation can be solved through the use of compatible materials such as rubbers with high fluorine content (Viton) [31]. Nylon can also prove resistant, but only at low temperatures (< 30°C) [32]. This polymer can therefore be used for the fuel supply line, provided that the fuel temperature remains low.

Galvanic corrosion is characterised by contact between 2 metals, generally increased by the presence an electrolyte conductor. Corrosion may occur, for example, when stainless steel and aluminium are in contact.

The electrical conductivity of standard gasolines (no additives) is in the region of 10^{-8} to 10^{-6} $\mu S \times cm^{-1}$. The electrical conductivity of ethanol is $1.35.10^{-3}$ $\mu S \times cm^{-1}$. In addition, the possible conversion of ethanol into acetic acid leads to a significant increase in electrical conductivity (41 $\mu S \times cm^{-1}$ for a 0.1 molar solution of acetic acid). The presence of acetic acid may therefore considerably increase galvanic corrosion and chemical attack. Carbon steel, stainless steel and bronze are recommended for use in engines running on ethanol. Metals such as magnesium, zinc, brass and copper must be banned [33]. The presence of water considerably worsens these phenomena [34]. It is therefore essential to use anhydrous ethanol, to avoid corrosion of some engine parts. This parameter remain the main factor limiting the development of ethanol, both as regards its supply and storage. European studies aimed at drawing up ethanol specifications are currently in progress.

2.6.1.7 Ethanol in spark ignition engines

In view of its characteristics, various approaches can be adopted to use ethanol in engines.

A. Ethanol at low content in gasolines

When ethanol is added to gasoline in small quantities, maximum contents of 5% or 10% are proposed, corresponding to oxygen contents of 1.85% and 3.7% by weight respectively. Incorporated at these contents, it is possible to switch between gasolines with or without ethanol. All European vehicles proposed by manufacturers can run on these E5 fuels. The same is true in the United States with E10 mixtures. The main adaptations concern the use of 100% compatible materials.

During combustion, addition of 5 to 10% ethanol in gasolines:

- reduces emissions of CO and unburnt hydrocarbons by up to 15%,
- tends to increase NOx emissions, by up to 5%. With some types of vehicle, however, a reduction of NO_X emissions has been observed. This could be due to the lower combustion chamber temperatures related to ethanol's high latent heat of vaporisation,
- significantly increases the emissions of aldehydes, in particular acetaldehyde,
- increases the risks of inlet valve clogging, which can easily be corrected through the use of a suitable additive in the mixture.

Lastly, note that the risks of increasing emissions by evaporation are very high due to the increases in vapour pressure described previously. This is a critical point since, in this case, emissions by evaporation concern light hydrocarbons, paraffins or olefins, olefins having one of the highest ozone formation potentials. It is therefore essential to control this problem for E5-E10 fuels.

B. Ethanol at high content in gasolines

The properties of ethanol described previously, in particular the octane numbers, demonstrate its potential as fuel in spark ignition engines. This allows use in a wide field of applications. In addition to use in mixtures at low contents, for instance, studies have often been conducted on use of the pure product. In this context, however, the following two points must be taken into account:

- due to some of the properties of ethanol, mixtures with very high ethanol contents are distributed (85% ethanol and 15% gasoline) instead of pure ethanol to overcome the problems of cold starting. The hydrocarbon fraction provides the mixtures with sufficient volatility to allow acceptable cold running operation,
- generally speaking, since the distribution of E85 type mixtures is geographically limited, it must be possible to fill up with either E85 or a traditional gasoline.

In this context, the widest possible range of characteristics must be taken into account to define an adaptation allowing vehicles to run either on E85, pure gasoline or an intermediate mixture after filling up with fuel. The main characteristics to be taken into account are the fuel NCV per unit volume and oxygen content, the combustion vaporisation and initiation characteristics and the interaction between fuel and materials.

To obtain a vehicle capable of running on E85 or gasoline while offering good driving comfort and compliance with statutory requirements concerning polluting emissions, it must be possible at all times to adapt the quantity injected and the combustion timing to maintain performance as well as to regulate the mixture equivalence ratio to stoichiometry (since ethanol NCV is 33% lower than that of gasoline, 35% more ethanol must be injected to maintain performance, for identical engine efficiency).

A specific system must therefore be developed and implemented, which will be capable of:

- detecting at all times the ethanol content of the mixture present in the tank. This information can be obtained by using a detector, which generally operates by measuring the conductivity and the dielectric constant. Looping via a specific probe could also be used,
- modifying the quantity injected, the injection advance and the ignition advance depending on the mixture present in the tank, i.e. depending on the information received via the measurement described previously. It is therefore essential to develop and install a specific computer.

Some of the materials on the vehicle will also have to be changed to ensure they are perfectly compatible with fuels of very high ethanol content (metallic materials and polymers, on the engine and on the vehicle fuel supply circuit).

The Fuel Flexible Vehicle (FFV) solution is quite attractive and is currently looked upon favourably throughout the world as a means of accompanying the development of ethanol fuel. However, since the fuel tank remains unchanged, the autonomy when running on E85 is considerably reduced compared with a tank full of gasoline, due to the differences in energy content between these two types of fuel.

Sale of E85 in France has been authorised since 1 January 2007, following in the footsteps of Brazil, the United States and Sweden. In December 2007, 4 000 Flexfuel vehicles were registered in France, with 191 service stations selling E85, at an average price of 0.815 euros/litre (French price recommendation by a "charter for the development of E85" signed by the French state and the various players involved in the production, distribution and use of this fuel), at a time when SP95 unleaded gasoline costs 1.27 euros/litre.

In 2006, there were about six million FFV vehicles throughout the world. Ford was the first manufacturer to sell this type of vehicle in 1996, with the Taurus FFV. Since then, virtually all the major car manufacturers have developed their own models.

Flexfuel vehicles are sold at roughly the same price as gasoline models of equivalent category.

C. Another possible application

Like the situation some twenty years ago with methanol, the development of a dedicated engine could offer attractive perspectives in terms of consumption and therefore in terms of CO_2 emissions. By increasing the engine compression rate via high octane numbers, it is quite feasible to obtain dedicated highly efficient engines.

2.6.1.8 Ethanol in Diesel engines

Ethanol is significantly different from conventional Diesel in terms of octane number, calorific value, auto-ignition temperature, vaporisation characteristics and boiling point.

The cetane number of pure ethanol, poorly soluble in Diesel, is just 8, while the European specification (EN 590) for Diesel requires a cetane number above 51.

In addition, the lubricating power of ethanol is too low to be compatible with current high-pressure Diesel injection pumps. The low boiling point of ethanol compared with a Diesel increases the risk of cavitation and its high conductivity increases the risks of corrosion.

A. Use of pure ethanol

In Sweden, Scania has manufactured buses running on ethanol mixed with a procetane additive and an additive to improve the lubricating power. This fuel is known as *Etamax D*. A number of technical modifications were also required, such as higher compression rate, use of a special injection system and use of a specially designed catalyst to control aldehyde emissions.

Etamax D consists of precisely 90% hydrated ethanol (to 95%), 7% procetane additive (Beraid 3540), denaturing agents (2.3% MTBE and 0.5% isobutanol) and 125 ppm of corrosion inhibitor additives.

Buses fuelled on ethanol are also running in Mexico, Australia and Denmark.

In all cases, heavy goods vehicles running on these mixtures with 95% ethanol (E95) release far less particulates and less NO_x than when running on Diesel. The results observed for CO and HC emissions, however, vary considerably depending on the application.

Emissions from Scania buses fuelled with *Etamax D* are lower than those obtained with pure Diesel concerning the regulated pollutants, but higher than those from buses running on natural gas. Compared with operation using Diesel, the Scania tests conducted with *Etamax D* give the following results:

- 92% reduction in CO emissions,
- 80% reduction in HC emissions,
- 80% reduction in particulate emissions,
- 28% reduction in NO_x emissions.

However, the use of additives significantly increases the cost of this solution, and technological solutions (spark plug and/or glow plug, exhaust gas recycling) must be found to overcome this disadvantage.

B. Use of an ethanol-Diesel (E-Diesel) mixture

Use of a mixture of Diesel with a low concentration of ethanol (or emulsion), together with specific additives guaranteeing the solubility of ethanol in Diesel and the stability of the emulsion, represents an alternative. This fuel, called E-Diesel, is a mixture of more than 15% ethanol and more than 5% additives in Diesel. The problems with this type of fuel concern safety and adaptation to Diesel engines (low cetane number, low lubricating power, stability).

Studies have been conducted to find additives capable of forming microemulsions of ethanol with Diesel. These emulsifying additives are required even for ethanol contents of about 5% in Diesel to prevent the ethanol from separating out of the Diesel at very cold temperature or if the mixture is contaminated by water. The ethanol/biodiesel mixture could therefore represent a fuel with very high potential, since ethanol can help reduce NO_x emissions and improve the cold properties, the two weak points of biodiesel. In addition, biodiesel can solve the problem of ethanol's low lubricating power, while both ethanol and biodiesel favour the reduction of particulates.

Apart from the problems of solubility and stability of the emulsion, the additives offer other gains in terms of lubricating power, detergence and cold properties, as well as a procetane additive.

A mixture of additives of this type, Puranol, has been developed and sold in the United States by Pure Energy Corporation. The results obtained with E-Diesel containing Puranol show a significant reduction in CO (– 27%), NO_x (– 5%) and particulates (– 41%) emissions. This emulsion consists of 83% to 94% Diesel No.2, 5% to 15% ethanol and 1% to 2% Puranol.

In terms of safety, it has been demonstrated that the risks associated with mixture of Diesel with 10-20% ethanol are the same as those of pure ethanol. Consequently, E-Diesel must be considered in the same risk class as a gasoline since the ignition temperatures are similar.

Furthermore, problems of resistance of materials have also been encountered, as well as problems of wear and corrosion in the injection pump and the injectors, due to ethanol's low lubricating power and risks of cavitation.

The presence of water increases the risk of ethanol separating out of the Diesel, but most manufacturers of emulsifying additives guarantee stability of the emulsion up to 3% water.

Numerous studies have been conducted in Brazil, a country which is extremely interested in the use of ethanol in Diesel engines due to its high alcohol production.

Currently, even in Europe, interest in this type of fuel is growing and several studies have been conducted in various countries.

2.6.2 ETBE-fuel

As we have just seen, some limitations remain concerning the use of fuels with low ethanol contents. Faced with the increased use, during the 1980s, of MTBE mixed in gasolines, the conversion of ethanol into ETBE by reaction with isobutene became an alternative worthy of interest.

Due to the similarity of the two production processes and its characteristics, listed in Table 2.13, ETBE appears as a feasible alternative.

Table 2.13 Physical properties of ETBE.

	ETBE	Ethanol	Gasoline (standard)
Molecular weight (g/mol)	102	46.07	102.5
C (wt%)	70.6	52.2	86.5
H (wt%)	13.7	13.1	13.5
O (wt%)	15.7	34.7	0
Relative density (kg/m^3)	750	794	735-760
Latent heat of vaporisation (kJ/kg)	321	854	289
Boiling point (°C)	72.8	78.4	30-190
Net calorific value per unit weight (kJ/kg)	35 880	26 805	42 690
Net calorific value per unit volume (kJ/l)	26 910	21 285	32 020
Stoichiometric ratio	12.1	8.95	14.4
RON	117	111	95
MON	101	92	85

Firstly, note that ETBE contains 45% by weight of chemically-combined ethanol. Incorporating 5% volume of ethanol in gasolines, as currently stipulated by specifications EN228, corresponds to 11% volume of ETBE, below the limit set at 15% volume.

For the formulator, addition of ETBE offers various advantages since the major problems generated by the use of ethanol, such as vapour pressure and tolerance to water, are removed: the vapour pressure of ETBE-based mixtures is maintained or even slightly reduced, demixing phenomena in the presence of water are non existent and there are no problems of compatibility with the materials.

Lastly, ETBE offers the same advantages as ethanol in terms of higher octane number.

Numerous studies have been published in the United States and Europe on discharge of pollutants from vehicles, and the studies conducted at IFP Energies nouvelles corroborate all the trends observed. Incorporation of ETBE in gasolines, in compliance with applicable European regulations on polluting emissions at the exhaust, leads to:

- 5 to 15% reduction in CO emissions,
- slight reduction (3 to 5%) of unburnt hydrocarbons,
- slight increase (2 to 5%) in NO_x emissions,
- as with ethanol, a significant increase in aldehyde emissions, especially acetaldehyde.

Emissions by evaporation are unaffected by introduction of ETBE, which represents an advantage and a significant added value for this product.

2.7 CHEMICAL CONVERSION OF ETHANOL

2.7.1 Production of hydrogen

Numerous studies are currently being conducted on bioethanol as fuel that could be used to supply a fuel cell and as a source of hydrogen.

Today, hydrogen is mainly produced with hydrocarbons obtained from natural gas, LPG and naphtha. It is used almost exclusively as reactant in the chemical industry (production of ammonia, amines, methanol, etc.) and in the refining industry (hydrotreatment of petroleum products).

In the future, hydrogen could find new applications as energy vector, in stationnary applications and transportation sectors, through its conversion into electricity and heat in fuel cells.

2.7.1.1 Production of hydrogen by ethanol steam reforming

Steam reforming is the industrial process most widely used to produce hydrogen. Ethanol reacts with steam over a specific catalyst to form syngas.

A. Chemistry and reaction mechanisms of ethanol steam reforming

The first step, decomposition of ethanol, leads to the simultaneous formation of acetaldehyde by dehydrogenation (reaction 1a) and ethylene by dehydration (reaction 1b).

Reaction 1a: $CH_3\text{-}CH_2OH \Leftrightarrow H_2 + CH_3\text{-}CH = O$

Reaction 1b: $CH_3\text{-}CH_2OH \Leftrightarrow H_2O + CH_2 = CH_2$

More or less ethylene is formed depending on the operating conditions chosen and the type of catalyst. A catalyst containing acid sites will lead to the formation of large quantities of ethylene. In a second step, acetaldehyde and ethylene are converted by steam reforming into syngas. The primary products of steam reforming are difficult to identify since the secondary reactions of CO conversion (reaction 4) and methanation (reaction 5) are very fast. However, it seems likely that methane and CO_2 are the primary products obtained from acetaldehyde (reaction 2a) and that methane and CO are the primary products obtained from ethylene (reaction 2b).

Reaction 2a: $CH_3\text{-}CH = O + H_2O \Leftrightarrow CH_4 + CO_2 + H_2$

Reaction 2b: $CH_2 = CH_2 + H_2O \Leftrightarrow CH_4 + CO + H_2$

Ethylene may also produce polymeric species which deposit on the catalyst and are responsible for the coke formation at low temperatures (< 600°C). To minimise the formation of coke, by polymerisation of ethylene, ethanol steam reforming must be carried out at high temperatures (> 600°C) on a catalyst containing few or no acid sites. Increasing the temperature improves the ethylene conversion selectivity towards the formation of hydrogen and CO via the steam reforming reaction, whereas reducing the number of acid sites on the catalyst limits the formation of ethylene.

B. Chemical thermodynamics

The main reactions are balanced:

Reaction 3: $CH_3\text{-}CH_2OH + H_2O \Leftrightarrow 2\ CO + 4\ H_2$ $\Delta H°_{298} = 255$ kJ/mol

Reaction 4: $CO + H_2O \Leftrightarrow CO_2 + H_2$ $\Delta H°_{298} = -41$ kJ/mol

Reaction 5: $CO + 3\ H_2 \Leftrightarrow CH_4 + H_2O$ $\Delta H°_{298} = -206$ kJ/mol

The overall ethanol steam reforming reaction becomes:

Reaction 6: $CH_3\text{-}CH_2OH + 3\ H_2O \Leftrightarrow 2\ CO_2 + 6\ H_2$

Ethanol steam reforming leads to the formation of carbon monoxide, which is then converted into CO_2 and H_2, and to the formation of methane according to the balanced reactions 4 and 5. Conversion of ethanol into H_2 and CO_2 is limited by the chemical equilibrium and the composition of the reformed gas depends on the operating conditions (temperature, pressure) and the quantity of steam used. Formation of carbon, via the balanced Boudouart reaction (reaction 7), can be avoided by operating at high temperatures (> 500°C) with steam injection at least equal to the stoichiometry of reaction 3 (H_2O/ethanol ≥ 1).

Reaction 7: $2\ CO \Leftrightarrow CO_2 + C$ $\Delta H°_{298} = -172$ kJ/mol

C. Catalysts and operating conditions

An alumina-supported nickel catalyst is generally used in methane steam reforming. This catalyst is not suitable for ethanol steam reforming, since it generates too much coke [35]. Studies concerning ethanol steam reforming are relatively recent and focus on development of the catalyst. These studies have demonstrated that most group VIII metals can be used to catalyse ethanol steam reforming. Most studies have been conducted on catalysts based on rhodium (the group VIII metal most active in ethanol steam reforming) or based on nickel or cobalt. The choice of support is extremely important. It must exhibit the acido-basic properties required firstly to avoid the formation of ethylene by dehydration of ethanol and secondly to activate the water molecule [36]. Basic supports like MgO, ZnO, CeO_2, etc. are preferred to supports with a certain degree of surface acidity such as aluminas or silica-aluminas which favour the formation of coke.

To achieve high hydrogen yields, the catalyst must work under severe operating conditions, similar to those of methane steam reforming. The compositions of the dry gases at equilibrium can be obtained from curves produced for temperatures between 500°C and 800°C (Figure 2.18), pressures between 1 and 10 bar (Figure 2.19), H_2O/ethanol ratios between 2 and 8 (Figure 2.20).

As with methane, ethanol steam reforming is globally endothermic and requires equipment such as burners or exchangers to compensate for this endothermicity.

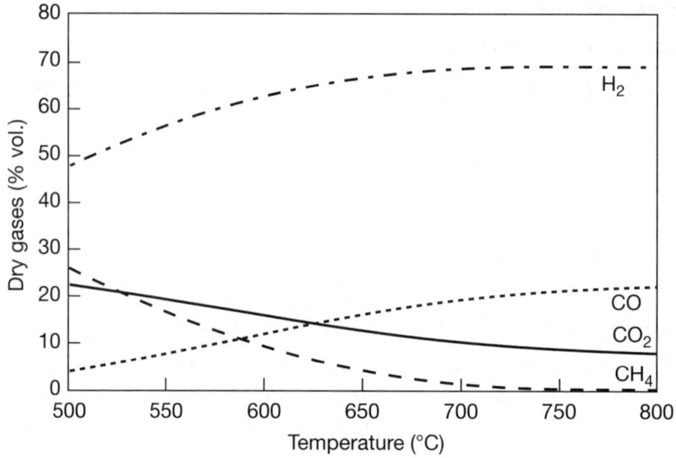

Figure 2.18

Influence of temperature on the composition of dry gases at thermodynamic equilibrium.

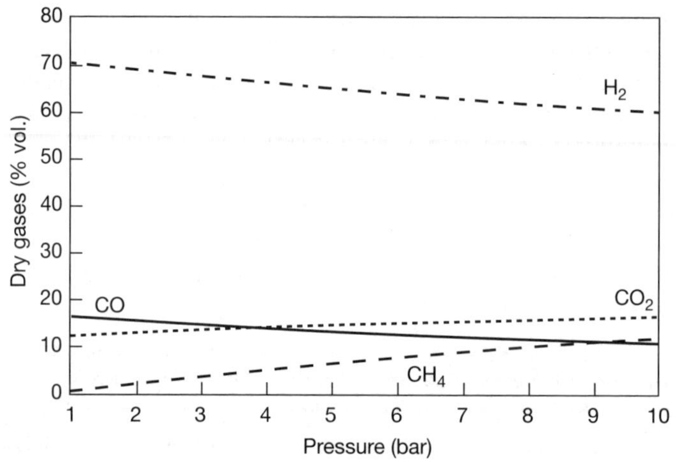

Figure 2.19

Influence of pressure on the composition of dry gases at thermodynamic equilibrium.

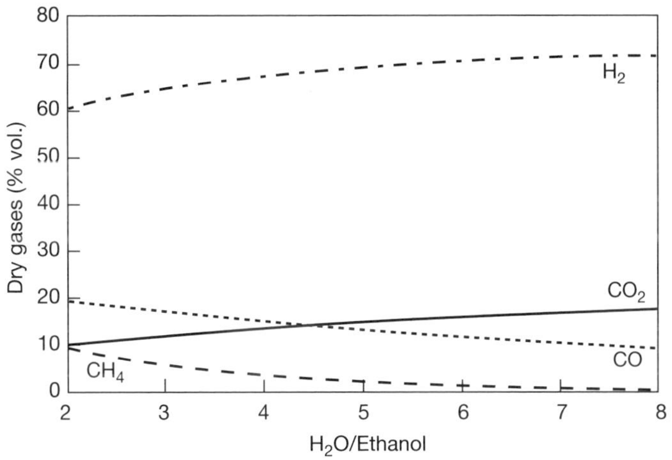

Figure 2.20

Influence of the H_2O/ethanol ratio in the feedstock on the composition of dry gases at thermodynamic equilibrium.

2.7.1.2 Purification of the hydrogen produced

A. Conversion of CO into hydrogen

Depending on the operating conditions chosen for ethanol steam reforming, the reformed gas contains between 10% and 20% molar of CO (fraction of dry gases). To improve the hydrogen yield, the carbon monoxide must be converted into hydrogen via the CO conversion reaction, also known as the water-gas shift reaction:

$$CO + H_2O \Leftrightarrow CO_2 + H_2 \qquad \Delta H^\circ_{298} = -41 \text{ kJ/mol}$$

This balanced, exothermic reaction is favoured by low temperatures and the presence of steam. CO conversion takes place in two steps: a conversion step at high temperature followed by a conversion step at low temperature. To favour the kinetics of the CO conversion reaction, the first step is conducted at high temperature. It is carried out on iron- and chromium-based catalysts which operate at between 300°C and 500°C. This step leads to residual CO contents of 1% to 3% by volume. The second step is carried out at low temperature, to favour the chemical thermodynamics and reach higher conversion levels: catalysts consisting of alumina-supported copper and zinc oxides are used for this finishing step which takes place at temperatures between 200°C and 250°C, resulting in residual CO contents of between 0.2% and 0.5% by volume.

These catalysts have been used for many years. They must be activated before use and inerted before exposure to air, to make them less pyrophoric. Numerous studies are in progress to find new catalytic formulations which do not suffer from this disadvantage [37]. Precious metal-based catalysts [38] meet this criterion while offering a level of activity at least as good as that of copper- and zinc-based catalysts. Due to their high cost, however, precious metal-based catalysts are less competitive on the market. Other formulations such as copper or nickel supported on ceria, cobalt on perovskite are also being investigated [39]. These formulations seem more active than iron/chromium catalysts at high temperature and comparable at low temperature to alumina-supported copper/zinc catalysts. Above all, they offer the advantage of not being pyrophoric.

B. Final purification of hydrogen

The hydrogen produced for use as chemical reactant requires a final purification step to eliminate the carbon oxides and methane present in the reformed gas. Final purification can be carried out by Pressure Swing Adsorption (PSA) or by separation on palladium/silver membrane (dense membrane 100% selective to hydrogen). Hydrogen of purity greater than 99% by volume can be obtained with these two purification processes.

If the hydrogen is produced to supply fuel cells, final purification may consist simply of preferential oxidation of the residual CO to reduce its content to the target value of 10 ppm necessary for Proton Exchange Membrane (PEM) cells. Final purification is not required for Solid Oxide Fuel Cells (SOFC). Numerous studies are currently being conducted on development of catalysts for preferential oxidation of CO rather than hydrogen, to achieve a residual CO content of 10 ppm. Amongst the catalytic systems being investigated, gold-based catalysts seem especially interesting and active at low temperatures (< 100°C).

2.7.1.3 Production of hydrogen by partial oxidation

Partial oxidation (POX) consists in treating a hydrocarbon with air in the presence of steam. This technology applies to all feedstocks, whether gaseous, liquid or solid. For ethanol, the overall reaction can be written in simplified form as follows:

$$C_2H_5OH + {}^1/_2\, O_2 \Leftrightarrow 2\, CO + 3\, H_2 \qquad \Delta H^\circ_{298} = +20 \text{ kJ/mol}$$

Few studies have been published on the production of hydrogen by POX of ethanol.

At the traditional steam reforming temperature (700°C), purely thermal POX of ethanol does not seem interesting for the production of ethanol since it produces yields of only 10% compared with the reaction stoichiometry. This poor selectivity is partly due to oxidation of the hydrogen formed [40]. Better yields can be obtained by operating at temperatures above 1 200°C, which correspond to the traditional thermal conditions of industrial POX processes treating hydrocarbon feedstocks.

2.7.1.4 Production of hydrogen by autothermal reforming

Catalytic POX, also known as oxidising steam reforming, gives better hydrogen yields. With a catalyst, the steam reforming steps can be carried out after partial oxidation of ethanol, according to the following reaction:

$$C_2H_5OH + {}^1/_2\, O_2 + 2\, H_2O \Leftrightarrow 2\, CO_2 + 5\, H_2 \qquad \Delta H^\circ_{298} = -\,50 \text{ kJ/mol}$$

Production of hydrogen in autothermal mode consists in adjusting the O_2/ethanol ratio, for a H_2O/ethanol ratio chosen so as to compensate, by combustion with air, for the endothermicity of the steam reforming reaction. The following overall equation corresponds to an autothermal mode for the production of H_2:

$$C_2H_5OH + 0.608\, O_2 + 1.784\, H_2O \Leftrightarrow 2\, CO_2 + 4.784\, H_2,\ \Delta H^\circ_{298} = 0 \text{ kJ/mol}$$

The reforming autothermal process seems to be a promising pathway for the production of hydrogen from ethanol. It gives better hydrogen yields and offers the advantage of being autonomous in terms of energy since the endothermicity of the reforming reactions is compensated in the reactor by the exothermicity of the oxidation reaction.

Compared with steam reforming, catalytic POX does not produce coke and the reformer can be started much more quickly. However, the hydrogen produced is diluted by nitrogen: purities of about 45% to 50% by volume in the fraction of dry gases are obtained, which penalises the downstream purification steps.

2.7.1.5 Hydrogen production systems and technologies

Numerous companies propose and commercialise hydrogen production systems. Most of these systems treat natural gas, LPG and methanol by steam reforming, or heavy feedstocks (coke, coal) by POX. Use of ethanol is often mentioned in patents and articles [41]. Haldor Topsoe, for example, filed a patent on a *steam reforming*/PSA purification sequence for the production of hydrogen from ethanol. CETH filed a patent on ethanol steam reforming combined with purification by membrane. Helbio SA also has a patent for the production of hydrogen by catalytic POX, to supply fuel cells, and has published on the subject with the University of Minnesota, with the development of a compact reformer to produce hydrogen from ethanol by oxidising steam reforming [42]. With monolith-supported rhodium-cerium oxide (Rh/CeO_2) catalyst, complete conversions can be achieved with residence times below 10 ms and H_2 selectivities greater than 95% (compared with the stoichiometry of the thermal POX reaction). Similarly, company H2Gen is working on autothermal reforming combined with membrane purification.

At least two prototypes of integrated process for the production of hydrogen from methanol have been developed. Under a multi-year contract, HyRadix and IFP Energies nouvelles are working in partnership to extend the capability of HyRadix's autothermal reforming technologies to liquid hydrocarbons and, initially, to ethanol [43]. A pilot generator producing 50 Nm^3/h of hydrogen has been built and validated. The generator is intended for the industrial hydrogen market for treatment of metals, hydrogenation of oils, the glass industry and electronics.

Similarly, N-GHY, in partnership with the Spanish company HYNERGREEN, has developed a generator to produce hydrogen from ethanol. The technology implemented is a steam reforming sequence followed by PSA purification. The target capacity is 200 Nm^3/h [44].

We can expect to see the first commercial facilities producing hydrogen from ethanol being built in the near future.

2.7.1.6 Conclusion

Ethanol appears as being the preferred alternative fuel to supply fuel cells: it is a liquid, clean, non-toxic biofuel, which can be converted quite easily into hydrogen, offering good thermal efficiency (> 80%). Thermodynamic analysis of the production of hydrogen by ethanol reforming for PEM type fuel cells shows that, at atmospheric pressure, the maximum yields that can be obtained with a POX and a steam reformer are fairly close. They are respectively 4.74 and 4.65 mol./mol. of ethanol, which corresponds to a thermal efficiency of 90-92% [45]. The choice between the steam reforming and catalytic POX technologies largely depends on the targeted application. POX technology, which is simpler and less expensive, will be preferred for an onboard application for which the system startup time is very fast (< 5 s) [42], while the technology used to produce hydrogen by steam reforming will be reserved for stationary applications. Over the last few years, numerous articles and patents have been published on the use of ethanol to produce hydrogen.

2.7.2 Other ways for the conversion of ethanol into fuel base

The most well-known process for chemical conversion of ethanol, apart from ETBE synthesis, is the conversion into acetaldehyde by catalytic oxidation or dehydrogenation. While acetaldehyde is a major petrochemical intermediate, used in particular to manufacture acetic acid, acetic anhydride and butanol, it cannot be considered as source of fuel. The pathways described below are examples of approaches used to manufacture biofuels from ethanol.

2.7.2.1 Direct oligomerisation of ethanol into hydrocarbons

Several studies have been conducted on direct oligomerisation of ethanol into hydrocarbons belonging to the range of gasolines. The authors have always tried to favour the production of aqueous ethanol, to avoid all the ethanol concentration steps which require considerable energy. Most studies have therefore focused on use of a ZSM-5 zeolite with no associated metal [46]. It has been demonstrated that ethylene is formed very rapidly by dehydration of ethanol and that this molecule represents the true reactant.

The products obtained are described in three groups:

- light olefins (propene and butene),
- light paraffins (mostly propane and butane),
- C_5+ compounds, belonging to the range of gasolines.

The latter products are obtained by oligomerisation-cracking and oligomerisation-cyclisation-aromatisation mechanisms. The C_5+ fraction is rich in aromatic compounds, comprising a ring more or less substituted by methyl and ethyl groups.

The difficulty with the reaction lies in the fact that catalyst deactivation by coking quickly reduces the yield in C_5+ compounds. Adding water to the feedstock delays deactivation by coking. However, the water introduced, added to that generated by the reaction, deactivates the catalyst by dealumination. For an MFI zeolite, dealumination occurs for water contents between 6 and 75% by weight if the reaction temperature exceeds 400°C.

The yields in C_5+ compounds obtained at atmospheric pressure reach about 55% by weight on an anhydrous base, for a water content of 6% by weight in the ethanol feedstock and a temperature of 350°C. These compounds enter the gasoline fraction.

Data in pressurised fixed bed have also been published [47], for feedstocks containing between 50% and 100% by weight of ethanol. At 30 bar and temperature between 400°C and 500°C, on ZSM-5 zeolite, the fuel yields lie in the following ranges:

- gasoline C_5^+ – 150°C 17 to 32% by weight,
- kerosene 150-250°C 4.5 to 12% by weight,
- Diesel 270-370°C 0.28 to 0.73% by weight.

The reaction generates a large quantity of light paraffins (between 34% and 72% by weight) which cannot be recycled, as well as light olefins (between 7% and 26% by weight). It seems possible to treat the loss of activity over time by regular oxidising regeneration of the catalyst.

With this process, we therefore see that ethanol conversion leads to a range of products, not just fuel bases. Note that the maximum production of kerosene base is 12% by weight under the conditions studied, whereas the production of Diesel base is negligible.

2.7.2.2 Ethanol conversion in two steps via ethylene

Since direct conversion of ethanol into fuel is only possible with poor selectivities, another solution would be to start by a first step of dehydration into ethylene. This is the opposite process to that currently used to produce synthetic ethanol. In this case, it is easier to convert the ethylene obtained into higher hydrocarbons.

A. Dehydration of ethanol into ethylene

Catalytic dehydration of ethanol into ethylene is an old, well-known process, but industrialised only under highly specific situations, since ethylene is manufactured in larger volumes and at lower costs in petrochemical complexes by steam cracking or pyrolysis of naphtha or ethane. The oldest catalytic processes for dehydration of ethanol into ethylene operate on activated alumina or supported phosphoric acid, at temperatures of between 300°C and 400°C, on fixed bed reactor [48]. Supported phosphoric acid has the advantage of producing 99.5% pure ethylene. The associated constraints in terms of corrosion and handling, however, make the activated alumina solution more attractive. With this process, the ethylene obtained is 97% pure. With the subsequent ethylene purification steps, over 99% purity can be obtained.

These processes require a feedstock of at least 95% vol. of ethanol and are carried out at temperatures between 300°C and 400°C. It is widely accepted that the main reactions involved in dehydration of ethanol into ethylene are as follows:

$$2\ C_2H_5OH \rightarrow (C_2H_5)_2O + H_2O$$

$$(C_2H_5)_2O \rightarrow 2\ C_2H_4 + H_2O$$

Diethylether is formed if the reaction temperature is less than 300°C. Above this temperature, ether is only a reaction intermediate. Ethylene is therefore produced with a yield of about 37% by weight, i.e. a molar yield of 94%.

ABB Lummus has developed an efficient fluidised bed process, using a silica-alumina type catalyst [49]. The ethylene yield reaches 60.5% by weight, i.e. 99.5% molar.

We must also mention a certain number of academic research studies concerning dehydration of ethanol produced by fermentation of sugars, i.e. from aqueous solutions containing between 2% and 20% by weight of ethanol. In this case, the catalysts studied are MFI zeolite-based, generally modified to improve their stability [50].

B. Oligomerisation of ethylene by organometallic catalysis

The Dimersol-E process developed by Axens-IFP Energies nouvelles [51] converts ethylene into a high octane olefinic gasoline cut. An industrial plant operated for 10 years in South Africa, from 1986. The Ziegler catalyst is a nickel hydride formed *in situ* by separate introduction in the reactor of a mixture containing a nickel salt and an alkylaluminium-based co-catalyst. However, this type of catalyst does not support water contents above a few ppm. The process is only possible if combined with highly efficient prior drying of the ethylene produced by dehydration of ethanol.

Other industrial processes are available for the synthesis of long-chain α-olefins by oligomerisation of ethylene with organometallic catalysis. These processes, developed by Ethyl, Chevron, IFP Energies nouvelles and Shell [52], implement Ziegler catalysis with alkylaluminium- or nickel-based catalysts. Apart from the type of catalyst, these processes differ in terms of whether or not they use a solvent and in terms of operating conditions. This leads to a difference in the type and distribution of the products.

Irrespective of the process, the oligomers obtained exhibit a very broad carbon distribution, around C_6-C_{18}, with a variable but low proportion of branched olefins. This very broad distribution prevents direct conversion into fuels. While light products are highly suited to the gasoline pool, hydroisomerisation would be required on the C_{10}+ cut to obtain good quality Diesel.

C. Oligomerisation of ethylene by heterogeneous catalysis

Numerous studies are still underway on oligomerisation of ethylene using metal-exchanged-based catalysts on solid acid. Nickel is very frequently used and there is an extensive range of associated acid solids (alumina, silica-alumina, ZSM-5 or USY zeolite, mesoporous solid, etc.). All studies tend to demonstrate that the active sites are the combination of a low-valence nickel species (Ni^+) and a weakly acid site [53], correct equilibrium between the acid and metallic phases being highly important with respect to the activity, stability and selectivity of the catalyst. The catalytic tests are generally carried out under operating conditions of temperature 20-200°C and pressure 1-50 bar.

In particular, interesting results have recently been described on Ni MCM-41 catalyst, with *n*-heptane solvent [54]. The oligomers produced contain between 4 and 12 carbon atoms, with a majority of C_4 at 150°C – 35 bar and a majority of C_8 at 200°C – 35 bar, the C_8 oligomers being formed *a priori* by hetero-oligomerisation reactions. The reaction is therefore relatively selective. The maximum quantity of C_8+ is obtained at 150°C and 50 bar. It is limited to 32% by weight. The hydrocarbon branching rate is low for C_6– compounds, increasing sharply for C_8+ compounds.

2.7.2.3 Conversion of ethanol via butadiene

Ethanol can be converted into butadiene in a single step on an $MgO\text{-}SiO_2$ type catalyst using a very old process [55]. The overall reaction, which includes dehydrogenation, dehydration and dimerisation steps, is written:

$$2\ C_2H_5OH \rightarrow H_2C = CH\text{–}CH = CH_2 + 2\ H_2O + H_2$$

70% butadiene selectivity can be achieved with this process.

A second butadiene telomerisation step is required to produce fuel bases. Companies Snamprogetti and Euron have demonstrated, for example, that by conducting the butadiene telomerisation reaction in the presence of methanol or ethanol, followed by a hydrogenation reaction, we obtain *n*-octyalkylethers which could be considered as additives to a Diesel base [56]. Telomerisation is conducted in homogeneous phase, using palladium complexes in the presence of phosphine ligands. Telomerisation leads to 1-methoxy-2,7-octadiene, or 1-ethoxy-2,7-octadiene, with a molar yield in primary ether of respectively 94% and 91%. Subsequent hydrogenation is carried out, for example, on noble metal- or nickel-based catalyst. This is once again a sequence of reactions, with formation of secondary products at each step.

2.7.2.4 Conversion of ethanol into butanol-1

Butanol-1 is one of the most important chemical compounds. It is used in particular in the manufacture of butyl acetate, phthalates (dibutylphthalate, butylbenzylphthalate) and various solvents. It could also be converted into acetal, by reaction with formol or acetaldehyde, into ether, e.g. dibutylether, or into butyl ester. The boiling points of these compounds are compatible with the Diesel distillation curve.

The very old Guerbet reaction, discovered at the end of the 19th century, is used to dimerise aliphatic alcohols with release of a water molecule to form branched alcohols [57]. The reaction mechanism is composed of numerous steps:

- catalytic dehydrogenation of alcohol into aldehyde,
- condensation of aldol from two aldehyde molecules previously formed followed by dehydration, resulting in the formation of an α,β-unsaturated aldehyde,
- hydrogenation of the condensation product formed previously, to give the branched dimer alcohol.

The reaction must be carried out with an alkaline type catalyst such as potash, generally doped by a transition metal (Ni, Pd, Cu, Rh, Ir, etc.) during the first hydrogenation step. The reaction takes place at high temperature and pressure.

Numerous fatty alcohols have been synthesised with this reaction. Unfortunately, ethanol is the primary alcohol which reacts with the lowest selectivity according to the Guerbet reaction.

Other catalytic systems were studied later for the synthesis of *n*-butanol from ethanol. Silica-supported mixed oxides composed of 65% MgO – 25% CuO – 10% MnO have proved to be amongst the most efficient [58]. The *n*-butanol selectivity is at most 18% molar compared with ethanol, i.e. 29% by weight, for an ethanol conversion of less than 70%. The secondary organic reaction products are acetaldehyde, ethylene and, to a lesser extent, crotonaldehyde and butene-1. Magnesium oxide alone has also been mentioned more recently in the literature [59]. It leads to an 18% molar yield of *n*-butanol for an ethanol conversion of 56%. Consequently, with the current state of knowledge, the process cannot be industrialised.

2.7.2.5 Conversion of ethanol by reaction with another molecule

Various reactions are mentioned sporadically in the literature concerning the conversion of ethanol by reaction with another molecule. We may mention, for example, the synthesis of acetal (1,1-diethoxyethane) by reaction of two moles of ethanol with one mole of acetaldehyde. Acetal could be interesting as an additive for a Diesel fuel.

The reaction operates well in homogeneous catalysis on a strong inorganic acid such as sulphuric or phosphoric acid. Amongst the heterogeneous catalytic systems studied, resins seem to be the most efficient [60]. Acetaldehyde itself can be obtained by oxidation of ethanol.

2.7.2.6 Conclusion

In the current state of knowledge on the possible reactions for the chemical conversion of ethanol, the processes which could be implemented to produce bases for gasoline or Diesel fuel seem to be complex. They have been discussed in more or less numerous publications depending on the approaches taken, but none of the pathways described has yet been industrialised. Their cost and/or lack of performance make them quite uncompetitive for the time being compared with the processes used to synthesise ETBE or vegetable oil ethyl esters.

Studies conducted in-house by IFP Energies nouvelles have demonstrated that the cost of ethylene produced by dehydration of ethanol is more or less double that of the initial ethanol, i.e. about €1 000/t using European ethanol and about €550/t using Brazilian ethanol. This must be compared with the price of ethylene manufactured with the traditional petrochemical processes, in the region of €750-800/t during the second half of 2008, when the average price of Brent type crude oil was €35 per barrel; at the present time, this price of ethylene is higher with a price of crude oil between €60 and €70. The use of this process to obtain ethylene from European ethanol does not currently appear to be worthwhile and could only be considered in case of a very difficult oil economic climate. Furthermore, the subsequent conversion of ethylene into fuel generates a non-negligible additional cost.

2.8 ENVIRONMENTAL ASPECTS

Numerous studies have been conducted to measure the impact of biofuel production and use on the environment. These "well to wheel" studies compare the greenhouse gas emissions and energy consumption over the entire fuel life cycle between the biofuel solutions and the traditional solutions using oil fuels. These are the two main impacts mentioned below. The results of two studies are shown in the following tables:

– the study conducted in France on behalf of the ADEME (French Environment and Energy Management Agency), the MAP (Ministry of Food, Agriculture and Fisheries) and the MEEDAT (Ministry of Ecology, Energy, Sustainable Development and the Sea), supported by several R&D institutes and French companies [61]. This study does not take into account the land use changes,

– the European study conducted jointly by the JRC (European Commission Joint Research Centre), the CONCAWE (the oil companies' European association for environment), and EUCAR (which coordinates the R&D actions of the European Automobile Manufacturers Association) [62].

The results of these two studies concerning reductions of fossil energy consumptions and greenhouse gas emissions, are provided in Tables 2.14 and 2.15 respectively.

Table 2.14 Reduction of non-renewable energy consumption.

	ADEME study (February 2010)	JRC/EUCAR/CONCAWE study (March 2007)		
	Reduction of consumptions compared with the petroleum reference (%)	**MJ ex/km**	**Reference MJ ex/km**	**Gain compared with the petroleum reference (%)**
Ethanol from sugar beet	52	1.65	2.16	24
Ethanol from wheat	49	1.68	2.16	22
Ethanol from corn	53			
Ethanol from sugar cane	85	0.2	2.16	91
ETBE from sugar beet	20			
ETBE from wheat	18			
ETBE from corn	22			
ETBE from sugar cane	54			

Table 2.15 GHG balance of biofuel sectors compared with the reference petroleum fuel.

	ADEME study (February 2010)	JRC/EUCAR/CONCAWE study (March 2007)		
	Reduction of emissions compared with the petroleum reference (%)	**gCO_2 eq/km**	**Reference gCO_2 eq/km**	**Gain compared with the reference petroleum fuel**
Ethanol from wheat	49	114	164	30
Ethanol from sugar beet	6	111	164	32
Ethanol from sugar cane	72	19	164	88
Ethanol from corn	56			
ETBE from sugar beet	42			
ETBE from wheat	24			
ETBE from corn	31			
ETBE from sugar cane	47			

After analysing these results, several comments can be made. Greenhouse gas emissions are in general significantly reduced by the use of biofuels compared with conventional solutions. This conclusion must nevertheless be tempered by the fact that biofuels are generally used in blends at relatively low contents (5% to 10%, with a maximum of 24% in Brazil, and sometimes 85%).

While from a qualitative point of view the use of biofuels offers a clear gain in terms of greenhouse gas emissions, this gain remains difficult to quantify accurately; it also depends on N_2O emissions which are extremely important since this gas is 300 times more noxious in terms of greenhouse effect than CO_2. These emissions depend themselves on the quantities of fertilizer used, weather conditions, soil quality, etc., and are very difficult to assess precisely. We must emphasise that the methodologies used in the two studies differ widely concerning the attribution of impacts when co-products are manufactured. The study conducted in France on behalf of ADEME/DIREM uses the traditional mass prorata. In contrast, the JRC/EUCAR/CONCAWE study adopts the so-called substitution method which consists in attributing all impacts to the biofuel alone and then deducting a "credit", once all the impacts have been evaluated. This credit is estimated by assuming that the co-product avoids greenhouse gas emissions due to its manufacture alone. When producing ethanol from wheat, for example, the co-products used as animal feed are the DDGS. Co-production avoids the greenhouse gas emissions which would have been generated if an equivalent quantity of animal feed had to be produced from wheat. This calculation method is less favourable than that used in the ADEME/DIREM study.

The studies conducted agree on the fact that the best result is obtained when ethanol is produced from sugar cane. Use of bagasse to generate the heat required for distillation largely contributes to this result.

For the production of ethanol from sugar beet and ethanol from corn, the results are intermediate, between those obtained for wheat and sugar cane. Concerning the use of ETBE, reductions of greenhouse gas emissions are even lower. This is mainly due to the fact that half the manufacture of this compound uses fossil origin raw materials.

Several conclusions can be drawn from the energy balances.

Firstly, the production of biofuels involves a non-negligible consumption of fossil energies. Consequently, the final gain expected with respect to the "petroleum fuel" solution will never be 100%. Depending on the assumptions made and the sectors studied, reduction in non-renewable energy consumption varies from 22% to 91%. Once again, ethanol produced from sugar cane leads to the largest reductions. In view of this broad interval, we must be cautious when discussing the gain in terms of energy consumption related to the use of biofuels. In addition, as with greenhouse gas emissions, the expected gain will depend on the methodology chosen. Note, however, that these balances are not fixed and that the "fossil energy" balances of the biofuel production units could be improved, either by energy saving programmes or even by using renewable energy resources.

The ADEME [61] study gives indications on other impacts. For the eutrophication potential indicator, expressed in kg eq. phosphates/MJ, the balances for ethanols are slightly unfavourable compared with gasolines. Once converted into ETBE, we observe a gain compared with their direct incorporation. The agricultural step, by leaching of nitrates and emissions of ammonia, is the largest contributor to this impact. However, the industrial conversion step is also accompanied by strong emissions. For the photochemical oxidation potential, expressed in kg eq. C_2H_4/MJ, ethanols, incorporated at 10% in gasolines, emit fewer oxidising molecules than the gasoline they are replacing. This reduction is largely due to the decrease in CO emissions when ethanols are used in vehicles. The impacts on the human toxicity indicator, expressed in kg eq. 1-4 dichlorobenzene (1-4 DB)/MJ are much higher for ethanols than for gasolines. The main contribution comes from the agricultural step.

2.9 ECONOMIC ASPECTS

Ethanol is by far the most extensively produced biofuel worldwide. The main producers of ethanol fuel are the United States, Brazil, China and Europe. Due to this hierarchy, the United States and Brazil currently benefit from the best scale effects and therefore the lowest costs in terms of ethanol production, in different contexts.

The manufacture of biofuels generates co-products whose conversion and price variations affect the biofuel economy. When producing ethanol from wheat, about 1.2 tonnes of protein-rich DDGS are obtained per tonne of ethanol and used in animal feed. From sugar beet, 0.75 tonne of pulp used as animal feed are obtained per tonne of ethanol.

2.9.1 Ethanol production cost in the United States

In the United States, ethanol is mainly produced from corn. Two types of process can be implemented depending on the type and the markets targeted for the co-products: wet milling and dry milling. The co-products and therefore the type of markets for these co-products are of paramount importance for the ethanol production economy: the volume of co-product generated represents about 55% by weight of raw material fed into the process and its sale may represent up to 50% of the price of the raw material. Each of these pathways has its own strong and weak points. The "dry" pathway involves lower initial investment and easier implementation; this is offset by the fact that the DDGS co-produced have less added value (mainly intended for the animal feed market). Inversely, the "wet" pathway, more difficult to implement, requires greater investment; the main co-product is a fructose-rich syrup mainly intended for the food market, especially for the production of soft drinks or cooked products. Until recently, the latter process was mainly used, in particular by Archer Daniels Midlands (leading player in this sector of activity in the United States). With the increased production of ethanol fuel observed in the United States since the start of the years 2000, however, the "dry" pathways are taking over on the American market. These processes prevail since they depend only on the sale of ethanol and DDGS in an expanding market, unlike the multitude of co-products generated by the "wet" pathway, whose fluctuating markets are unable to absorb the surpluses produced by the new large plants set up.

Over the last few years, all new plants have production capacities in excess of 100 000 t/year, resulting in a scale effect leading to lower production costs.

The cost breakdown shown on Figure 2.21 is substantially the same for the two processes [63]. It highlights the importance of the raw material, even though it is attenuated by the income generated by the sale of co-products. Finally, while the cost of the raw material is higher with wet milling, more income is generated by the sale of the co-products.

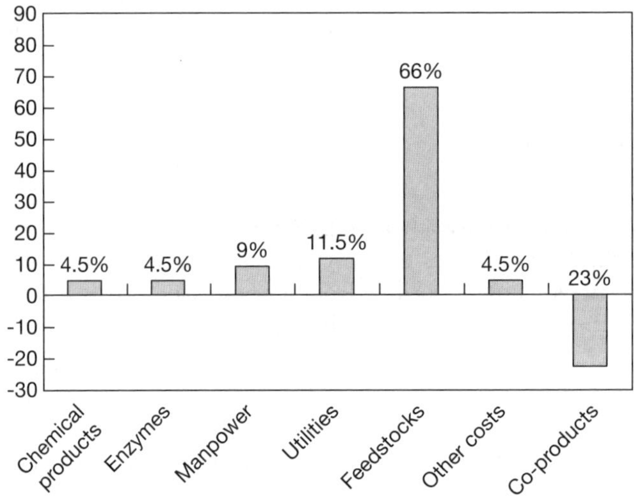

Figure 2.21

Cost breakdown for the production of ethanol from corn.

Taking this income into account, the net costs of producing ethanol from corn are roughly the same for the two processes. At the start of the last quarter in 2009, the costs were about \$550/m^3 for 92% ethanol and \$580/m^3 for anhydrous ethanol [64].

2.9.2 Ethanol production cost in Brazil

The lowest ethanol production cost is obtained in Brazil, to a lesser extent in India and potentially in all tropical regions capable of producing large quantities of sugar cane. In Brazil, for example, the price per unit volume of 92% ethanol, which can only be used to fuel dedicated vehicles, dropped below the price of gasoline in 2002 and 2003 (about \$0.2/l). Similarly, in India, the price per unit volume of ethanol came close to that of gasoline in 2002. The prices currently observed are highly dependent on variations in the price of sugar. The average ethanol production costs in Brazil are detailed in Table 2.16.

The trend in ethanol costs, between 2000 and 2009, per unit energy, in euros per gigajoule, is compared with the trend in oil costs. The values indicated are given before tax (Figure 2.22), [65]. Major variations may be observed over time, mainly caused by variations in the price of raw materials. The revenues obtained on the sugar market, which were

Table 2.16 Breakdown of ethanol production costs in Brazil in 1990 ($/l).

Cost of raw material	**0.13**
Operating costs	**0.04**
Labour/Administrative/Maintenance	0.01
Chemical products	0.002
Energy	0.002
Other	0.024
Fixed costs	**0.062**
Return on investment	0.05
Other	0.011
Total production costs	**0.23**

Source: IEA

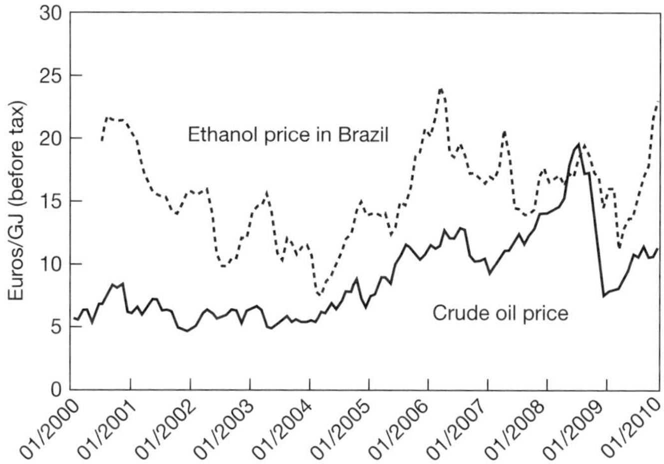

Figure 2.22

Graph of prices (euros/GJ).

close to those from the ethanol market, have been higher since the start of 2009. The prices of ethanol in Brazil, per unit energy, are generally always higher than those of oil. In 2008, however, they were very close.

Since the 1990s, significant progress has been made in the production of ethanol. Bagasse, for example, is used as raw material in electricity and heat cogeneration plants. In 2004-2005, the production costs for 92% ethanol dropped to about $0.15/l, which is competitive compared with the price for gasoline.

This cost becomes competitive compared with that of gasoline, per unit energy, the energy content of ethanol being about 2/3 that of gasoline.

2.9.3 Ethanol production cost in Europe

Ethanol production costs are higher in Europe, due to much smaller plants, a sector less structured than in the United States or Brazil and more expensive raw materials. In France, one of the leading producers of ethanol fuel in Europe, the average capacity of the production plants was about 20 000 t/year until 2006. It is also important to note that the raw materials, wheat and sugar beet, are not the same as those used in Brazil and the United States. The process with wheat is similar to that of corn already described for the United States. With sugar beet, however, the situation is closer to that observed in Brazil. Table 2.17 below gives an estimation of the average cost breakdown for the production of ethanol. These costs are only provided as an indication and cannot reflect the diversity of the situations encountered in Europe.

Table 2.17 Estimation of the average ethanol production cost in Europe (€/l).

	Wheat	Sugar beet
Raw material	0.22-0.34	0.2-0.32
Production cost	0.28	0.22
Co-product credit	(0.15)	0
Total cost	0.36-0.48	0.42-0.52

In Europe, ethanol costs in October 2009 oscillate between 50 and 55 euros/hl, while they were 10% more expensive one year before.

Taxes levied on fuels remain the responsibility of each individual country. To date, numerous member states have obtained waivers for biofuels. In France, for example, the tax cut which was 21 euros/hl in 2009, was reduced to 18 euros/hl in 2010, 11 euros/hl in 2011 and is expected to be cancelled in 2012.

2.9.4 ETBE production cost

The production cost of ETBE, calculated in 2005, with a price of raw materials (ethanol and isobutene) of €295/t, amounted to €397/t. This evaluation includes fixed costs of €46/t and other variable costs (utilities, catalysts, labour, etc.) of €56/t.

The cost of raw materials represents 75% of the ETBE cost. Consequently, the slightest fluctuations in the cost of raw materials change the cost price of ETBE.

REFERENCES

[1] Swain EJ (1999) US MTBE Production at a Record High in 1998. Oil and Gas Journal **97**, 24, pp 99-101.

[2] Lamberth R (2004) 2003 Was a Year of Transition for the MTBE, Fuels Industry. Oil and Gas Journal **102**, 2, pp 52-58.

[3] UOP sells first Licence for DIPE Production, European Chemical News (1996) **6**, pp 1710.

[4a] Koskinen M, Jarvelin H and Lundqvist P (1997) NExETHERS – New Way of Producing Ethers. Oil-Gas European Magazine 23, (3), pp 31-34.

[4b] http://www.sriconsulting.com/CEH/Public/Reports, 7 octobre 1998.

[4c] Caillaud V (2007) Les biocarburants en perspective. GP Arts et métiers, 10 septembre 2007.

[5a] Fayolle F, Hernandez-Perez G and Vandecasteele JP (2001) Biodegradation of Ethyl t-butyl ether (ETBE), Methyl t-butyl ether (MTBE) and t-amyl methyl ether (TAME) by *Gordonia Terrae*. Microbiology and Biotechnology **55**, 1, pp 117-121.

[5b] Allard M (2005) Methanol Industry Overview, GTLtec Doha, Novembre 2005.

[6] Journal Officiel de la République Française, 30 décembre 1991.

[7] Forestière A. Rapports internes IFP Energies nouvelles 1991et 1996.

[8] Quiroga M, Capeletti MR, Figoli N and Sedran U (1999) Catalyst Deactivation by Acetonitrile in MTBE Synthesis. Applied Catalysis A-General **177**, 1, pp 37-42.

[9] Capeletti MR, Figoli N and Sedran U (2002) Propionitrile Hydrolysis over sulfonic Resins. Catalysis Letters **83**, 1-2, pp 15-18.

[10] Marion MC, Forestière A and Chodorge JA (1996) ETBE Production, the French Experience. ISAF XI International Symposium on Alcohol Fuels, Sun City: Proceedings 2, pp 520-527.

[11] Nocca JL, Forestière A et Cosyns J (1994) Nouvelles technologies IFP Energies nouvelles pour la reformulation des essences. Revue de l'Institut Français du Pétrole **49**, 5, pp 461-473.

[12] Chodorge JA, Forestière A, Hennico A and Nocca JL (1995) ETBE: The Renewable Oxygenate of the Future. AICHE – Summer National Meeting of the American Institute of Chemical Engineers – Extented Abstract, Paper 44F.

[13] Hunszinger P, Jarvelin H, Nurminen M and Birkhoff R (2003) Case History: Converting an MTBE to Isooctane Operation. Hydrocarbon Processing **82**, 9, p 53.

[14] Process Developped to Convert MTBE Plants (2003) Material Performance **42**, 9, p 53.

[15] Barger P (2004) Communication orale de l'UOP à l'AICHE.

[16] Anonyme (2000) UOP Process Selected for Japan MTBE Revamp. Chemical Week **162**, 14, p 27.

[17] Di Girolamo M and Tagliabue L (1999) MTBE and Alkylate Coproduction: Fundamentals and Operating Experience. Catalysis Today **52**, 2-3, pp 307-319.

[18] Nocca JL (1995) Proceedings: IFP Energies nouvelles Customer Seminar: IFP Energies nouvelles's Etherification Technologies. Newport Beach, Sept 1994.

[19a] Krishna R and Taylor R (2000) Modelling Reactive Distillation. Chem Eng. Sci. **55**, pp 5183-5229.

[19b] Krishna R and Baur R, Taylor R (2001) Influence of Column Hardware on the Performance of Reactive Distillation Columns. Catal. Today **66**, pp 225-232.

[20] Mikitenko P (1986) La distillation réactive : principe, applications et perspectives. Pétrole et Techniques 329, pp 34-38.

[21] Nocca JL, Leonard J, Gaillard JF and Amigues P (1989) US Patent 4 847 431, Process for Manufacturing a Tertiary Alkyl Ether By Reactive Distillation.

[22] Chodorge JA, Hennico A, Leonard J and Nocca JL (1990) IFP Etherification Technology: The MTBE and TAME Processes. Hydrocarbon Technology International 1989-1990, pp 127-129.

[23a] Nocca JL, Leonard J, Gaillard JF and Amigues P (1991) US Patent 5 013 407: Apparatus for Reactive Distillation.

[23b] Asselineau L, Mikitenko P, Viltard JC and Zuliani M (1994) US Patent 5 368 691: Reactive Distillation Process and Apparatus for Carrying it Out.

[24] Forestière A, Nocca JL, Torck B and Leprince P (1991) Proceedings of the International Symposium on Alcohol Fuels.

[25] Pucci A, Mikitenko P et Zuliani M (1991) Brevet français Fr 2 683 523 : Procédé de séparation d'éthyl tertiobutyl éther et d'éthanol; US Patent 5 158 652: Process for the Separation of Tert. Butyl Ethyl Ether from Mixtures; US Patent 5 250 156: Method of Separating Ethyl Tertiobutyl Ether from Mixtures with Ethanol; US Patent 5 348 624: Process for Separating Ethyl Tert. Butyl Ether and Ethanol.

[26a] Prévost I, Mauci E, Chau C, Ballaguet JP et Forestière A (2000) GFGP-Intégration des membranes dans les procédés. Mise en œuvre des nouvelles membranes dans les procédés. Comptes rendus du Séminaire du Groupe Français de Génie des Procédés 14, pp 155-161.

[26b] Ballaguet JP, Forestière A, Maucci E, Prévost I and Streicher Ch (2000) New Membrane Technology to Produce ETBE. ISAF XIII-International Symposium on Alcohol Fuels-Proceedings, Part I, p 10.

[27a] Asselineau L, Streicher Ch and Forestière A (1995) Separation of Alcohol/hydrocarbon Mixtures in Industrial Etherification Processes for Gasoline Production. Pure and Applied Chemistry **67**, 6, pp 985-992.

[27b] Baromètre Biocarburants. Etude réalisée par EurObserver (2009) Systèmes solaires. Le Journal des Énergies Renouvelables, 192, pp 55-77.

[28] Supply an Cost of Alternatives to MTBE in Gasoline (December 1998). Technical Appendices – Ethanol Blending Properties for Task 3 Modeling Work – Staff Report – California Energy Commission.

[29] Alternative Fuel in the Automotive Market (1995) Concawe report 2/95.

[30] Hammel-Smith C, Fang J, Powders M and Aabakken J (2002) Issues Associated with the Use of Higher Ethanol Blends (E17-E24), NREL Technical Report NREL/TP-510-32206.

[31] Nersasian A (E. I. du Pont de Nemours & Co., Inc.) The Vol. Increase of Fuel Handling Rubbers in Gasoline/Alcohol Blends, SAE Paper n° 800789, June 1980.

[32] Dunn JR and Pfisterer HA (Polystar Limited) Resistance of NBR-Based Fuel Hose Tube to Fuel-Alcohol Blends. SAE Paper n° 800856, June 1980.

[33] Bechtold RL (1997) Alternative Fuels Handbook, SAE Press, ISBN 0-7680-0052-1.

[34] Mueller Associates, Inc., Status of Alcohol Fuels Utilization Technology for Highway Transportation: a 1986 Perspective, in: Volume 1 – Spark-Ignition Engines, Report n°ORNL/Sub/85-22007/4, National Technical Information Service, Springfield, Va.

[35] Fatsikostas AN and Verykios XE (2004) Reaction Network of Steam Reforming of Ethanol over Ni-based Catalysts. Journal of Catalysis **225**, pp 439-452.

[36] Auprêtre F, Descorme C, Duprez D, Casanave D and Uzio D (2004) New Multifunctional Catalysts for the Steam Reforming of Bioethanol. 13th International Congress on Catalysis, 11-16 July 2004, pp 4-155.

[37] Haryanto A, Fernando S, Murali N and Adhikari S (2005) Current Status of Hydrogen Production Techniques by Steam Reforming of Ethanol: A Review. Energy & Fuels, **19**, 5, pp 2098-2106.

[38] Panagiotopoulou P and Kondarides DI (2004) Effect of Morphological Characteristics of TiO2-Supported Noble Metal Catalysts on their Activity for the Water-Gas Shift Reaction. J. Catal., 225, pp 327-336.

[39] Ladebek JR and Wagner JP (2003) Catalyst Development for Water-Gas Shift Handbook of Fuel Cells. Fundamentals, Technology and Applications, 3, Part 2, pp 190-201.

[40] Christensen D, Silveston P, Croiset E and Hudgins R (2004) Production of Hydrogen from the Noncatalytic Partial Oxydation of Ethanol. American Chemical Society **43**, 11, pp 2636-2642.

[41] Rapport Alphea (2006) Etat de l'art des systèmes et technologies de production d'hydrogène, 13 juillet 2006.

[42] Deluga G, Salge J, Schmidt L and Verykios X (2004) Renewable Hydrogen from Ethanol by Autothermal Reforming. Science 303, pp 993-997.

[43] Giroudière F, Sanger R, His S, Boyer C, Doshi K and Xu J (2006) Production of Hydrogen from Bio-ethanol, WHEC 16/13-16 June 2006, Lyon, France.

[44] Brochure N-GHY/HYNERGREEN (2008) H2 Expo, Hambourg, Allemagne.

[45] Ioannides T (2001) Thermodynamic Analysis of Ethanol Processors for Fuel Cell Applications. Journal of Power Sources 92, pp 17-25.

[46] Aguayo AT, Gayubo AG, Tarrío AM, Atutxa A and Bilbao J (2002) Study of Operating Variables in the Transformation of Aqueous Ethanol into Hydrocarbons on an ZSM-5 zeolit. Journal of Chemical Technology and Biotechnology 77, pp 211-216.

[47] Bhattacharyya KG, Talukdar AK and Sivasanker S (1994) Production of Hydrocarbons from Aqueous Ethanol over HZM-5 under High Pressure. Journal of Assam Science Society **36**, 3, pp 177-188.

[48] Winter O and Eng M-T (1976) Make Ethylene from Ethanol. Hydrocarbon Processing 1976, 11, pp 125-133.

[49] Tsao U and Reilly JW (1978) Dehydrate Ethanol to Ethylene. Hydrocarbon Processing, **57**, 2, pp 133-136.

[50] Hillips CB and Datta R (1997) Production of Ethylene from Hydrous Ethanol on ZSM-5 under Mild Conditions. Industrial Engineering Chemistry Research **36**, pp 4466-4475.

[51] Olivier-Bourbigou H, Chodorge JA and Travers P (1999) Olefin Oligomerisation with Homogeneous Catalysis. Petroleum Technology Quarter (PTQ), Autumn 1999, pp 141-149.

[52] Skupińska J (1991) Oligomerization of α-olefins to Higher Oligomers. Chemical Review 91, pp 613-648.

[53] Sohn JR and Park WC (2003) The Roles of Active Sites of Nickel Sulfate Supported on γ-Al_2O_3 for Ethylene Dimerization. Applied Catalysis A: 239, pp 269-278.

[54] Hulea V and Fajula F (2004) Ni-exchanged AlMCM-41-An Efficient Bifunctionnal Catalyst for Ethylene Oligomerization. Journal of Catalysis 225, pp 213-222.

[55] Weissermel K and Arpe HJ (1997) Industrial Organic Chemistry. 3rd Ed. VCH, Weinheim. pp 106-107.

[56] Marchionna M, Patrini R, Giavazzi F and Pecci GC (1996) Symposium on Removal of Aromatics, Sulfur and Olefins from Gasoline and Diesel. Presented before the Division of Petroleum Chemistry, Inc. 212th National Meeting, American Chemical Society. Orlando, FL, August 25-29, 1996, pp 585-589.

[57] Knothe G (2002) Synthesis, Applications, and Characterization of Guerbet Compounds and their Derivatives. Lipid Technology, pp 101-104.

[58] Nagarajan V (1971) Kinetics of a Complex Reaction System – Preparation of n-butanol from Ethanol in one Step. Indian Journal of Technology 9, pp 380-386.

[59] Ndou AS, Plint N and Coville NJ (2003) Dimerisation of Ethanol to Butanol over Solid-base Catalysts. Applied Catalysis A: 251, pp 337-345.

[60] Capeletti MR, Balzano L, de la Puente G, Laborde M and Sedran U (2000) Synthesis of Acetal (1,1-diethoxyethane) from Ethanol and Acetaldehyde over Acidic Catalysts. Applied Catalysis A: 198, L1-L4.

[61] Analyses de cycle de vie appliquées aux biocarburants de première génération consommés en France. ADEME, Synthèse de février 2010.

[62] Concawe, Eucar, JRC (2007) Well-to-Wheels Analysis of Future Automotive Fuels and Powertrains in the European Context, Well-to-Wheels Report, March, http://ies.jrc.ec.europa.eu/WTW

[63] Shapouri H (2003) Ethanol Production Economics. USDA/OCE First World Summit on Ethanol for Transportation. Quebec (Canada) 02-04 Nov. 2003.

[64] F.O.Lichts (2009) World Ethanol Price Report, October 26, **5**, 29, pp 1-5.

[65] Chaud B (2009) IFP Energies nouvelles Sessions – Which Technologies to Diversify Transportation Fuels? Rueil-Malmaison (France), December 9.

3

First generation biofuels for Diesel engines

D. Ballerini, T. Chapus, G. Hillion, X. Montagne

First generation biofuels for Diesel engines are produced from vegetable oils. After more or less deep purification, they can be used directly as fuel in Diesel engines and are still currently used in some limited applications. Improvements in the technology and performance of Diesel engines has led to stricter requirements concerning the quality of the fuels used, however, which the crude vegetable oils are unable to meet (see 3.10).

For use in recent engines, vegetable oils are transesterified with an alcohol, generally methanol, but also possibly ethanol to produce a vegetable oil methyl or ethyl ester (VOME or VOEE). The same type of reaction can be applied to animal fats and used cooking oils. Known for over a century, the transesterification reaction was adapted during the 1980s, resulting in industrial processes for manufacture of VOME, generally known as biodiesel or sold in France under Diester® brand.

A second type of biodiesel, consisting of a mixture of hydrocarbons, is obtained by catalytic hydrotreatment of vegetable oils or animal fats. Vegetable oils may also act as raw materials for oil refineries, being mixed with various diesels before hydrotreatment, which increases the production of good quality fuels for Diesel engines.

This chapter describes the entire vegetable oil and – to a lesser extent – animal fat fuel production sector. After describing the raw materials and the process used to obtain the oils and methanol required for the transesterification reaction, the various processes involved in the manufacture of vegetable oil methyl and ethyl esters and hydrocarbons are presented. The characteristics of the products manufactured are then given. Manufacture of biodiesel is described in the main countries committed to the development of first generation biofuels as replacement or complement to Diesel in Diesel engines.

The glycerine market is then discussed. Conversion of glycerine, a by-product of the transesterification reaction, plays a role in the global economic equilibrium of the sector. Use of vegetable oils and VOME in Diesel engines is then described, followed by the environmental and financial aspects.

In terms of protection of the environment, use of these products reduces the consumption of fossil fuels, by replacing Diesel, and secondly it leads to a significant reduction in emissions of pollutants and greenhouse gases (see 3.13).

3.1 RAW MATERIALS OF VEGETABLE ORIGIN

3.1.1 World production of vegetable oils

World production of vegetable oils was estimated at 130 million tonnes (Mt) in 2007, reaching only 86 Mt in 1999 and 101 Mt in 2003. This sharp upturn is expected to continue over the next few years since the first estimations indicate production figures well above 130 Mt in 2010. About 20 to 25 Mt animal fats and 1 and 2 Mt of tall oil, a by-product of kraft paper pulp processing, must be added to these sources of fat.

Palm oil is the largest contributor to world production of vegetable oils (about 30% of the total output), ahead of soybean oil. Rapeseed and sunflower oils come next, with volumes of about half of the first two oils mentioned (Figure 3.1). Over the last few years, palm oil production has recorded the largest increase.

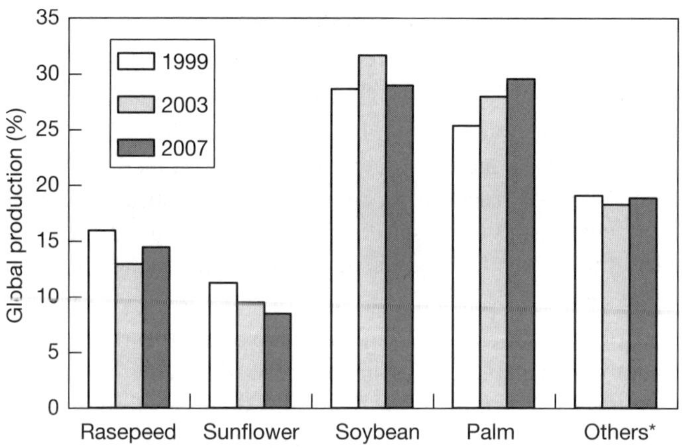

Figure 3.1

Production of the various vegetable oils (with respect to world output).

Other vegetable species, capable of producing oils, have been tested and are under development. The most important is probably *Jatropha curcas*, also known as "purghere" or medicinal nut. Native to Latin America, this plant is grown over several thousand hectares throughout the world, mainly Asia, South America and Africa, in a geographic region located between 30 degrees latitude north and 35 degrees latitude south. India and China are making considerable effort to develop these crops.

Jatropha curcas is a shrub whose fruit is rich in oil, with a content of about 30% to 35%. The distribution of fatty acids is as follows: palmitic acid 15%, stearic acid 6%, oleic acid 42%, linoleic acid and linolenic acid 35%, other acids 2%. This species can live up to 50 years. The first harvest can be made 2 to 3 years after plantation. The fruit is toxic for animal and human consumption.

Since this plant grows well and displays excellent resistance to difficult weather conditions (high temperature and periods of drought), it could play a role in fighting soil erosion and desertification and also act as raw material to produce energy (production of electricity and biofuels for tractors), for example in the rural areas of India, China and some African countries.

Jatropha curcas can be planted in several ways: in hedges, open fields or intercropping, for example with cotton, still a cash crop, as complement to food crops, but declining due to its currently poor profitability. Other annual crops such as peanut can be planted between the rows of *Jatropha curcas*.

The best species are being selected, while other points deserve further investigation, for example:

- real nitrogen, phosphorus and potassium requirements,
- change in fruit and oil yields, depending on the fertilisers and water provided. Seed yields vary considerably, from 2 to 5 t of seeds/ha/year, i.e. 0.7 to 1.7 t of oil/ha/year,
- resistance of the plant to drought and pests,
- shortening of the time for the fruit to reach maturity to reduce the harvesting duration,
- conversion of cakes.

Numerous R&D studies must be conducted before this plant is recognised for its oil production potential, in geographic areas where weather conditions and soil quality considerably limit the development of other plants.

In north-eastern Brazil, for example, plantations of babassu oil palm are developing, with an average oil production of 1.6 t/ha/year.

3.1.2 Production of oilseeds outside Europe

In 2007-2008, the main soybean producers were the United States (70 Mt) followed by Brazil (58 Mt), Argentina (45 Mt) and China (16 Mt) [1]. Together, they supply 90% of world production. US production has nevertheless dropped by about 20% compared with 2006, since some of the areas dedicated to soya have been turned over to corn. In 2008, Brazil became the leading exporter of soybeans, with 29 Mt, ahead of the United States and Argentina.

Indonesia is the No. 1 world producer of palm oil, the main source of vegetable oil, with 78 Mt of palm fruit, followed by Malaysia with 77 Mt. Several countries in West Africa such as Nigeria, Ghana and Ivory Coast come well behind. Six million hectares are dedicated to the cultivation of oil palm worldwide, with 80% in South East Asia.

3.1.3 European production of vegetable oils

In Europe, the production of rapeseed prevails, ahead of sunflower, with soybean lagging far behind (Table 3.1). The European Union is almost 90% self-sufficient for rapeseed, 35% for sunflower and only 1% for soya (67% of European consumption of protein cakes consists of imported soya).

Table 3.1 European production of vegetable oils (seeds).

	Country	Cultivated area (10^3 ha)			Yield (t/ha)			Production (10^3t)		
		2003	2007	2008	2003	2007	2008	2003	2007	2008
Rapeseed	Germany	1 269	1 548	1 373	2.87	3.4	3.76	3 644	5 320	5 160
	France	1 077	1 616	1 442	3.08	2.9	3.32	3 321	4 684	4 783
	Great Britain	542	603		3.27	3.1		1 771	1 896	
	EU	3 216	6 511		2.82	2.8		9 080	18 231	
	Czech Republic	250	338	357	1.56	3.1	2.96	390	1 032	1 055
	Poland	426	797	771	1.77	2.7	2.70	754	2 130	2 084
Sunflower	Spain	787	601	710	0.91	1.2	1.27	716	703	899
	France	686	519	626	2.08	2.5	2.53	1 424	1 308	1 585
	Italy	129	126		1.47	2.2		190	277	
	EU	1 712			1.45			2 482		
	Hungary	505	512	560	1.92	2.1	2.52	970	1 060	1 409
	Rumania		836	846		0.65	1.63		547	1 379
	Bulgaria		602	724		0.94	1.64		564	1 184

The values given for the European Union are those of the 15 countries in 2003 and of the 27 countries in 2007.
Sources: 2003: COPA/COGECA, 2007-2008: http://epp.eurostat.ec.europa.eu

Production of rapeseed, harvested in the EU-15, decreased between 1999 and 2003, dropping from 11.4 Mt to 9.1 Mt. In 2007 and 2008, however, the harvest increased in numerous countries, reaching a total of 18.2 Mt in EU-27.

The main producers are Germany and France, well ahead of Great Britain, Poland, Bulgaria and Rumania.

The agricultural yields vary from one year to the next. The average in EU-15 oscillates between 3.2 t/ha in 1999 and 2.8 t/ha in 2003, reaching a figure of 2.8 t/ha in EU-27. The best yields obtained in the main producing countries exceed 3 t/ha. They are significantly lower in the Czech Republic and Poland, Bulgaria and Rumania.

Generally speaking, the agricultural areas dedicated to rapeseed dropped in the EU-15 (3.2 Mha in 2003 *versus* 3.6 Mha in 1999) especially in France (1.08 Mha in 2003 *versus* 1.37 Mha in 1999), before picking up in 2007 and 2008 to reach a total value of 6.5 Mha in 2007 in the 27 European countries.

Sunflower seed production in EU-15 also dropped dramatically, from 3.3 Mt in 1999 to 2.5 Mt in 2003, due to a reduction in cultivated areas together with lower yields, before recording an upturn further to the involvement of countries such as Bulgaria and Rumania. Yields vary from one country to another. In 2007, for example, the two largest European producers, Spain and France, recorded respectively 1.2 t/ha and 2.5 t/ha, against 2.2 t/ha in Italy, 0.94 t/ha in Bulgaria and 0.65 t/ha in Rumania.

The areas dedicated to soya in EU-27, mainly Italy, Rumania and France, represent 5% of the areas set aside for rapeseed. Yields are highly variable from one year to the next (2.89 t/ha in 2002, compared with 1.89 t/ha in 2003 for the average yields obtained in France).

3.1.4 Agricultural yields (seeds, oils, cakes)

Table 3.2 summarises the various data.

Production of palm oil offers the best yield per hectare, well ahead of rapeseed. About 0.6 t of palm kernel oil, extracted from the fruit seeds, must be added to the 6.3 tonnes of oil produced per hectare. The rate of oil extraction from palm bunches is about 20%, roughly the same as that obtained with soya. Soya is mainly grown to obtain protein-rich cakes, intended for animal fodder.

The rapeseed oil production yield lies between 1.3 and 1.5 t/ha. The oil extraction rates, in the region of 40%, are substantially the same for rapeseed and sunflower.

Table 3.2 Agricultural yields of various oilseed plants (average values in 2002).

Oilseed plant	Seeds (t/ha)	Oil (t/ha)	Cakes (t/ha)
Soya	3.4	0.7	2.7
Palm	30 [a]	6.3	–
Rapeseed	3.1	1.2-1.3	1.8-1.9
Sunflower	1.7	0.7	0.6-0.7 [b]

[a] *Palm bunches.*

[b] *Hulled cakes.*

3.1.5 Composition of vegetable oils

Fats and oils, also called lipids, consist of 95-98% triglycerides. Minor constituents present in oils include free fatty acids, mono- and di-glycerides, phospholipids, tocopherols, sterols, natural colouring agents as well as more or less volatile odorous compounds.

Triglycerides are composed of a glycerol molecule esterified with 3 similar or different fatty acid molecules.

Some 20 fatty acids are found in nature and their numerous possible combinations with the 3 alcohol functions of the glycerol produce a wide variety of triglycerides and therefore of oils.

Table 3.3 illustrates the fatty acid composition of the most common vegetable oils [1].

Table 3.3 Composition in fatty acids of various oils (% by weight).

Oil \ Fatty acids	C6:0	C8:0	C10:0	C12:0	C14:0	C16:0	C18:0	C18:1	C18:2	C18:3	C20:0	C20:1	C22:0	C22:1	Iodine value
Rapeseed						5	2.5	59	21	9	< 0.5	1	< 0.5	1	111
Sunflower						6	5	18	69	< 0.5	< 0.5	< 0.5	1		135
Oleic sunflower						4	4	80	10	< 0.5	1	< 0.5	1		87
Soya						10	4	23	53	8	< 0.5	< 0.5	< 0.5		133
Palm					1	44	6	38	10	< 0.5	< 0.5				50
Palm kernel	< 1	4	3	47	16	8	3	16	3						19
Copra	1	7	6	46	18	9	3	8	2						10
Flax						6	3	19	17	54		< 0.5	< 0.5		187

3.1.6 Manufacture of vegetable oils

The processing of rapeseeds is described first.

3.1.6.1 The example of rapeseed

Production of rapeseed oil from seeds, whose composition is given in Table 3.4, consists of 3 steps: trituration, extraction and refining.

1. To illustrate the symbolic representation of fatty acids, oleic acid, for example, is represented by C18:1, where C18 is the number of carbon atoms and:1 is the number of double bonds of the fatty chain, i.e. one double bond for oleic acid.

Table 3.4 Mean composition of rapeseeds weighted by the areas (2003 harvest).

Characteristics	Mean weighted by areas
Impurities (%)	0.9
Water content (%)	7.2
Oil (%)	41.6
Proteins (% fat free dry matter)	39.9
Glucosinolates (µmoles/g. of seeds with 9% humidity)	14.1

Source: Info-Prolea, No. 62, April 2004

A. Trituration

To release the oil, the rapeseeds are crushed, cooked, then pressed in a screw press. After these steps, we obtain a crude oil and a cake which still contains between 12% and 15% by weight of oil.

B. Extraction

During this second step, the oil contained in the cakes is extracted. The operation is carried out using a non-polar solvent of boiling point less than 70°C. *n*-hexane is the most generally used solvent.

The deoiled cakes, whose composition is given in Table 3.5, represent a protein-rich food for animals. Soya cakes with high protein content but also a balanced distribution of essential amino acids, including high lysine content, form a good source of protein for animal feed and fish meal.

Table 3.5 Composition of cakes produced from various oilseeds (expressed in wt%).

Characteristics	Rapeseed	Sunflower (hulled)	Soya
Humidity	6-10	7.5-13.8	8-12
Protein matter	34	30-53	41-47
Residual fats	4-9	0.8-13.8	1-6
Cellulose	10-15	7-15	4-7
Ash	2.4-8.5	4.3-7.7	4-6
Phosphorus	0.9	1.0	0.6
Lysine [a]	–	3.8	6.2

[a] *As a percentage of the sum of the amino acids, constituents of proteins.*
Source: ITERG

Removing animal meal from feedstock should increase the demand for proteins of vegetable origin, in the coming years. This increase should primarily benefit soya cakes, whose world consumptions reached 159 Mt in 2008, including 35 Mt in Europe, 32 Mt in the United States and 30 Mt in China. However, use of rapeseed cakes in Europe, especially in dairy herds, should reduce these consumptions of soya cakes slightly. Recent studies with new formulations where rapeseed cakes replace soya cakes, with a substitution rate of 1.3 to 1, prove highly encouraging [1].

An average of about 2% oil is lost during the trituration and extraction steps: 100 kg of seeds with 42% oil give 41 kg of crude oil and 59 kg of cakes.

C. Refining

The main refining steps, described in Figure 3.2, lead to the production of an edible oil [2].

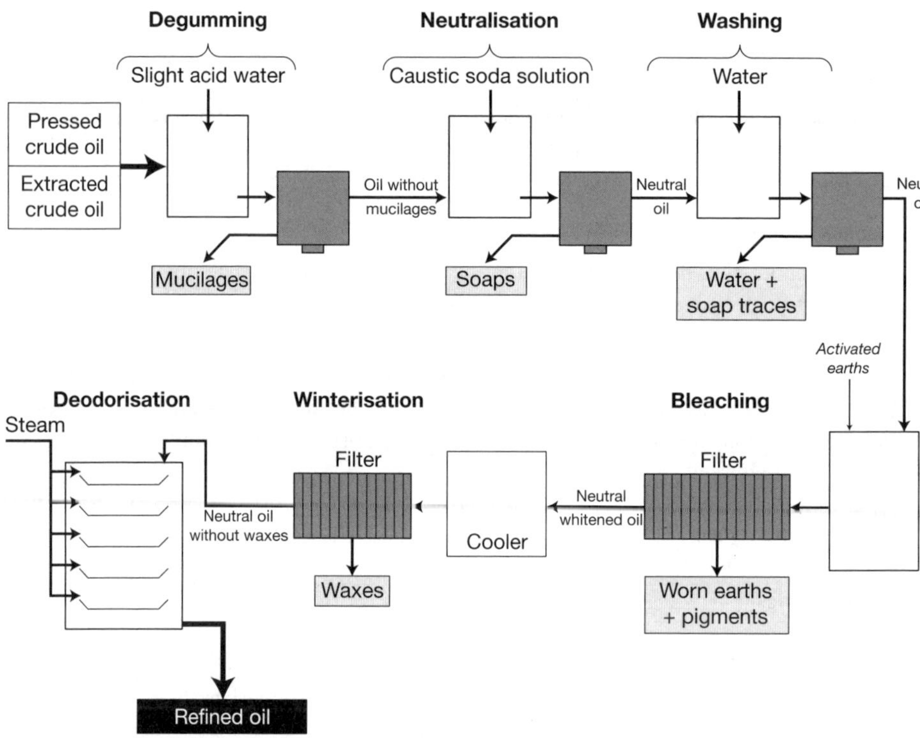

Figure 3.2

Main vegetable oil refining steps.

In the production of biofuels, refining stops after the oil washing then drying step.

Degumming consists in removing the phospholipids or mucilaginous products which precipitate in the presence of water. Precipitation is accelerated by adding phosphoric acid at 70-80°C. The hydrated mucilaginous products are removed by centrifuging before or after the neutralisation step with the sodium soaps formed.

1. Source: CEREOPA, Centre d'Études et de Recherche sur l'Économie et l'Organisation des Productions Animales (Centre for Research on the Economics and Management of Animal Production).

The free fatty acids, whose crude oil content rarely exceeds 1.5% by weight, cause instability and bad taste. When synthesising biofuels, they neutralise stoichiometrically the basic catalyst added during the transesterification reaction. Consequently, it is absolutely essential to remove them. By mixing the crude oil with a solution of caustic soda at 80-90°C, insoluble sodium soaps form, which can then be separated by centrifuging. These soap stocks carry along some of the pigments and impurities contained in the oil.

The traces of sodium salt still remaining in the neutral oil are washed out with water. 41 kg of crude oil give 40.2 kg of neutral oil.

To obtain an edible oil, two additional operations are required:

- bleaching by adding activated earth at 90°C, followed by filtration,
- vacuum steam deodorisation at 220°C to eliminate the volatile compounds which give the oil unpleasant smells and flavours.

D. Energy consumptions

The energy consumptions for the two steps, trituration and extraction, per tonne of seeds, amount to 680 MJ (steam) and 40 kWh.

For refining, the energy consumptions per tonne of crude oil amount to 150 MJ (steam) and 10 kWh.

3.1.6.2 Other oils

For corn and sunflower oils, an additional step called winterisation is required. During this operation, the oil is progressively cooled down to +5°C, when the natural waxes become insoluble. They are esters of fatty acids and alcohols with between 42 and 58 carbon atoms, present in the oils at a content of less than 0.1%.

For soya oils, the lecithin content (phospholipid with excellent surfactant properties of value in the food industry) is high enough to justify extraction before the neutralisation step.

Palm oil must be obtained very quickly after harvesting the palm tree fruits, which do not keep and are rapidly damaged, unlike oilseeds. The treatment facilities must therefore be built very close to the plantations.

The palm oil production process includes the following steps [3]:

- continuous steam sterilisation after crushing the bunches,
- threshing, to separate the fruit from the bunches,
- mixing and pressurising,
- clarification of the crude juice by settling, to recover about 90% of oil,
- deoiling of the oily sludges by centrifuging or settling in several steps,
- dehydration under partial vacuum,
- storage of the palm oils.

In the palm crushing plant the nuts and fibres are separated in columns. The nuts are then polished to remove any remaining fibres, then cracked. The shells and kernels are then dried and stored before extraction of the palm kernel oil.

The palm oil mill produces much more energy than it consumes. Using only some of the solid waste (fibres contained in the fruit, palm shells, bunches) resulting from oil production, it is sufficient to generate the steam and electricity required to run the process. For a well-designed oil mill, combustion of the fibres alone covers all the energy requirements of the industrial plant. In addition, the other effluents as sludges have a high organic matter content (COD of about 60 000 mg/l). Large quantities of biogas could be obtained by methanisation of the sludges, for use in boilers or heat engines.

For oil with a high natural acidity such as palm oil, the free fatty acids are removed by vacuum steam distillation at 220°C, under reduced pressure, during the deodorisation step. This type of treatment, called physical refining, is quite suitable when the acid number of the crude oils is greater than 4-5, unlike chemical refining with soda which generates significant oil losses in the soap stocks.

3.1.7 Applications of vegetable oils

Competition between food and energy applications for oilseeds? Concerning rapeseed, it is important to note that, apart from the straw of this plant which can be used as a complement to other plant raw materials, such as cereal straw and wood to generate electricity and heat (cogeneration), the cakes are used as animal feed. They partially replace soya cakes, with a production of about 1 to 1.5 tonnes per tonne of VOME synthesised. In addition, glycerine (0.1 t/t VOME) can be partly used in the food industry in addition to its other potential applications. Only the oil fraction, representing 40% by weight of the seed, is used to manufacture biofuels. Note also that this oil is used in the food sector, as well as for other industrial applications.

In 2008, the European Union was the second user of vegetable oils, behind China, with 22.3 Mt (*source*: USDA), with a consumption in the food sector of 13.1 Mt and 9.2 Mt for industrial applications, mainly production of biodiesel. The USDA envisages a world consumption of 130 Mt of vegetable oils, including 105 Mt for food applications and 25 Mt for industrial requirements.

In France, the requirements for land dedicated to oilseeds (rapeseed and sunflower) are estimated at 2.1 Mha for 2010 [4]. With a target of introducing 7% biodiesel in Diesel, 1.45 Mha, i.e. 70% of the oilseed land, would be dedicated to this supply. This clearly reflects the trend observed in the use of rapeseed and sunflower oils in France.

In 2003, about 70% to 85% of rapeseed oil was used in food-related applications in Europe. The figure reaches an average of 95% for sunflower and almost 100% for soya (Table 3.6).

The rest is intended for oleochemical syntheses (fatty alcohols, fatty amines), the manufacture of lubricants, surfactants, solvents, soaps, care products. The main industrial applications of the lipochemical derivatives of fatty acids are shown on Figure 3.3.

Chemists consider biolubricants as being the most promising sector for oilseeds, especially in fields of application such as lost oils or those giving rise to repetitive accidental losses.

Table 3.6 Applications of vegetable oils (seeds) in Europe (2003).

Plant	Country	Food application				Non-food application			
		S [a]		P [b]		S [a]		P [b]	
		10^3 ha	%	10^3 t	%	10^3 ha	%	10^3 t	%
Rapeseed	Germany	944	74.4	2 733	75.0	325	25.6	911	25.0
	France	784	72.8	2 419	72.8	293	27.2	902	27.2
	Great Britain	460	84.9	1 548	87.4	82	15.1	223	12.6
	EU-15	2 475	77.0	7 009	77.2	740	23.0	2 071	22.8
	Czech Republic	125	50.0	195	50.0	125	50.0	195	50.0
	Poland	426	100.0	754	100.0	0	0	0	0
Sunflower	Spain	787	100.0	716	100.0	0	0	0	0
	France	639	93.1	1 326	93.1	47	6.9	98	6.9
	Italy	116	89.9	172	90.5	13	10.1	18	9.5
	EU-15	1 646	96.2	2 355	94.9	66	3.8	127	5.1
	Hungary	505	100.0	970	100.0	0	0	0	0
Soya	France	81	100.0	153	100.0	0	0	0	0
	Italy	140	100.0	450	100.0	0	0	0	0
	EU-15	236	100.0	635	100.0	0	0	0	0

[a] *Areas cultivated in thousands of hectares and as a percentage of total areas.*
[b] *Production in thousands of tonnes of seeds and percentages of the various applications.*

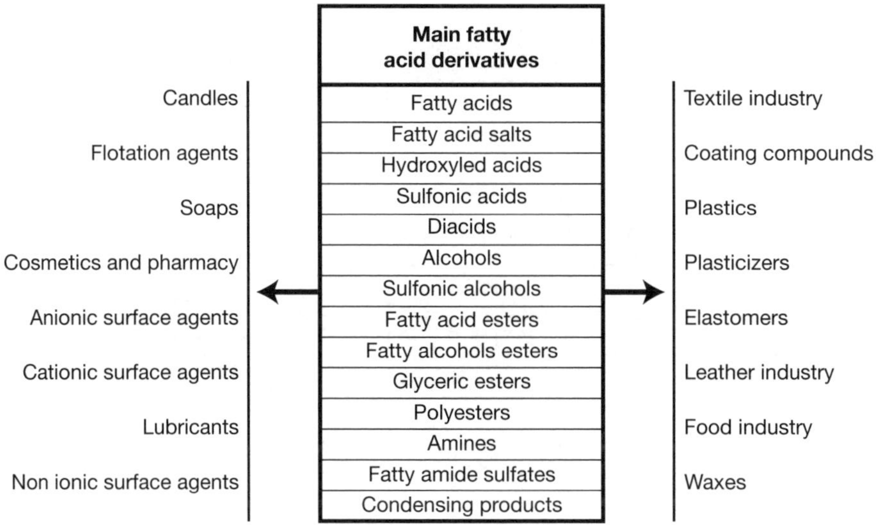

Figure 3.3

Industrial applications of the main derivatives of fatty acids.

Surfactants represent the main non-food outlet for oils, produced mainly outside Europe. Together with oil used for lighting, soap was one of the first known non-food technical applications of vegetable oils [1].

Lastly, VOME-based biosolvents represent an ecological alternative to the solvents targeted by the new European directives, in particular those aimed at limiting the emissions of Volatile Organic Compounds (VOC). Paints, lacquers, varnishes and inks have always contained vegetable oils.

In 1998, the European chemical industry consumed 470 thousand tonnes of vegetable oils.

3.2 METHANOL

Methanol is one of the most widely manufactured organic chemicals worldwide, with a production of about 21 Mt/year, including 5.1 Mt in the United States, 2.3 Mt in Canada, 2.2 Mt in Saudi Arabia and 1.6 Mt in Germany.

1. A clay tablet, of Sumerian origin, found at Babylon and dating from 2800 BC describes a soap used to clean fabrics.

3.2.1 Methanol synthesis

Methanol [5] is mainly manufactured from syngas ($CO + H_2$), itself obtained by partial oxidation with oxygen of methane or gasification of heavier hydrocarbons and in particular by steam reforming of methane or naphtha, according to the following reactions:

$$CH_4 + H_2O \rightleftharpoons CO + 3\,H_2$$
$$CO + H_2O \rightleftharpoons CO_2 + H_2$$

Methanol synthesis is based on the reaction:

$$CO + 2\,H_2 \rightleftharpoons CH_3OH$$

And, to a lesser extent, on conversion of CO_2:

$$CO_2 + 3\,H_2 \rightleftharpoons CH_3OH + H_2O$$

The gaseous mixture requires a hydrogen/carbon molar ratio of between 2 and 3.

Production of methanol is favoured by higher pressure and lower temperature; this last point means that catalysts must be used. Copper-based catalysts have now replaced zinc-chromium catalysts.

3.2.2 Methanol production processes

The processes generally operate between 50 and 100 bar at temperatures between 240°C and 270°C, in vertical reactors.

The main technologies are those of the following companies: Johnson Matthey (JM) (former ICI, Imperial Chemical Industries), Lurgi, Haldor Topsoe.

The reactors can be extrapolated up to unit production capacities of 3000 to 5000 t/d.

The main chemical applications of methanol are the syntheses of formol, MTBE, methyl terephthalate and acetic acid.

3.3 VOME PRODUCTION PROCESSES

3.3.1 The transesterification reaction

Methyl esters are obtained by transesterification of triglycerides with methanol according to the following diagram. It is a balanced reaction, carried out in the presence of a catalyst.

Oil triglycerides are glycerol esters (glycerol sometimes known as glycerine) and fatty acid R-COOH.

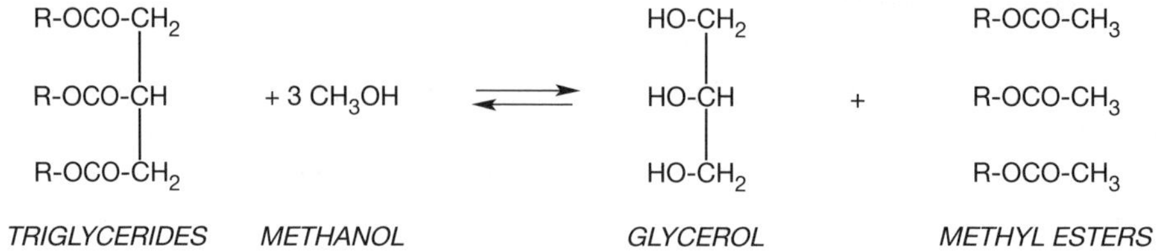

The reaction diagram in Figure 3.4 clearly shows that the release of glycerol is only complete after the consecutive reactions of transesterification of the triglycerides into diglycerides, then into monoglycerides and the formation of the third molecule of fatty acid methyl ester.

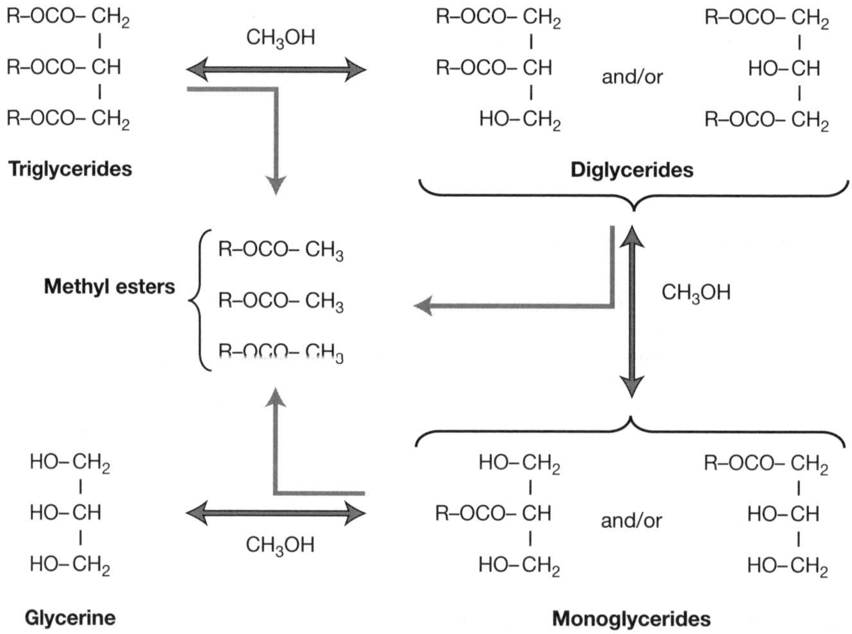

Figure 3.4

Transesterification reaction diagram.

This reaction can be carried out by homogeneous catalysis with catalysts soluble in the reaction environment, or by heterogeneous catalysis, with catalysts totally insoluble in the reactants.

This reaction can be conducted with other techniques using innovating technologies such as microwave heating [6] (with considerably reduced catalysis times) or by enzyme catalysis (with in this case relatively long residence times). To date, these technologies are not mature enough to allow industrialisation in the near future.

3.3.2 Homogeneous catalysis processes

Homogeneous catalysis is currently the most widely used technique in biodiesel production processes.

Transesterification can be carried out by basic or acid catalysis. Greater reactivity is generally obtained in basic solution.

3.3.2.1 Catalysts

There are three broad classes of catalysts:

- basic catalysts
  - hydroxides, alkoxides or soaps of alkali or alkaline-earth metals (Li, Na, K, Ca, Ba, Cs, etc.)
  - amines of the guanidine family, for example.
- acid catalysts
  - mineral acids: HCl, H_2SO_4
  - sulphonic acids
  - ion-exchange resins (strong acid)
  - zeolites
- other catalysts
  - titanium alkoxides: Ti (OBu) 4, Ti $(OiPr)_4$, etc.
  - oxides of various metals such as Sn, Mg, Zn, Ti, Pb, etc.

Acid catalysts are rarely used due to their lower activity and high risks of corrosion of the industrial facilities. The metal alkoxides or oxides are mainly used to synthesise heavy alcohol esters from various cuts of fatty acid methyl esters.

The catalysts generally chosen for biodiesel production are soda in methanolic solution or sodium methylate.

3.3.2.2 Raw material specifications

The specifications of the raw materials in the transesterification reaction, vegetable oils and methanol, are described below.

A. *Vegetable oils*

Acid number AN (expressed in mg KOH/g oil) < 1

Phosphorus content ≤ 10 ppm

Water content (wt%) ≤ 0.1

B. *Methanol*

Methanol content (wt%) ≥ 99.85

Water content (wt%) ≤ 0.1

3.3.2.3 Discontinuous processes

An example of schematic diagram is shown on Figure 3.5.

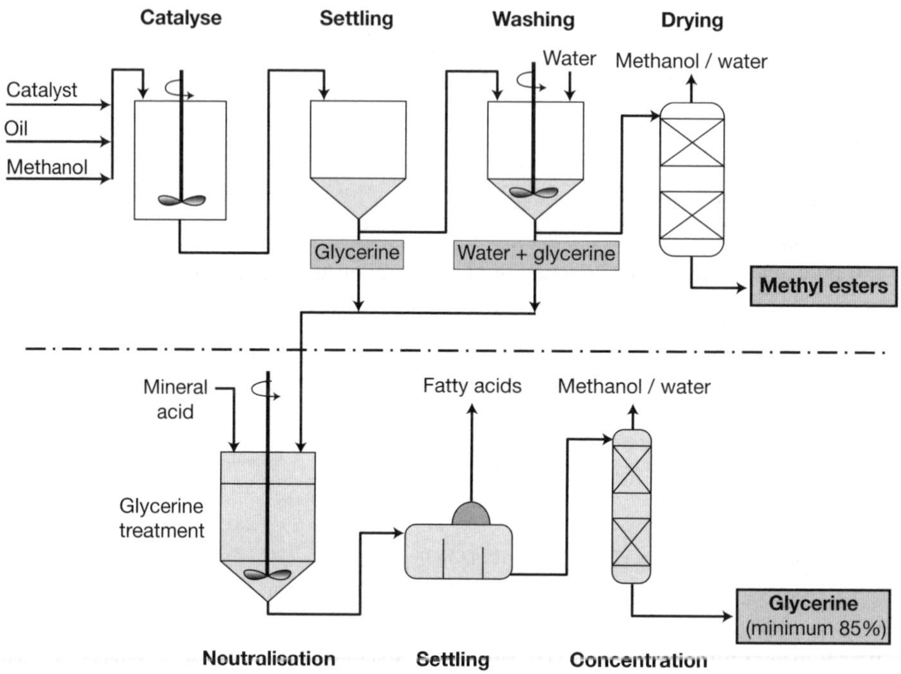

Figure 3.5

Schematic diagram of a discontinuous process (Esterfip process – IFP Energies nouvelles and Axens).

Oil and methanol are introduced in an reactor equipped with a stirring system and the mixture raised to a temperature of 45°C to 85°C, under a maximum absolute pressure of 2.5 bar, before adding the catalytic solution.

The residence time is about one hour to reach the reaction thermodynamic equilibrium. As an example and allowing for the emptying and filling times, a reactor of 15 m^3 useful capacity can achieve an annual production of 80 000 tonnes of VOME, generally considered as the maximum for discontinuous processes with a single reaction chain.

Unlike the transesterification reaction which is carried out discontinuously, all the downstream ester purification operations are carried out continuously.

Methyl esters are separated from the glycerine in a static settling tank before being introduced into a second stirred reactor where two water washing steps are carried out to eliminate the salts and residual glycerine. To improve the quality of the esters, acidulated water can be added in a third stirred reactor. Any traces of methanol and water still contained in the esters are eliminated by distillation.

After the reaction, most of the basic catalyst and the soaps are contained in the glycerine fraction. This fraction and the washing water are combined then neutralised with hydrochloric acid. They are reconcentrated by distillation, after separation by settling of the free fatty acids released during neutralisation. Note that salts are formed during the neutralisation step.

The excess of methanol recovered is dried before being recycled.

Several variants are possible:

- static settling tanks replaced by centrifuges,
- distillation columns replaced by falling-film evaporators.

Deep purification of the esters, to totally eliminate the soluble metal ions, can be carried out by passage over a bed of ion-exchange resins.

Depending on the technologies and catalyst chosen, the yields by weight of the methyl esters with respect to the oil vary from 98.5% to 99.4%. The losses correspond to the fatty acids contained in the oil which react with the catalyst to form alkaline soaps (saponification).

The yields therefore vary with the acid number of the oil used (the higher the value, the lower the yield), and with the water content of the raw materials, which poisons the catalysts and favours saponification.

Respecting the raw material specifications (paragraph 3.3.2.2), the catalyst consumptions per tonne of oil lie between 3.6 and 4.4 kg of soda or between 3.8 and 4.9 kg for sodium methylate (available in 30% solution in methanol).

The yields could potentially be improved by recycling the fatty acids produced when neutralising the basic glycerine [7]. Glycerolysis of the fatty acids with the basic glycerine results in the formation of a mixture of mono-, di- and tri-glycerides which can be reinjected in the transesterification reactor for conversion into methyl esters.

The operating conditions necessary during glycerolysis to obtain an acid number < 3 are as follows:

- glycerine/fatty acids stoichiometry = 1.25,
- the crude glycerine produced during the process must still contain sufficient basic catalyst to meet the requirements of the glycerolysis reaction,
- temperature between 200°C and 210°C,
- reaction time 3 to 4 hours.

After recycling the fatty acids, the overall yield of the transesterification may reach almost 100%.

3.3.2.4 Continuous processes

Since the capacity of the units built in recent years often exceeds 100 000 t/year, continuous processes are now regularly chosen. The main advantages lie firstly in the significant reduction in the size of the equipment, and consequently in investments, and secondly in the lower operating costs, including labour. The following illustration is only an example of this type of process. Several others are in fact available, using the same physico-chemical and chemical bases.

The schematic diagram (Figure 3.6) illustrates the main steps in a continuous process.

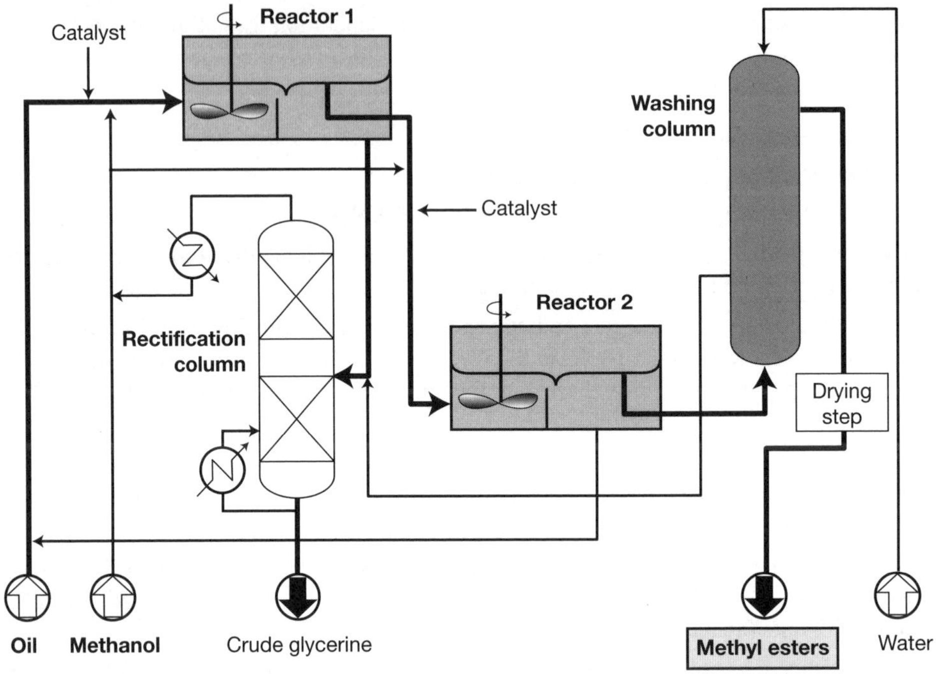

Figure 3.6

Schematic diagram of a continuous VOME production process (process proposed by Lurgi).

The transesterification reaction is generally carried out in two reactors in series, of the same capacity, with optimised residence times to obtain maximum conversion rate and esters whose quality corresponds as closely as possible to the specifications imposed on fuel esters. This quality is further improved by countercurrent water washing, which eliminates traces of catalyst, glycerine and methanol. Lastly, the ester is dried under reduced pressure, between 40 and 60 mbar at 140°C.

Note that process developed by Cognis includes no water washing steps: the ester is purified by vacuum flash distillation.

The basic salts from the catalyst and contained in the crude glycerine and the washing water are neutralised by a mineral acid (HCl, H_2SO_4, H_3PO_4). The excess methanol is recovered by distillation, bringing the glycerine content to a value between 80 and 90% by eliminating the water.

As with the discontinuous processes, methyl ester yields of between 98.5% and 99.4% are obtained. These yields could be further increased by subsequent glycerolysis of the fatty acids.

The characteristics of the products manufactured are substantially the same as those obtained with the discontinuous processes.

Still respecting the raw material specifications (paragraph 3.3.2.2), the catalyst consumptions per tonne of oil are roughly equivalent to those indicated for the discontinuous processes.

3.3.2.5 Energy consumptions

Since the operating conditions (temperature, pressure) of the discontinuous and continuous processes are comparable, the energy consumptions per tonne of oil treated are similar: 900 MJ (use of natural gas) for steam production, plus electricity consumption of 30 to 35 kWh.

3.3.3 Continuous heterogeneous catalysis processes

Heterogeneous catalysis offers significant advantages regarding respect of the environment. It meets the criteria associated with the new "green chemistry" concepts since, due to the purity of the products obtained together with high synthesis yields, there are virtually no polluting discharges. In addition, unlike homogeneous catalysis, the fact that there are no salts in the reaction products removes the need for expensive purification treatments and increases the range of industrial outlets for the glycerine, without prior purification.

As mentioned previously, the processes currently using homogeneous catalysis in basic environment operate on lost catalyst. These choices make purification of the two products, VOME and glycerine, more complicated and increase the cost of added compounds, such as hydrochloric acid used for glycerine neutralisation.

3.3.3.1 The catalyst

The catalyst developed by IFP Energies nouvelles [8], which consists of extrudates (Photo 3.1), is a spinel type zinc aluminate of general formula $ZnAl_2O_4$.

This catalyst is highly stable over time, no loss of activity being observed after one year's operation, and exhibits good resistance to attrition.

3.3.3.2 Raw material specifications

The raw material specifications are more or less the same as those required for homogeneous catalysis.

A. Vegetable oils

Acid number AN (expressed in mg KOH/g oil) ≤ 10

Water content (wt%) ≤ 0.1

Phosphorus content ≤ 10 ppm

With this type of process, even acid oils such as palm oil and palm kernel oil can be converted into esters without prior treatment. In addition, partially refined oils (degumming without neutralisation) are accepted as feedstocks.

Photo 3.1

Zinc aluminate extrudates.

B. Methanol

Methanol content (wt%) ≥ 99.85

Water content (wt%) ≤ 0.1

3.3.3.3 Continuous process

The schematic diagram of the process on Figure 3.7 and the photograph of the Sète Diester unit (Photo 3.2) illustrate this continuous process.

The two transesterification reactors, installed in series to obtain theoretical yields close to 100%, are continuous fixed-bed catalytic reactors. Since the catalyst activity is relatively low, reaction temperatures of between 180°C and 220°C are required, much greater than those for homogeneous catalysis, and pressures of between 40 and 60 bar.

The methanol/oil ratio is between 35% and 50% by weight. The mixture introduced in the first reactor has a residence time of about one hour. The effluent, whose methyl ester content is in the region of 95%, is subjected to partial evaporation to eliminate the excess alcohol. This methods improves separation of the glycerine supplied and favourably shifts the reaction balance to achieve nearly 100% transesterification efficiency in the second reactor.

The pressure drop of the effluent as it leaves the first reactor, and the heat generated by the reaction, are used in the excess alcohol evaporation step between the two transesterification steps.

After the second transesterification reaction, all excess methanol is eliminated by distillation and recycled. The ester content is greater than 98% by weight. A final methyl ester purification step may consist in eliminating the soluble glycerine by passage through a column filled with a selective adsorbent. The glycerine is separated by settling; its purity is above 98%, the main impurity being water, present initially in the raw materials used.

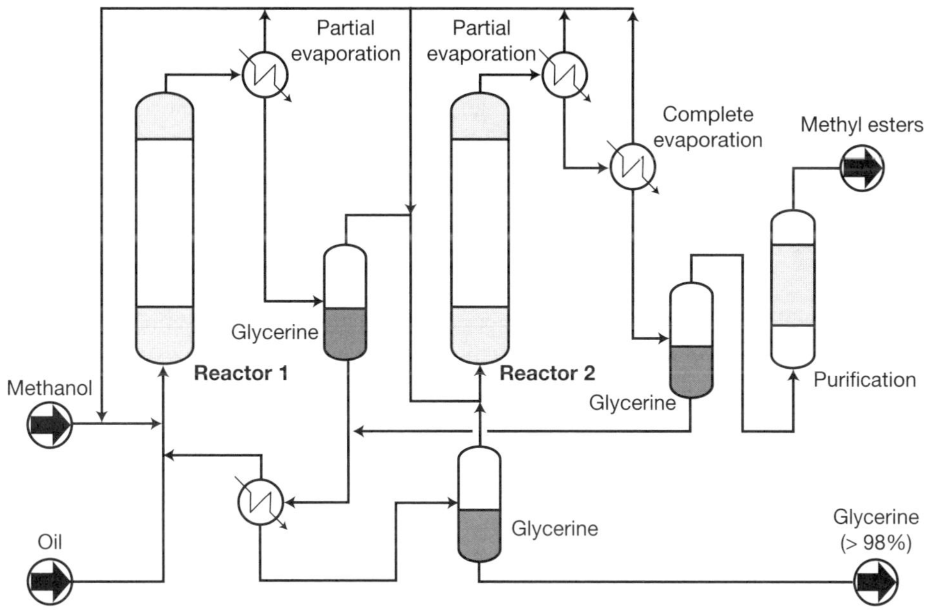

Figure 3.7

Schematic diagram of a continuous VOME production process by heterogeneous catalysis (Esterfip H process, IFP Energies nouvelles and Axens).

Photo 3.2

VOME continuous production unit (Esterfip H process, IFP Energies nouvelles and Axens). *Photograph credit*: Déclic Éditions.

The purity of the glycerine obtained is greater than that obtained by the homogeneous catalysis processes, with in particular the complete absence of soluble salts. It can be used in numerous chemical applications without prior purification treatment, for example glycerophthalic paints, polyols for polyurethane, synthon for various chemical syntheses of ethers, acetals and carbonates.

3.4 PRODUCTION OF VEGETABLE OIL ETHYL ESTERS (VOEE)

Since more and more Diesel vehicles are being sold in Europe, introduction of ethanol as VOEE in Diesel pool represents an opportunity for ethanol producers to extend the alcohol outlets in the field of biofuels.

The possibility of manufacturing vegetable oil ethyl esters, replacing methanol by ethanol, is under development. It represents an indirect pathway for the introduction of ethanol in the Diesel pool, since the characteristics of VOEE as fuel are at least as good as those of VOME. It must nevertheless be pointed out that ethanolysis via homogeneous basic catalysis is very difficult. High conversion values cannot be achieved in a single step since glycerine, being completely soluble in the reaction medium, modifies the reaction equilibrium. A second transesterification step is therefore required. It is carried out after distillation of the excess alcohol which results in demixing of the glycerine formed and allows its separation. This distillation operation is difficult to implement in homogeneous catalysis, since the catalyst still active and present in the medium may favour the back reaction with the glycerol released and reform glycerides.

The heterogeneous catalysis process removes the problem of back reaction, since glycerol separation is not carried out in the presence of the catalyst.

Ethanol being generally less reactive than methanol, ethyl ester productivities are always lower than those of methyl esters.

The following specifications are required for ethanol:

- ethanol content (wt%) $\geq$ 99,
- water content (wt%) $\leq$ 0.1.

Water is a poison common to the methanolysis and ethanolysis processes using homogeneous or heterogeneous catalysis. In homogeneous catalysis, water is a precursor of the saponification reaction, leading to excess catalyst consumption and reduced yield. In heterogeneous catalysis, this results in temporary and reversible inhibition of catalytic activity, which may significantly reduce the conversion rate.

Two processes are under development at Petrobras in Brazil, one by esterification of oil with ethanol, the other by direct pathway from seeds, without extracting the oil, still in the presence of ethanol. Two demonstration units are implementing these two pathways at Guamaré in Brazil, prior to extrapolation to industrial stage.

A demonstration unit with an annual production capacity of 10 000 t is operating at La Rochelle (France), implementing a technology developed by Bioenergy Pilot. One feature of this process concerns the esterification reaction, conducted directly from oilseeds, not vegetable oils.

3.5 OTHER RAW MATERIALS

The processes described previously are also suitable, provided that the alcohol/oil ratios and operating conditions are adapted, for the conversion of sunflower, soya, palm, palm kernel and copra (coconut) oils into methyl esters.

In all cases, the ester qualities obtained with heterogeneous catalysis are superior to those obtained with homogeneous catalysis in basic medium.

The physical properties of castor oil, composed of nearly 85% 12-hydroxyoleic acid (ricinoleic acid), are different from those of other oils. It can be transesterified with no major problems, but the esters obtained cannot be used as biofuels since, at high temperature, conjugated dienes forming by dehydration at the hydroxyl function tend to polymerise and form gums.

In 2001, some 40 000 t of VOME were produced in the United States, including a large proportion from recycled cooking oils, the rest from soya oils. This use of recycled cooking oils reduces the cost of raw materials but increases the treatment cost. They generally consist of hydrogenated oils with a high pour point, or peanut and/or sunflower oils; in addition to a high acid number, which is a handicap in transesterification by homogeneous basic catalysis, they generate polymers during cooking. These compounds may cause serious problems in engines (piston gumming). Vacuum distillation of the esters obtained is virtually essential for units dedicated to transesterification of these oils. Another solution, after filtration and esterification and to reduce their free acidity, would be to introduce a few percent of these oils into existing biodiesel production units. After validation on engine tests, the dilution effect could avoid the need for final distillation of the esters manufactured. This new approach to recycle cooking oils should considerably offset their production cost.

Given the characteristics of animal fats, there are no real difficulties in transesterifying them with methanol or ethanol, provided that they contain no water, free fatty acids, sugars or soluble proteins. Refining of the raw material must be adapted to each type of fat.

There are five types of animal fat: lard (pig fat), beef and sheep tallow, poultry fats, bone fats and fish oils. Fish oils cannot be used to produce biofuels, however, due to their high iodine value.

The properties of the esters obtained comply with the biodiesel specifications. However, their cold properties resulting from their high content in saturated fatty acid esters are penalising for use in winter, if these esters alone are mixed with Diesel. Due to the high pour points of saturated fatty acid methyl esters, the fuel may in fact crystallise and/or freeze.

In contrast, use of animal fat esters mixed with sunflower esters, for example, delays or prevents crystallisation of the animal fat esters at low temperatures and lowers the iodine value of sunflower esters, thereby making them jointly acceptable as biofuels.

Lastly, animal fats are inexpensive compared with vegetable oils, although availability remains limited. As an indication, it is still below 400 000 t/year in France.

3.6 OTHER PROCESSES, INCLUDING HYDROTREATMENT AND HYDROISOMERISATION OF RENEWABLE FATS

A new process has been developed to produce a type of synfuel from tall oil, at the Saskatchewan Research Council, sponsored by the Canadian Mineral and Energy Technology Centre (CANMET). Tall oil is composed mainly of unsaturated C18 fatty acids, resin acids and unsaponifiable compounds such as terpene alcohols and aldehydes. Tars are removed from the crude tall oil before it is hydrogenated by catalytic pathway and cracked. Three distinct fractions can be separated: an initial fraction (initial boiling point 160°C), a Diesel fuel fraction (not methyl esters in this case) with a cetane number of 64, and a heavy fuel [9].

Vegetable oils are favoured resources to supply the Diesel fuel pool, since the chemical structure of the fatty chains is similar to that of hydrocarbons in Diesel cuts. However, the methyl or ethyl esters produced retain oxygen in their molecules and their properties, especially cetane number and pour point, depend on the feedstocks used for their production, which could limit their use as fuel.

To overcome these limitations, the equivalent of a Diesel from vegetable oils, animal fats or cooking oils and algae can be obtained via another pathway. It consists in direct conversion of the oils by total hydrogenation of triglycerides, to produce Diesel bases completely free from oxygen and fully compatible with diesels of fossil origin. The unsaturated carbons of the fatty acid chains and the carboxyl groups can be fully hydrogenated by catalytic hydrotreatment of the oils at 300°C to obtain a final production of *n*-paraffins, propane, water and carbon dioxide. A hydroisomerisation step is required to respect the specifications imposed for use as fuel, in terms of cloud and pour points of the mixture of hydrocarbons obtained, which has a cetane number of about 70.

3.6.1 Hydrotreatment

Conversion of vegetable oils by hydrotreatment consists in treatment under hydrogen pressure in the presence of a suitable catalyst, resulting in total hydrodeoxygenation of the initial fat, forming mainly linear paraffins. Two conversion pathways coexist:

- hydrogenation/hydrogenolysis, which preserves the number of carbons in the fatty chains. This reaction is accompanied by the formation of water,
- decarboxylation, which is accompanied by the formation of CO_2 and which leads to hydrocarbons with one carbon atom less than the initial fatty chain.

In all cases, the unsaturated carbons of the fatty chains are totally hydrogenated and the glyceric group is converted into propane. In addition, secondary reactions may lead to the formation of CO and methane. The main reactions for the conversion of triglycerides by hydrotreatment are indicated schematically below.

A. Hydrogenation/hydrogenolysis

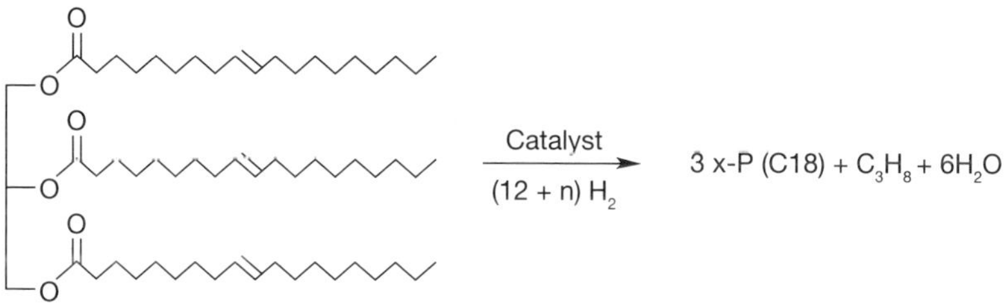

n is the number of unsaturations per triglyceride mole.

B. Decarboxylation

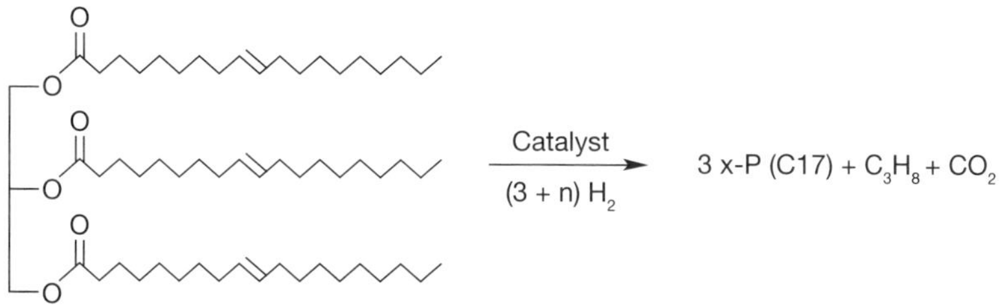

n is the number of unsaturations per triglyceride mole.

C. Shift and methanation reactions

$$CO_2 + 3\,H_2 \rightleftharpoons CO + 3\,H_2O$$

$$CO_2 + 4\,H_2 \rightleftharpoons CH_4 + 2\,H_2O$$

Hydrotreatment of vegetable oils consists of two main steps:

- a hydrotreatment section, which hydrogenates the triglycerides, to form mainly linear paraffins. The two conversion reactions mentioned previously lead to the formation of water, CO_2, CO and propane,
- a hydroisomerisation section, designed to improve the cold properties of the Diesel bases produced. As an illustration, the Cold Filter-Plugging Point (CFPP) in Europe is 0°C in summer and – 15°C in winter.

For the hydrotreatment section, sulphide type catalysts used in traditional hydrotreatment of diesels are active to obtain complete hydrodeoxygenation of the fats. Hydrogen consumptions depend on the type of initial oil or animal fats, in particular the number of unsaturations per chain and the length of the hydrocarbon fatty chains.

The following table (Table 3.7) shows the orders of magnitude of hydrogen consumptions obtained with the two pathways, hydrogenation and decarboxylation, as well as the corresponding theoretical yields by weight.

For example, catalytic hydrotreatment of rapeseed oil would require the introduction of 3.6% by weight of hydrogen. With maximum selectivity, 100 kg of oil give 86.3 kg of Diesel equivalent, 5 kg of *n*-propane and 12.3 kg of water. A production unit of 100 000 tonnes/year

Table 3.7 Yields by weight and hydrogen consumptions according to the two conversion pathways, hydrogenolysis (hydro) or decarboxylation (decarbo), depending on the initial oil type.

Oil type	Rapeseed		Palm		Soya	
	Hydro	Decarbo	Hydro	Decarbo	Hydro	Decarbo
H_2 consumption (weight%/feedstock)	3.6	1.5	3.2	1.2	3.8	1.7
Paraffin yield (weight%/feedstock)	86.3	81.5	85.3	80.4	86.4	81.6
Propane yield (weight%/feedstock)	5.0	5.0	5.2	5.2	5.0	5.0
H_2O yield (weight%/feedstock)	12.3	–	12.7	–	12.4	–
CO_2 yield (weight%/feedstock)	–	15.0	–	15.6	–	15.1

Diesel equivalent would require a feedstock of 116 000 t of rapeseed oil, 4 300 t of hydrogen and would generate 5 800 t of propane.

Due to the low density of the paraffins obtained, the yields per unit volume (with respect to the feedstock) would be less penalising. With rapeseed, for example, the theoretical yields per unit volume are respectively 101.9% and 96.3% for the hydrogenolysis and decarboxylation pathways.

The hydrogen consumption for palm oil, due to its more saturated structure, is lower than those measured for rapeseed and soya oils.

Hydrotreatment of vegetable oils and animal fats offers a certain number of advantages compared with other pathways such as transesterification (Table 3.8):

- the bases produced have a higher cetane number,
- the process is compatible with existing refining infrastructures,
- their qualities make the products compatible with current engines and offer the flexibility required for blending with convention Diesel (sulphur content, cetane number, calorific value).

Table 3.8 Comparison of fuel properties between a conventional Diesel and the products obtained with vegetable oil (transesterification or hydrotreatment).

Properties	Conventional Diesel	Biodiesel (transesterification)	Diesel produced by hydrotreatment
O content (wt%)	0	11	0
S content (ppm wt)	< 50 (10)	< 10	< 10
Relative density	0.82-0.845	0.883	0.78
NCV (MJ/kg)	44	38	44
Cetane number	51	50-65	70-90
Aromatic content (wt%)	< 11	0	0
CFPP (°C)	– 5	– 5 to +15	– 10 to – 20 [a]

[a] *Cold properties after a hydroisomerisation step further to hydrotreatment.*

3.6.2 Hydroisomerisation

The linear paraffins obtained by hydrotreatment of fats, as described above, are molecules with high cetane number but high melting point. Consequently, they can only be used as fuel (Diesel) base when blended in small proportions.

This constraint can be removed by converting these linear paraffins into molecules of identical composition (isomers) with slightly branched carbon chain.

Isomerisation of paraffins is a well-known reaction frequently used in refining and petrochemistry. Generally conducted under pressure and in the presence of hydrogen (in this case we speak of hydroisomerisation), it is favoured by several catalyst families.

The ability of a paraffin to remain liquid at low temperature increases with the degree of branching (we also speak of "branching rate"). However, branching a paraffin lowers its cetane number. For most isomerisation catalysts, the search for a high branching rate is generally accompanied by risk of breaking the carbon chain (hydrocracking).

The choice of catalyst and the hydroisomerisation operating conditions must therefore be optimised to reach the best compromise between yield and the targeted properties.

For Diesel, the cold properties must obviously be improved while retaining the best cetane number.

If the absence of a cetane constraint, due to their extremely low freezing points (Figure 3.8) some highly branched isoparaffins could be used as bases for the formulation of fuels suitable for extreme environments (polar regions) and in aviation. Note also that, for this application, hydrocracking associated with the high isomerisation rates may in fact prove to be an advantage, since the fuels obtained are composed of lighter molecules than those forming automotive Diesel [10].

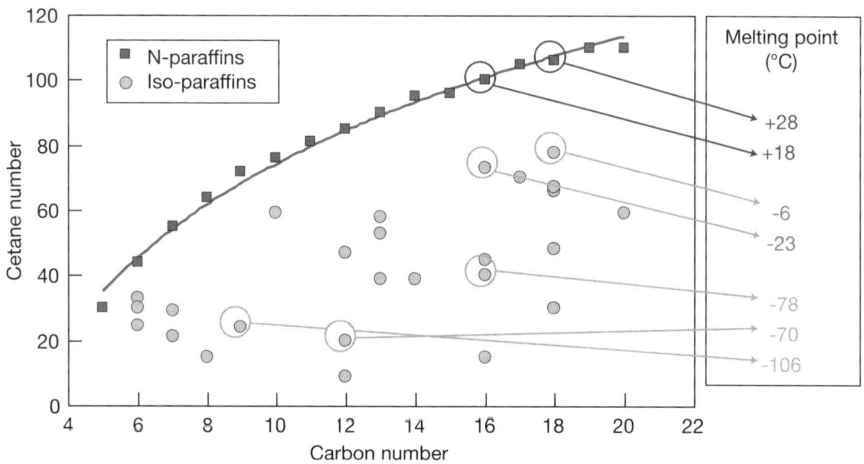

Figure 3.8

Properties of *n*-paraffins and isoparaffins.

3.6.3 Processes

When implementing this type of hydrotreatment process, followed optionally by a hydroisomerisation section, a certain number of problems must be taken into account, including:

- controlling the content in various contaminants which may be present in the feedstock, such as alkali metals, proteins and sugars. One solution consists in refining the feedstock before sending it to the hydrotreatment section,
- maintaining the catalyst activity, in particular tolerance to the presence of some reaction products containing oxygen, especially water,
- controlling the heat given off by the reactions involved (hydrodeoxygenation and hydrogenation of the unsaturations in the hydrocarbon fatty chains). This point is essential, in particular to avoid unwanted polymerisation reactions,
- controlling corrosion phenomena which may result from the presence of CO_2 and water in the reaction medium.

Industrial units implementing this type of process will have to be installed on sites with very large quantities of hydrogen available. This could be the case of crude oil refineries, selected also depending on their ability to upgrade propane. Nevertheless, by privileging the conversion of saturated oils and fats, such as palm, palm kernel and copra oils (although obviously not produced in Europe), as well as cooking oils and animal fats, hydrogen consumption should be significantly reduced.

While this pathway for the synthesis of biofuels from vegetable oils totally removes the problems related to upgrading glycerine, there is a problem concerning the small quantity of the co-product generated, propane.

By carefully selecting catalysts, often well-known in the hydrocarbon industry, and specific operating conditions, it will be possible to consider the production of "bio-paraffins" that could be used in the formulation of Diesel and jet fuel from a wide variety of renewable fats, some only capable of producing a poor quality ester.

A number of industrial projects have emerged or have been announced over the last few years. Two broad approaches can be identified.

3.6.3.1 Co-treatment with fossil feedstocks

The renewable fat is mixed in the feedstock (in very small proportions) in a unit intended for hydrodesulphurisation of a petroleum cut (Diesel, VGO). *A priori*, this approach will limit (by dilution) problems specific to the hydrotreatment of fats. The main advantage, however, lies in the possibility of using the existing infrastructures.

The following companies are involved in industrial projects:

- Petrobras (Brazil) and ConocoPhillips (Ireland): co-treatment of Diesel and soya oil,
- BP (Australia): co-treatment of Diesel with animal fats,
- Total: hydrotreatment,
- ExxonMobil: co-treatment of VGO with renewable fats,
- Preem (Sweden): co-treatment of Diesel with tall oil (residue from the paper industry).

3.6.3.2 Treatment in dedicated unit

While co-treatment may allow (partial) re-use of existing facilities, and therefore reduction of the necessary investments, precise optimisation of the operating conditions is more difficult. Desulphurisation of petroleum cuts generally requires more severe operating conditions than the conversion of fats into bio-paraffins, which has a negative impact on yields. This phenomenon is increased by the fact that the presence of these feedstocks inhibits the desulphurising function of the catalysts used, which may mean that the operating conditions have to be made more severe. Lastly, note that isomerisation of paraffins in a mixture is more difficult to implement.

In view of all these reasons, numerous companies have considered developing specific processes for implementation in units processing only renewable fats. The isoparaffins produced will then be used pure or in mixture, in Diesel or jet fuel.

Firstly, since mid-2007, the company Neste Oil (NExBTL internal process), a pioneer in field, has been operating a 170 000 t/year unit on the site of the Porvoo refinery (Finland) and is conducting three other projects (Porvoo2, Singapore, Rotterdam) for a total capacity of 1 770 000 t/year.

Then, the commercial processes of the companies UOP (Ecofining, co-developed with ENI) and Axens (Vegan process, developed by IFP Energies nouvelles) may produce, depending on the version, either biodiesel or bio-kerosene.

3.7 MAIN INDUSTRIAL PLAYERS

The main technology suppliers and manufacturers are Desmet Ballestra, Lurgi, BioDiesel International (BDI), Technip, Axens, Agrar-Technik (AT), Renewable Energy Group (REG), as well as other companies. These companies are active in numerous geographical areas, including Europe, United States, Latin America and Asia. They propose different technologies, suitable for a range of raw materials which can therefore be processed on the same industrial site.

The two products manufactured are vegetable oil methyl esters and glycerine. Brazil, a large ethanol producer, focuses primarily on the manufacture of vegetable oil ethyl esters.

The processes used are generally based on homogeneous catalysis. Axens is the only company proposing two technologies, one based on homogeneous catalysis and the other on heterogeneous catalysis.

In Europe, the largest producing company is Diester Industrie which inaugurated three new production plants in France during 2008, at Grand-Couronne, Capelle-la-Grande and Bordeaux-Bassens, and the new Mériot unit at the start of 2009. In France, Diester Industrie has a total production capacity of 2 Mt/year.

Diester Industrie International, a group composed of Diester Industrie and Bunge, manages production units in Germany, Italy, Austria and Belgium.

The other leading producers of biodiesel in Europe are ADM Biodiesel, a subsidiary of the American group Archer Daniels Midland, which operates several units in Germany,

Biopetrol Industries, a Swiss group based in Germany and Holland, the German company Verbio AG and the Spanish company Acciona Energia.

Dynol, Shaw Group, ADM and Renewable Energy Group are well established in the United States, while in Malaysia and India, companies such as BioEnergy International and Gujarat Oleo Chem are highly involved in the production of biodiesel. Numerous other companies producing biofuels on a lower scale are set up around the world.

3.8 GLOBAL PRODUCTION OF BIODIESEL

Europe is the largest producer and consumer of biodiesel. Production of VOME increased significantly between 1992 and 2001 in Europe, with an average annual growth rate of 33%. In 2005 for example, Germany, which had granted major tax incentives, produced more than half of the biodiesel for the European Union, i.e. about 1.7 Mt out of a total of 3.2 Mt. France ranked second, with 0.5 Mt.

Since then, as shown on the following table (Table 3.9), European production has considerably increased, reaching 7.75 Mt in 2008, i.e. 143% growth compared with 2005.

In 2009, the production capacities in the 27 countries of the European Union were estimated at 20.9 Mt (*source*: EBB, European Biodiesel Board), which is more than the current production and consumption and than those corresponding to incorporation of 10% biodiesel in Diesel in Europe.

Table 3.9 European production of biodiesel (thousands of tonnes).

Year	2005	2006	2007	2008
Germany	1 669	2 662	2 436	2 819
France	492	743	1 269	1 815
Italy	396	447	508	595
European Union	3 184	4 890 (53%)	5 995 (23%)	7 755 (30%)

The production increases from one year to the next are shown between parentheses.
Source: European Biodiesel Board

In 2010, France reached the target of 7% incorporation of biofuels, with 2.75 Mt/year of biodiesel. Germany still has a target of 10% incorporation of biofuels by 2020.

In 2008, European production accounted for about 60% of world production, estimated at 13.7 Mt (based on consumption of raw materials, this value was 15 Mt in 2008 and 9.7 Mt in 2007), of which about 80% (11 Mt) were consumed in Europe.

In 2008, Europe imported 1.5 Mt from the United States, as compared with just slightly over 1 Mt in 2007 and less than 100 000 t in 2006 (*source*: EBB). These American exports were favoured by the attractive prices obtained in the United States.

Production in the United States tripled between 2004 and 2005, reaching 370 000 t/year. Then, due to the federal grants allocated by the American government of 1 \$/gallon (0.27 \$/l) on B99 biodiesel (99% biodiesel and 1% Diesel) produced in the United States, from vegetable oils and animal fats, and of 0.5 \$/gallon (0.13 \$/l) for the blend produced from spent oils or fats, irrespective of the origin of the biodiesel (Brazil, Argentina or Malaysia), production increased dramatically. Some American states offered more modest grants, in the region of 0.1 \$/gallon. B99 was therefore massively exported to Europe, creating an imbalance in the European market.

To stop European biofuel companies from being disadvantaged by this situation, a European penalty ranging from 213 to 409 euros/t, composed of anti-dumping and anti-subsidy taxes, was levied from July 2009 on biodiesel imported from the United States, for a period of 5 years. Applying to American exports only, just how effective will these measures be? If the biodiesel transits via a country not subject to this penalty, the taxes will not apply.

US production, manufactured mainly from soya oil (70% to 80%), rose sharply between 2005 and 2008, increasing to 740 000 t in 2006, 1.65 Mt in 2007, 2.3 Mt in 2008, before dropping to 1.6 Mt in 2009 further to the export difficulties. The RFS-2 (Renewable Fuel Standard-2) is nevertheless considering the possibility of doubling production to 3.2 Mt (3.6 Mm^3) in 2012, which would correspond to 1.75% of the American Diesel demand of 225 Mm^3.

Other countries, which have become large biodiesel producers in recent years, include Argentina, Brazil, Malaysia and Indonesia. While their outputs vary depending on the year, the production per country generally exceeds 1 Mt from soya in Latin America and from palm in South East Asia. The Philippines, Thailand and Australia are also developing their production.

3.9 THE GLYCERINE MARKET

Glycerine is a by-product of the transesterification reaction.

For the past years 2008-2009, world production of glycerine has reached more than 1.5 Mt (quantity measured with 100% pure glycerine), (Table 3.10), with a sharp increase of the share from the manufacture of biodiesel, amounting to 1 Mt in 2009, i.e. about 60% of world glycerol production.

The glycerine market has now been increasing for several years. This trend is the result of several events:

- general growth of the world economy, especially at the end of the 1990s,
- significant development of new applications such as personal healthcare products, dental hygiene and the more widespread use of glycerine in the food industry.

The current trend in the growth of glycerine applications is expected to continue, mainly supported by the markets of healthcare, body hygiene and dental products, the agribusiness and pharmaceutical industries (Table 3.11).

Table 3.10 Glycerine productions, in thousands of tonnes (measured at 100% purity).

Year / Origin of the glycerine	2007	2008	2009
Saponification	155	130	105
Fatty acids	470	490	460
Biodiesel	690	950	1 000
Fatty alcohols	180	190	180
Other	25	25	20
Total	1 520	1 785	1 765

Source: HBI

Table 3.11 Glycerine applications (measured at 100% purity).

Year / Applications	2007	2008	2009	2010 Estimation
Cosmetics/pharmacy	800	830	750	780
Technical	420	400	300	330
Glycol substitutes	0	30	60	70
Sorbitol substitutes	0	10	10	10
Animal feed	170	100	130	150
Methanisation	50	100	300	300
Combustion	134	66	70	20
New uses	40	80	190	330
Total	1 614	1 616	1 810	1 990

Consumptions of refined glycerine are shown in bold, the other figures correspond to the use of crude glycerine.
Source: HBI

Glycerol is used for its emollient characteristics and as wetting agent due to its hygroscopic properties. It is also used as antifreeze. It is used in the synthesis of resins and explosives. It is a good solvent for the preparation of dyes and inks and a major compound of the agribusiness industry (food additive E422). New applications are under development, with the synthesis of products replacing certain compounds derived from the petrochemistry industry. A sharp rise is expected in these two domains for 2010 and the coming years. Currently however, glycerol surpluses are used mainly in methanisation and only small quantities for combustion purposes.

Glycerol can also be used as an intermediate for the chemical synthesis of polyglycerols and/or polyglycerol esters, frequently used as emulsifiers in food, cosmetics, lubricants in the textile industry and plastifiers. Glycerol may compete with other polyols (sorbitol, ethylene glycol, propylene glycol, neopentylglycol, pentaerythritol, etc.), provided however that it is cheaper and that its performance is at least as good, not forgetting the factor of acceptability by the end user, i.e. the customer.

Glycerol can also act as fermentation substrate, for example in the biosynthesis of 1-3 propanediol, a monomer used in the manufacture of polytrimethylene terephthalate (PTT), whose main applications concern the textile industry.

Lastly, studies have been conducted on the possibility of incorporating glycerol in the fuel pool after conversion. One of these pathways could consist in the synthesis of GTBE (glycerol-terbutyl ether) by reaction with isobutene. Glycerine could also be the oxygenated molecule used in the composition of fuel additives to improve the quality of Diesel engine emissions by reducing the particulate content.

Many of these new applications of glycerine depend above all on its cost price. As indicated in Table 3.12, the European prices of various grades of purified and crude glycerines, of animal or plant origin (kosher quality) have steadily declined, reaching in 2009 the lowest values ever recorded apart from those observed in 2006. This trend should definitely favour the development of new syntheses from glycerol, and new applications.

Purification of glycerine consists in distillation which mainly affects energy expenditure (77 euros/t), which is slightly higher than investment expenditure (54 euros/t) for a 10 000 t/year purification unit. The total cost of this operation was estimated at 170-175 euros/t in 2005 (*source*: Lurgi).

Table 3.12 Glycerine prices in Europe.

Year	Tallow glycerine 99.5% purity (€/t)	Kosher glycerine 99.7% purity (€/t)	Crude glycerine 80% purity ($/t)
2001	795	945	340
2002	870	1 015	390
2003	790	910	350
2004	660	745	335
2005	450	475	170
2006	400	450	70
2007	640	735	345
2008	650	755	365
2009	350	405	125

Source: HBI

Until 2003, the developed countries exported very little glycerine, which was mainly intended for their own consumption. Since 2004 however, exports from the European countries (Europe is the largest biodiesel producer) increased, reaching more than 30% of their 2008 production (Table 3.13).

In the United States, where production has increased significantly since 2005 due to the rise in biodiesel production, imports and exports varied very little, apart from in 2008 when exports doubled (Table 3.14).

Table 3.13 Production and consumption of glycerine in Europe (thousands of tonnes).

Year	Production	Total consumption	Consumption of refined glycerine	Exports	Imports
1995	215			13	27
1996	224			15	25
1997	250		230	14	24
1998	239		241	16	25
1999	268		267	16	20
2000	296		295	20	20
2001	297		280	24	18
2002	300		300	22	18
2003	340		325	25	12
2004	380	357	337	50	10
2005	545	411	351	75	20
2006	709	553	423	91	15
2007	675	562	418	100	10
2008	796	581	406	260	20
2009* est.	750	623	380	140	15

Source: HBI

Table 3.14 Production and consumption of glycerine in the United States (thousands of tonnes).

Years	Production	Total consumption	Consumption of refined glycerine	Imports	Exports
1995	157	187		61	19
1996	151	191		63	18
1997	163	194		48	21
1998	149	184		61	24
1999	163	214		54	17
2000	157	244		115	16
2001	159	229		94	18
2002	155	223		95	20
2003	141	184		75	24
2004	132	201		111	22
2005	101	194	193	114	24
2006	190	240	205	104	55
2007	270	343	205	96	37
2008	376	423	216	123	74
2009* est.	276	377	174	103	21

Source: HBI

Biodiesel production has risen sharply over the last few years in South America, mainly in Argentina, resulting in an annual glycerine production of about 100 000 t. We observe a similar situation in Malaysia and Indonesia where glycerine production, which varies depending on the year, is currently slightly less than 100 000 t/year.

3.10 DIRECT USE OF VEGETABLE OILS IN DIESEL ENGINES

3.10.1 Characteristics of vegetable oils

Vegetable oils were first developed as fuel for direct use, after more or less deep purification. However, the first experiments revealed the drawbacks of direct incorporation, which grew worse as engine technologies improved. In the early 80s, however, following tests conducted on vehicles and engine test benches, the Brazilian government decided that, from July 1982, consumers would be allowed to use blends with 30% vegetable oil in Diesel, for vehicles equipped with prechamber Diesel engines. Numerous experiments were also conducted in Europe, especially in Austria where they were initiated and where about a hundred agricultural tractors currently run on pure vegetable oils, with close technical monitoring.

The length of the fatty acid chain, the distribution and number of unsaturations give each oil its own particular properties as fuel.

The oils most frequently produced in Europe, rapeseed and sunflower, are composed of fatty acids with carbon chains (18 carbon atoms) including 3 chains globally longer than those of the hydrocarbons found in Diesel. These oils have high molecular weights, about 0.88 kg/mole, density above 910 kg/m^3, and low volatility. On heating, they generally crack at temperatures in the region of 300°C.

Other important properties, such as viscosity and resistance to cold, self-ignition delay, oxidation resistance and stability over time are highly dependent on the oil composition (Table 3.15) [11].

3.10.2 Impact on combustion

On reading the extensive literature dedicated to the subject, we quickly discover the limitations of direct use of vegetable oils in Diesel engines [12], [13], [14], [15]. Vegetable oils cause serious damage to combustion systems, due to the formation of deposits, and downgrade the performance and emission quality, especially in direct injection engines. In prechamber engines, the behaviour of the oils seems less problematic. Even in this case, however, formation of deposits in the longer term leads to complications.

This rather discouraging observation can be attributed to the physico-chemical characteristics of the oils. During the fuel vaporisation and combustion process, due to the high molecular weights and low volatility of the oils, the molecules crack, resulting in the formation of deposits. The high viscosity of the oils radically modifies the phenomena associated with spraying of the fuel and therefore the combusting timing, already disturbed by the low

Table 3.15 Main characteristics of vegetable oils

Vegetable oil	Viscosity at 40°C (mm^2/s)	Carbon residue (% w)	Cetane number	GCV (kJ/kg)	Ash content (% w)	Sulphur content (% w)	Iodine value (I/g oil)	Saponification value (mg KOH/g oil)	CFPP (°C)
Cotton	33.7	0.25	33.7	39.4	0.02	0.01	113.2	207.71	
Poppy	42.4	0.25	36.7	39.6	0.02	0.01	116.83	196.82	
Rapeseed	37.3	0.31	37.5	39.7	0.006	0.01	108.05	197.07	+ 20 [a]
Sunflower	34.4	0.28	36.7	39.6	0.01	0.01	132.32	191.7	+ 15 [a]
Sesame	36	0.25	40.4	39.4	0.002	0.01	91.76	210.34	
Flax	28	0.24	27.6	39.3	0.01	0.01	156.74	188.71	
Palm	63.6 (30°C)		42				35-65		
Jatropha	49.9 (38°C)		40-45						
Castor	29.7	0.21	42.3	37.4	0.01	0.01	88.72	202.71	
Soya	33.1	0.24	38.1	39.6	0.006	0.01	69.82	220.78	+ 11 [b]
Peanut	40	0.22	34.6	39.5	0.02	0.01	119.55	199.8	
Hazelnut	24	0.21	52.9	39.8	0.01	0.02	98.62	197.63	
Walnut	36.8	0.24	33.6	39.6	0.02	0.02	135.24	190.82	
Almond	34.2	0.22	34.5	39.8	0.01	0.01	102.35	197.56	
Olive	29.4	0.23	49.3	39.7	0.008	0.02	100.16	196.83	
Wheat	32.6	0.23	35.2	39.3	0.02	0.02	120.96	205.68	
Corn	35.1	0.22	37.5	39.6	0.01	0.01	119.41	194.14	
Diesel	2-4.5		47						0 to -20

[a] *Source:* Valenergol
[b] *Source:* USDA

cetane number of these oils (about 30 to 40), well below the 51 limit imposed by Diesel standard EN 590.

These last two characteristics disturb fuel combustion by delaying the start and end, resulting in higher emissions and partial conversion of the molecules, thereby contributing to the creation of deposits. In addition, any phenomenon delaying combustion also increases sound emissions. These disadvantages can be partially compensated by modifying the ignition advance, but there are no ways of solving the problems related to cracking of the molecules.

In addition, the cold properties of the oils, especially their Cold Filter-Plugging Point (CFPP) represent a further handicap. At low temperature, the resulting higher viscosity makes them virtually impossible to use and development of a special cold-starting technology is essential. Solutions include starting the vehicle with Diesel, or methods to warm up the fuel.

Due to their distillation characteristics, use of oils as fuel may dilute the engine lubricant, especially during idling, resulting in increased engine wear, piston gumming and scuffing on the walls of the combustion chamber.

Lastly, unsaturation of the fatty acids forming the vegetable oils also leads to problems during storage related to oxidation reactions. Some of these handicaps, if not fully overcome, can at least be reduced by technological artefacts. These modifications will lead to greater costs in two areas, investment and energy consumption during operation, especially with the installation of special injection and heating systems under cold operating conditions.

In addition, some of the difficulties mentioned above will be further amplified with improvements in Diesel engine technologies (smaller injector nozzles, more injectors resulting in higher risk of clogging). Direct use of vegetable oils with the technologies currently being developed will become increasingly critical [16].

3.10.3 Impact on emissions

Due to the need to adapt the combustion timing, use of vegetable oils in Diesel engines generally leads to higher CO, HC and particulate emissions.

In contrast, due to their slower combustion and lower temperatures in the combustion chamber, vegetable oils reduce NO_x emissions.

The following observations were made with a modern Diesel engine running on pure rapeseed oil [17]:

- CO emissions increased by a factor of up to 2,
- HC emissions increased by a factor of up to 4,
- NO_x discharges reduced by 25%,
- emissions of particulates increased by 90 to 140%, aldehydes and ketones by 30 to 330% and PAHs by about 20%.

Tests conducted by the company Ricardo [18] on two vehicles (a Euro 2 engine with direct injection and a 2001 engine with indirect injection), specially equipped with systems to preheat the fuel to 80°C, reveal that use of vegetable oils has a negative impact on pollutant emissions (Table 3.16).

Table 3.16 Impact of pure vegetable oils on pollutant emissions (g/km, NEDC [a]).

Emissions (g/km)	Direct injection 1.9 l			Prechamber 1.5 l		
	Reference fuel	Rapeseed oil	Impact	Reference fuel	Rapeseed oil	Impact
CO	0.125	0.665	432%	0.073	0.204	179%
HC	0.068	0.206	203%	0.025	0.037	48%
NO_x	0.727	0.673	– 7%	0.409	0.412	1%
Particulates	0.073	0.153	110%	0.025	0.021	– 16%
CO_2	139.2	151.7	9%	124.3	126.7	2%

[a] *NEDC: New European Driving Cycle*

The emissions vary depending on the condition of the vehicle. The differences may increase with the mileage, the age of the engine technology and the degree of engine clogging.

These results could nevertheless be improved with the development of new injection systems and studies on the combustion chamber.

Experiments have also been conducted on use of vegetable oils in mixtures [19], [20], [21]. These tests were performed with 25/75 mixtures of sunflower or safflower oil in Diesel. Performance with the fuel formulated using sunflower oil rapidly deteriorated due to deposits on injectors and piston and to gumming of the piston rings. Mixtures based on safflower oil did not generate any special problems.

Other isolated and endurance tests have been conducted [22] with various fuel formulations based on cotton oil and Diesel, containing between 30% and 65% vegetable oil and with a 50/50 mixture of cotton oil and cotton oil ester. Although beneficial effects may be observed in the short term, serious problems of deposits, ash, wear and gumming appeared during endurance tests, in the longer term.

3.10.4 Special case of agricultural tractors

For the last decade or so, tractor and farm machinery manufacturers have been faced with increasingly demanding customer requirements. Apart from making the vehicles more pleasant to use (comfort, sound level, no dust), as with cars and heavy goods vehicles, customers want models with longer lifetime, less maintenance and lower servicing costs, and longer service intervals. In addition, anti-pollution standards for off-road vehicles (Directive 97/68/EC) will soon apply to agricultural and forestry tractors (Directive 2000/25/EC). This will result in a significant change in the technologies, combined with stricter use and maintenance rules.

In-depth analysis of the characteristics of the tractors now available on the market shows that new techniques currently applied to this sector include, for example, direct injection under high pressure (between 1000 and 1600 bar), with electronic management or greater control of lubricant consumption by better tighter piston ring tolerances. They lead to the same requirements as those encountered for cars and heavy goods vehicles.

Today, the leading agricultural tractor manufacturers such as John Deere, Deutz-Fahr, Massey-Ferguson, propose similar technological innovations in order to anticipate future emission laws. John Deere, for example, was the first manufacturer to introduce common rail injection and a multi-valve cylinder head. Implementing these innovations makes the engines more sensitive to fuel quality. In this context, proposals to modify the traditional fuel DFO (Domestic Fuel Oil) must be risk-free.

3.10.5 Conclusion

While vegetable oils represent an alternative, they nevertheless continue to present risks related to their intrinsic characteristics, which neither car nor agricultural tractor and machinery manufacturers are willing to assume. In a press release published in 2006, the French Automotive Manufacturers Committee (CCFA) condemned the use of pure or mixed vegetable oils in car engines.

3.11 USE OF VOME IN DIESEL ENGINES

As mentioned previously, due to their characteristics vegetable oils cannot be used directly at large scale in a context of major technological changes to engines and increasingly strict anti-pollution standards.

In the early 80s, IFP Energies nouvelles demonstrated that transesterification of vegetable oils led to the production of VOME. In view of their properties, these esters can be used as fuel in Diesel engines, thereby removing most of the disadvantages encountered with direct use of vegetable oils. Their potential as fuel was demonstrated during the second half of the 1980s, then confirmed in the 1990s through numerous experiments conducted in Europe, especially in Austria, Germany, France and Italy. In France, as part of a large-scale experimental programme, associating car and HGV manufacturers, refiners, players in the oilseed sector, IFP Energies nouvelles, ADEME, the Ministries of Agriculture and Industry, a standard blend of 5% VOME in Diesel has been produced, without the need for any information to be displayed on service station pumps.

At the same time, France published specifications guaranteeing the quality of rapeseed methyl esters which would be incorporated in Diesel. This standardisation approach was followed at European level by the CEN (European Committee for Standardisation). Other applications have been developed. In Germany, for example, the pure product was proposed for adapted vehicles; in France, VOME was incorporated in Diesel at contents of up to 30% for use on captive vehicle fleets (urban public transport vehicles), under a tax exoneration scheme granted by the Ministry of Industry and the Customs.

In the United States, biodiesel can be used in 20% mixture with Diesel (B20), while in Brazil, a mandate promulgated in July 2008 authorised the addition of 3% biodiesel (B3), i.e. 1.2 Mm^3, with a plan to add 5% biodiesel in 2013.

3.11.1 Characteristics of the products manufactured

The specifications applying to fuel methyl esters in Europe are derived from standards EN 14 214. They recommend a certain number of maximum values concerning the contents in esters, partially or unconverted oils, unsaturated structures (iodine value), and the oxidation stability. The latter point is still being studied since there is disagreement on the representativeness of the methods proposed. The maximum contents for water and the impurities produced when refining the oils are also specified. Products created in compliance with these specifications are perfectly compatible with Diesel and present no major disadvantages in terms of engine operation (clogging, performance of the post-treatment systems, etc.). The specifications are given in Table 3.17.

The methanol content must be limited, to avoid a significant reduction in the flash point and problems of metal corrosion. Limiting the residual metal and phosphorus contents avoids filter plugging caused by possible formation of soaps and sediments. This also avoids corrosion of the injection system and piston ring gumming. In addition, the water content must be controlled to eliminate the risks of ester hydrolysis and problems due to corrosion.

Amongst the main differences recorded between the countries, note that no iodine value (expressed in grams of iodine fixed per 100 g of fat, this value being proportional to the rate of unsaturation of the fatty acid chains involved in the oil composition) is mentioned in the Italian, Sweden or American regulations, according to standard ASTM D6751, drawn up to allow correct use of 20% biodiesel mixed in Diesel (blend B20) (Table 3.18). This standard does not impose any density constraints nor a maximum methanol content. The methanol content must always be low, however, generally less than 1 ‰, since the flash point of the esters must always be above 100°C. In the United States, this means that soya oil esters can be used as biofuels, since their iodine values (about 135) are very high, exceeding the French and German specifications which impose a maximum value of 120.

There are currently no specific standards concerning vegetable oil ethyl esters.

The characteristics of the methyl esters produced are shown in Tables 3.19 and 3.20. They comply with the above specifications.

The main data included in Table 3.19 show that the density, viscosity at 40°C and cetane number of methyl esters compare with the values for Diesel, which is not the case for the oils.

Esters produced by heterogeneous catalysis are slightly purer than those obtained by homogeneous catalysis, with lower mono-, di-, triglyceride and methanol contents (Table 3.20).

As a comparison, Table 3.21 lists the main characteristics of methyl esters produced from various vegetable oils. The respective values of cetane number, iodine value and pour point depend mainly on the fatty acid composition of each oil. The cetane number increases with carbon chain length. The cetane number and the pour point vary with the degree of unsaturation of the fatty acids, determined by measuring the iodine value. The lower the value, the lower the fatty acid unsaturation rate and the higher the cetane number. Consequently, methyl esters synthesised from palm oil, which contains over 50% saturated fatty acids (iodine value = 50), have the highest cetane number, about 54. The pour point increases when the proportion of saturated fatty acids, mainly responsible for the poor

Table 3.17 VOME specifications in various European countries, and European specifications (Standard EN 14214).

Standard	Conditions	Unit	Austria	France	Germany	Italy	Sweden	Europe	
			ON C1191	Official Journal	DIN E 51606	UNI 10635	SS 15 54 36	EN 14214	
								Min	Max
Density	15°C	kg/m^3	850-890	870-900	875-900	860-900	870-900	860	900
Viscosity	40°C	mm^2/s	3.5-5.0	3.5-5.0	3.5-5.0	3.5-5.0	3.5-5.0	3.5	5.0
Distillation	I.B.P.	°C	–	–	–	≥ 300	–		
Distillation	95%	°C	–	≤ 360	–	≤ 360	–		
Flash point		°C	≥ 100	≥ 100	≥ 100	≥ 100	≥ 100	101	
CFPP		°C	≤ 0/– 15	–	≤ 0/– 10/– 20	≤ 0/– 15	≤ – 5		
Fluidity point	Summer	°C	–	≤ 10	–	–	–		
Total sulphur		wt%	≤ 0.02	–	≤ 0.01	≤ 0.01	≤ 0.001		10 mg/kg
CCR	100%	wt%	≤ 0.05	–	≤ 0.05	–	–		
CCR	10%	wt%	–	≤ 0.3	–	≤ 0.5	–		
Sulphated ash		wt%	≤ 0.02	–	≤ 0.03	–	–		0.02
Ash (oxid.)		wt%	–	–	–	≤ 0.01	≤ 0.01		
Water content		mg/kg	–	≤ 200	≤ 300	≤ 700	≤ 300		500
Total impurities		mg/kg	–	–	≤ 20	–	≤ 20		
Cetane number		–	≥ 49	≥ 49	≥ 49	–	≥ 48	51	
Neutrality value		mg KOH/g	≤ 0.8	≤ 0.5	≤ 0.5	≤ 0.5	≤ 0.6		
Methanol content		wt%	≤ 0.20	≤ 0.1	≤ 0.3	≤ 0.2	≤ 0.2		0.2
Ester content		wt%	–	≥ 96.5	–	≥ 98	≥ 98	96.5	
Monoglycerides		wt%	–	≤ 0.8	≤ 0.8	≤ 0.8	≤ 0.8		0.8
Diglycerides		wt%	–	≤ 0.2	≤ 0.4	≤ 0.2	≤ 0.1		0.2
Triglycerides		wt%	–	≤ 0.2	≤ 0.4	≤ 0.1	≤ 0.1		0.2
Free glycerol		wt%	≤ 0.02	≤ 0.02	≤ 0.02	≤ 0.05	≤ 0.02		0.02
Total glycerol		wt%	≤ 0.24	≤ 0.25	≤ 0.25	–	–		0.25
Iodine value		gI2/100 g	≤ 120	≤ 115	≤ 115	–	≤ 125		120
Phosphorus		mg/kg	≤ 20	≤ 10	≤ 10	≤ 10	≤ 10		4
Alkali metals Na – K		mg/kg	–	≤ 5	≤ 5	–	≤ 10		5
Carbon residue		wt%							0.3
Total contamination		mg/kg							24
Oxidation stability (110°C)		hour						6	
Acid number		mg KOH/g							0.5
Linolenic acid methyl ester		wt%							12
Polyunsaturated methyl esters (≥ 4 double bonds)									1
Group II alkali metals Ca + Mg		mg/kg							5

Table 3.18 VOME specifications in the United States (American standard ASTM D6751).

Properties	Limits	Units
Flash point	≥ 100	°C
Water and sediment contents	< 0.05	vol%
Kinematic viscosity at 40°C	1.9-6	mm^2/s
Sulphated ash	< 0.02	wt%
Sulphur content	< 0.0015	wt%
Copper strip corrosion	No. 3	
Cetane number	≥ 47	
Carbon residue	< 0.05	wt%
Acid number	< 0.8	mg KOH/g
Free glycerine	< 0.02	wt%
Total glycerol	< 0.24	wt%
Phosphorus content	< 0.001	wt%
Distillation temperature	< 360	°C

Table 3.19 Main physico-chemical characteristics of Diesel, rapeseed oil and VOME.

Characteristics	Diesel	Rapeseed oil	Rapeseed methyl esters
Specific density (kg/m^3)	820-860	920	880-885
Viscosity at 40°C (mm^2/s)	2-4.5	30	4.5
Cetane number	> 49	35	50
Flash point °C	100	Decomposition after 320°C	170-180

resistance to cold, increases. As a result, palm oil esters, with a pour point of + 13°C, cloud point + 23°C and cold filter-plugging point + 11°C, can only be used in Europe and the United States when mixed with rapeseed, sunflower and soya oil esters, and only during the summer, to avoid possible clogging problems likely to occur when the palm oil esters freeze as ambient temperatures drop. 5% mixtures of palm oil esters are generally used in Europe during the summer months. These characteristics can nevertheless be improved by performing 2 successive crystallisation steps on the palm oil. The resulting palm olein contains fewer saturated fatty acids, which have precipitated out.

Soya and sunflower esters, whose iodine values exceed the specifications (maximum 120 in Europe), can only be used when mixed with other esters. For example, the iodine value specification of 120 can be reached with a mixture of rapeseed and sunflower esters in the ratio 60/40 by weight. In 13% mixture by weight with soya methyl esters, palm methyl esters lower the iodine value to 120 and lead to a cold filter-plugging point of + 1°C. At 5% content in palm methyl esters, the iodine value is 126 and the cold filter-plugging point drops down to – 1°C.

Table 3.20 Characteristics of rapeseed methyl esters obtained by homogeneous and heterogeneous catalysis.

Characteristics	Units	Specifications	Methods EN 14214 (2003)	Homogeneous catalysis product	Heterogeneous catalysis product
Ester content	% m/m	> 96.5	EN 14103	97.5	98.5
Monoglyceride content	% m/m	< 0.8	EN 14105	0.7	0.5
Diglyceride content	% m/m	< 0.2	EN 14105	0.15	0.02
Triglyceride content	% m/m	< 0.2	EN 14105	0.1	0.01
Methanol content	% m/m	< 0.1	EN 14110	0.1	< 0.03
Water content	mg/kg	< 500	EN ISO 12937	200	200
Acid number	mg KOH/g	< 0.5	EN 14104	0.4	0.3
Free glycerine	% m/m	< 0.2	EN 14105	0.018	< 0.02
Total glycerine (calcul.)	% m/m	< 0.24	EN 14105	0.23	0.15
Alkali content	mg/kg	< 3	EN 14108	3	
Metal content	mg/kg				< 3
Phosphorus content	mg/kg	< 10	EN 14107	< 10	< 8
Sulphur content	mg/kg	< 10	EN ISO 20846	< 10	< 10

Table 3.21 Characteristics of methyl esters of various vegetable oils.

Characteristics	Rapeseed methyl esters	Soya methyl esters	Sunflower methyl esters	Palm methyl esters	Specifications EN 14214
Cetane number	50	48.1	49	54	51
Iodine value	111	133	134	45	≤ 120
Pour point	– 9	– 3	– 7	+ 13	

The iodine value is also a good indicator of the oxidation stability of these products. Higher values make the ester more sensitive to oxidation and increase the need for stabilisation additives when storing these esters. The additives used are generally aromatics such as phenolic and aminated molecules. The presence of these additives in the esters at contents of 100 to 1 000 ppm stabilises the fuel against oxidation during storage over a period of about 6 months.

Table 3.22 lists the results of the toxicity and biodegradability tests conducted with rapeseed and sunflower methyl esters and Diesel.

In the light of the values given, the environmental risks of esters, due to their total ultimate biodegradability and low toxicity to animal species and microorganisms, are low or zero in case of accidental spillage in nature (ground and water contamination). Their innocuousness is recognised.

Table 3.22 Toxicity and biodegradability of rapeseed and sunflower esters, and Diesel.

Test conducted	Units	Rapeseed methyl esters	Sunflower methyl esters	Diesel
Oral toxicity in mammals [a] OECD 401	LD 50 (mg/kg)	> 5 000	> 5 000	> 5 000
Toxicity in fish [b] OECD 203	LC 50 (mg/l)	> 100 000	> 100 000	134
Toxicity in daphnia [c] OECD 202	EC 50 (mg/l)	2 500	1700	86
Toxicity in algae [d] OECD 201	EC 50 (mg/l)	73 700	25 600	55
Toxicity in bacteria [e] ISO 10 712	EC 0 (mg/l)	5 250	1 020	< 10
Biodegradability [f] OECD 301B	%	87.4	89.9	38.7

[a] *Concentration by weight of chemical per kilogram of body weight of the animal required to kill half the animals (rats) in the population tested.*
[b] *Initial concentration which kills half of the freshwater fish Brachydanio rerio in 48 hours.*
[c] *Effective concentration which immobilises 50% of the daphnia present (Daphnia magna) after 48 hours.*
[d] *Effective concentration which causes 50% growth inhibition in Selenastrum capricornum, after an incubation time of 72 hours.*
[e] *Maximum effective concentration giving 0% growth inhibition in Pseudomonas putida after 16 hours incubation.*
[f] *Ultimate biodegradability (modified Sturm method) by measuring the CO_2 produced in 28 days.*
Source: ONIDOL

For example, rapeseed methyl esters were used as flux when pumping the heavy fuel contained in the tanks of the Erika (which sank off the Brittany coast at the end of December 1999) and when cleaning the rocks of the coastlines polluted by this fuel.

3.11.2 VOME properties

Table 3.23 lists the properties of the various VOME [23], [24].

The cetane number, a parameter used to determine the self-ignition delay of Diesel fuels (for conventional Diesel engines, a high cetane number is recommended so that the self-ignition delay is as short as possible), is similar to values typically encountered for diesels, which must now be greater than 51.

The distillation interval of the VOME traditionally found in Europe lies between 320°C and 350°C, placing them in the same category as the heavy fractions of the Diesel cut.

Absence of sulphur and aromatic structures: the changes in Diesel specifications will eventually lead to sulphur levels below 10 ppm and control of polyaromatic compound contents. Any base not increasing these compounds is of interest.

Improved lubricating power of highly desulphurised diesels: this added value of esters is extremely important in the current climate, since deep desulphurisation of diesels significantly reduces their lubricating power. This disadvantage can now be overcome by incorporating additives. Nevertheless, by incorporating VOME, the lubricating power of diesels with very low sulphur content is considerably improved. Table 3.24 indicates the performance obtained with this method. Determined by wear tests [25], they show that the addition of rapeseed methyl esters at contents of between 2% and 5% significantly reduces the wear of metal parts in contact with the fuel.

Viscosity: viscosity is an important characteristic which has an impact on both wear and performance. Diesel fuel is in fact designed to lubricate the injection parts and viscosity is one of the factors affecting this performance. In addition, viscosity may have an induced impact on the spraying, injection timing and injected quantity, the latter effect being much more noticeable on the injection technologies of the 1980s.

Miscibility in all proportions in Diesel: due to the difference in densities between Diesel and VOME, thorough stirring is necessary to guarantee a homogeneous mixture.

Nevertheless, the intrinsic chemical nature of vegetable oil methyl esters induces certain properties whose consequences must be controlled. In particular:

- stability over time. This is a central point. Depending on the origin of the oils, the distribution and respective quantities of unsaturated chains (distribution between the oleic, linoleic and linolenic chains, respectively C18-1, C18-2 and C18-3) may be quite different. These unsaturations have an impact on the stability, whether oxidation or thermal, of esters during storage,
- cold characteristics. They are very similar to those of diesels but remain inferior to those of winter diesels and far inferior to those of extreme cold weather diesels, whose cold filter-plugging point is set in France between – 12°C and – 15°C.

Table 3.23 Main characteristics of various vegetable oil and animal fat esters.

	Density at 15°C (kg/m^3)	Viscosity at 40°C mm^2/s	Flash point °C	NCV kJ/kg	NCV kJ/l	GCV kJ/kg	C wt%	H wt%	O wt%	Cetane number	Cloud point °C	Pour point °C	Iodine value g I/100 g
Rapeseed ME	882	4.5	172	37 400	32 900	40 300	78.2	12.5	9.3	55	– 3	– 12	106
Rapeseed EE	875	6.5	160	38 000	33 250	40 600	76.8	11.8	11.4	66	– 1	– 12	102
Soya ME	884	4.1	178	37 200	32 900	40 000	76.2	11.2	12.6	47	1	– 4	133
Soya EE	881	4.4	175			40 000				48	– 1	– 4	123
Cotton ME	880	3.7	110			40 600				48		– 4	
Flax ME	884	3.5	172	38 700	34 150	39 750	77.5	11.6	10.9	52		– 15	183
Flax EE	884	3.6				39 650							
Sunflower ME	878	4.3	146	38 250	33 500	39 130	77.2	11.9	10.9	54.3	1	– 4	129
Sunflower EE	876	4.4				39 800							
Oleic sunflower ME	879	4.7	177	37 000	32 600	40 200				52.6			85
Palm ME	875	4.9	164	37 600	32 900	39 200	76.3	12.4	11.3	62	23	13	52
Palm EE		4.5				39 070				56.2	8	6	
Peanut ME	883	4.9	176	38 050	33 600					54	5		
Corn ME		4.5	111			38 480				65	– 3	– 3	
Tallow ME	876	4.6	150	37 500	32 900	40 000				59.5	13	10	43
Tallow EE	871					41 675							
Lard ME				36 800		39 700	75.0	11.8	13.2	63.6			

Table 3.24 Results of the wear tests with two types of rapeseed oil methyl esters.

Fuel composition			Wear scar diameter (μm)	
			RME [a]	RME [b]
Diesel		(Swedish City Diesel fuel) [c]	580	580
Diesel	+	0.5% vol of RME	550	580
Diesel	+	0.8% vol of RME	500	580
Diesel	+	1% vol of RME	300	580
Diesel	+	2% vol of RME	180	450
Diesel	+	5% vol of RME	180	280

[a] *RME: esters purified by flash distillation.*
[b] *RME: esters not subjected to flash distillation.*
[c] *Diesel containing less than 10 ppm of sulphur.*

It must be pointed out that the net calorific value of vegetable oil methyl esters per unit weight and per unit volume is respectively 11% and 7% less than that of diesels. Consequently, the consumption (expressed in litres/100 km travelled) is about 7% higher. Also note that esters are more hygroscopic than diesels, which may modify the behaviour of mixtures in the presence of water.

Lastly, extreme care must be paid when mixing in Diesel, due to the differences in density of the two products. The final Diesel fuel commercialised must comply with EN590 Diesel specifications.

3.11.3 Performance of VOME as fuel

When an alternative solution to conventional fuels is proposed, we must obviously make sure that it is risk-free and identify the potential areas of technological development to optimise its use. These points concern in particular the injection characteristics, combustion process, operation of post-treatment systems such as oxidation catalysts and particle filters, endurance performance, behaviour with respect to the lubricant, and lastly the risks of clogging.

Due to their perfect miscibility in Diesel and their characteristics, they can be used in various ways:

- VOME7: standard blend containing up to 7% VOME in Diesel, distributed in compliance with specification EN590,
- VOME30: blend composed of 70% Diesel cut and 30% VOME. This mixture is distributed for captive fleets operating under a tax exoneration scheme. In France, numerous local communities, urban transport companies and even private companies have adopted this fuel,

– VOME100: VOME used pure in adapted vehicles. This fuel is sold in some European countries such as Germany, which now seems to be considering its use in blended fuels, however.

It is important to examine how each of these blends will affect engine performance.

An extensive experimental programme, set up and coordinated by IFP Energies nouvelles during the 1990s, was conducted on VOME5 blends in fleets of light and heavy goods vehicles. The total mileages associated with engine bench tests demonstrated that use of VOME5 blend does not modify the combustion and injection characteristics, the endurance performance or the operation of the post-treatment systems. It does not significantly dilute the lubricant or lead to increased engine wear.

In addition, good compatibility between the materials used and the blend was observed.

Furthermore, all other things being equal, the fuel consumption increases in the presence of VOME due to the lower calorific value. Incorporating 5% VOME only increases consumption by about 0.3%, however, which is hardly noticeable for the user.

Numerous studies have been conducted on the use of blends with higher contents, in particular VOME30, in various types of engine and vehicle. They conclude that there is a good match between engine technologies and VOME30 fuels. For example, they demonstrated perfect compatibility of the injection systems, slight or zero modification to the combustion process, no impact on operation of the post-treatment systems, very little increase in sensitivity to injector clogging, good compatibility of materials and consumption changes in line with variations in the NCV.

Great attention must be paid to two points, however: risks of the fuel diluting the lubricant and the oxidation stability of the blends.

Due to the high boiling point of VOME, their distillation characteristics may induce risks of incomplete fuel vaporisation when the engine is cold or when driving at low temperatures, especially during prolonged periods at tickover. The result is dilution of the lubricant, reducing the lubricating power and possibly creating varnishes. The experiments conducted have not demonstrated increased engine wear and the conclusion comes down in favour of the use of these blends. Special attention must be paid to this aspect, however, by closely monitoring the service conditions of the vehicles (especially when unladen or running at low temperature), in order to adapt the oil change frequencies, if necessary.

Lastly, when using pure VOME, vehicles operate without any problems. The materials used for gaskets and hoses, such as natural rubbers, polypropylene and polyvinyl, are highly vulnerable to B100 which, in contrast, is compatible with nylon and teflon. In the latter case, however, manufacturers propose adapted vehicles, especially by selecting other polymers and by recalibrating the injection pump to allow for the lower NCV.

3.11.4 Impact of VOME on emissions

The environmental balance of VOME is positive. The main results are included in Table 3.25, for two fuel formulations, 5% and 20% VOME in Diesel.

Table 3.25 Impact of the use of VOME on emissions

	VOME (5%)	VOME (20%)
CO	0 (↘)	0 ↗ (≅ 5%)
HC	0 (↘)	0
NO_X	0 (↗)	0
Particulates	↘	↘ 15%
Insoluble particulates [a]	0 (↘)	↘ 20-25%
Soluble particulates [a]	0 (↗)	↗ 30-50%

[a] *Solvent: dichloromethane.*

In terms of local pollution, almost all the studies conducted with different types of biodiesel record a 10% to 30% reduction in emissions of CO, HC and particulates depending on the pollutants and reduced emissions of gaseous and particulate PAHs. The reduction in emissions of unburnt hydrocarbons can be explained by the oxygen content of the biodiesel, which allows more complete combustion, and by its lower volatility. More complete combustion also helps reduce carbon monoxide emissions, lowered even further by the higher cetane number. Reductions of up to 50% in the weight of particulates emitted have been measured when using pure biodiesel. This gain may nevertheless become zero or negative while the vehicle is warming up.

The effects are more pronounced as the VOME content increases.

In contrast, NO_X emissions generally tend to increase by a few points (< 5%). Optimising the injection parameters will limit this increase.

The impact of biodiesel on emission of pollutants such as PAHs, aldehydes and ketones has not yet been determined precisely. Results are limited, poorly repeatable and difficult to justify.

No significant increase has been recorded on aldehyde emissions. This point is directly related to compliance with the criteria specified in standard NFEN14214. Excessive residual oil content could lead to higher acrolein discharges, while excessive residual methanol content could lead to higher formaldehyde discharges.

In addition, appreciable gains are observed in terms of greenhouse gas emissions, in particular CO_2 (see 3.13).

3.12 USE OF BIOFUELS IN AIR TRANSPORT

As with road transportation, the aeronautical sector must adapt to the new energy and environmental context and reduce its emissions and fuel consumptions.

Air transport is currently responsible for 8% of total oil consumption. Traffic can be expected to increase further over the next few years (forecasts estimate that the number of passengers carried could increase from 2 billion in 2009 to 9-10 billion in 2050), and have a greater impact on the environment, in particular in areas located near airports.

New engine technologies designed to reduce fuel consumption are being considered, as well as new fuels, including some biofuels. Amongst the biofuels, hydrocarbons derived from natural gas or biomass, and obtained using the Fischer-Tropsch synthesis, should be adapted to air transport, as well as biofuels obtained by hydrotreatment followed by hydroisomerisation of vegetable oils and other fats from animals or algae. Choices are limited, since the constraints in terms of safety are very strict and the specifications for jet fuels are very restricting as regards the energy content, which governs the range of the aircraft, resistance to cold, thermal and storage stabilities, lubrication, explosiveness, etc.

In the longer term, natural gas and hydrogen are potential candidates.

Several experiments on the use of biofuels in air transport have already been conducted or are in progress, with promising results for the future.

3.13 ENVIRONMENTAL ASPECTS

The results of the two studies mentioned earlier (Chapter 2), [26], [27], concerning greenhouse gas emissions, are given in Table 3.26.

Table 3.26 GHG balances of biodiesel sectors, compared with the reference Diesel.

	ADEME study (February 2010)	**JRC/EUCAR/CONCAWE study (March 2007)**		
	Reduction compared with Diesel (%)	**g CO_2 eq/km**	**Reference g CO_2 eq/km**	**Gain compared with Diesel (%)**
Rapeseed VOME	59	73	156	53
Sunflower VOME	73	34	156	78
Soya VOME	77			
Palm VOME	76			
Spent cooking oil ME	90			
Animal fat ME	91			
Vegetable oils	65			

After analysing these results, several comments can be made:

- greenhouse gas emissions are in general significantly reduced by the use of biofuels compared with conventional solutions. The reduction ranges from 59% for rapeseed esters to 76% for palm oil esters [26]. This conclusion must nevertheless be tempered by the fact that biofuels are generally used in blends at relatively low contents (5% to 10%). The overall benefit to be expected in terms of greenhouse effect will therefore be less than 5% if only traditional biofuels are used. However, even though these figures seem low at first sight, it must be pointed out that they can be obtained in a sector whose growth is very difficult to control and that few options offering the same benefit within a relatively short time are available.

Based on the results of Table 3.27, showing the energy balances, several conclusions can be drawn.

Table 3.27 Consumptions of non-renewable energies.

	ADEME study (February 2010)	JRC/EUCAR/CONCAWE study (March 2007)		
	Gain compared with Diesel (%)	MJ ex/km	Reference MJ ex/km	Gain compared with Diesel (%)
Rapeseed VOME	65	0.73	2.05	64
Sunflower VOME	68	0.54	2.05	74
Soya VOME	69			
Palm VOME	78			
Spent cooking oil ME	81			
Animal fat ME	81			
Vegetable oils	82			

Firstly, the production of biofuel involves a non-negligible consumption of fossil energies. Consequently, the final gain expected with respect to the "petroleum fuel" solution will never be 100%. Depending on the assumptions made and the sectors studied, reduction in non-renewable energy consumption compared with fossil energy consumption varies from 65% for rapeseed esters to 78% for palm esters [26]. This gain may reach about 80% for spent oils, animal fats and pure vegetable oil. Note, however, that these balances are not fixed and that the "fossil energy" balances of the biofuel production units could be improved, either by energy saving programmes or even by using renewable energy resources.

Fewer studies are conducted on the three other impacts since the uncertainties are greater and these impacts are highly dependent on local conditions where the pollutants are discharged.

The ADEME [26] study provides indications, however:

- for the eutrophication potential indicator, the balances for biodiesels are unfavourable compared with Diesel. The agricultural step, by leaching of nitrates and emissions of ammonia, is the largest contributor to this impact. This indicator, expressed in kg eq. phosphates/MJ, is $4{\cdot}10^{-5}$ for Diesel and 3.5 to $4{\cdot}10^{-4}$ for rapeseed and sunflower methyl esters.

- for the photochemical oxidation potential, expressed in kg eq. C_2H_4/MJ, rapeseed and soya esters (1.5 and $1.2{\cdot}10^{-5}$) appear to emit slightly more oxidising molecules compared with Diesel ($1.1{\cdot}10^{-5}$), whereas spent oil esters emit less ($4.5{\cdot}10^{-6}$). For rapeseed, the value depends to 70% on hexane emissions during the oil extraction step,
- impacts on the human toxicity indicator, expressed in kg eq. 1-4 dichlorobenzene (1-4 DB) are much higher for Diesel than for methyl esters. The reduction in Polycyclic Aromatic Hydrocarbon (PAH) emissions accounts almost completely for the difference between the fossil and biofuel sectors.

3.14 ECONOMIC ASPECTS

The costs of manufacturing biodiesel by homogeneous catalysis are shown in Table 3.28 for annual productions of 100 000 t. The first column, in euros/t, is based on an average rapeseed oil cost calculated during 2003-2004 (€460/t), the second column on a soya oil price of $660/t, at the start of 2007 [28]. At the start of 2007, the prices of the other oils reached respectively $800/t for rapeseed, $500 $/t for palm and $330/t for the spent oils (€1 = $1.4).

Table 3.28 Biodiesel production costs.

	Cost (€/t) [a]	Cost ($/gallon) [b]
Oil and chemical products	502.1	2.45
Utilities	6.9	0.02
Labour	5.3	0.08
Maintenance, administration	13.6	0.11
Investments (amortisation over 10 years)	15.9	0.12
Other		0.12
Sale of glycerine	– 46.0	– 0.05
Total cost	497.8	2.85

[a] *Source: Lurgi (years 2003-2004).*
[b] *Source: SRI Consulting (early 2007).*

The following figures (Figures 3.9 and 3.10) show the trend in oil prices in euros/t and $/t, where significant cost variations can be seen between end 2005 and end 2008. In particular, from May 2007 to May 2008, we observe a sharp increase in prices, reaching €1 000/t for rapeseed and soya and values greater than €1 200/t for sunflower. The prices of palm oil always remain lower, but they nevertheless record a sharp increase between early 2007 and May 2008, doubling from €400/t to €800/t. Although the price variation of these raw

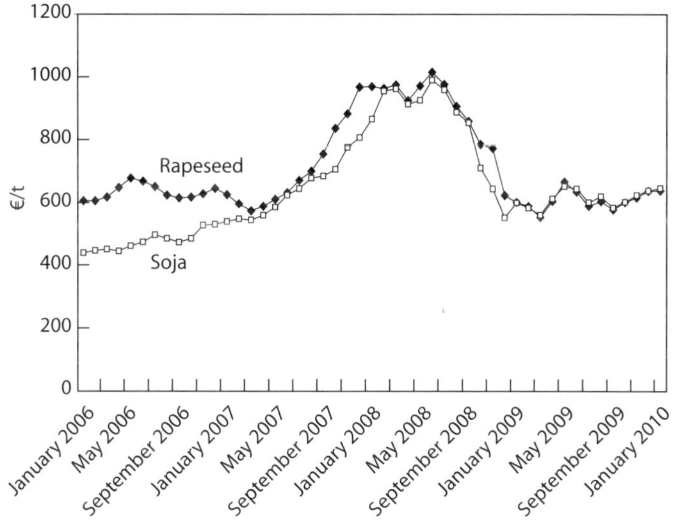

Figure 3.9

Graph of rapeseed and soja oil prices.

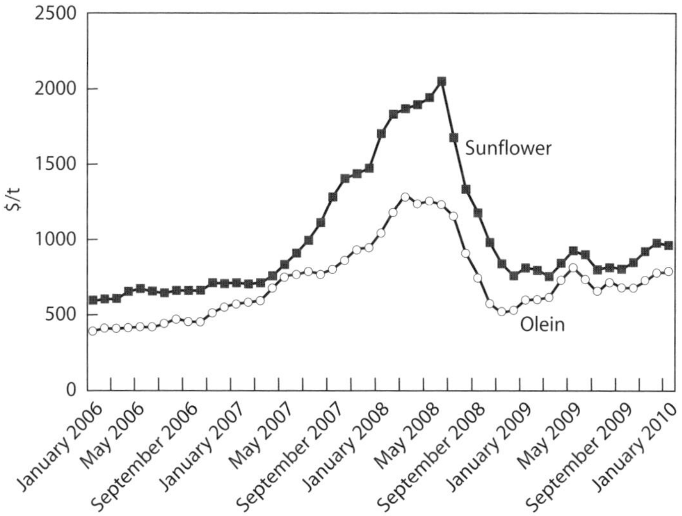

Figure 3.10

Graph of sunflower oil and olein prices.

materials is always large over time, it is tending to decrease, resulting in high variations in the prices of biodiesel manufactured from vegetable oils.

In 2003-2004, the price of raw materials, in particular rapeseed oil, represented 85% of the cost of producing vegetable oil methyl esters, whereas soya oil and methanol accounted for 85% of the cost of producing biodiesel.

In 2003-2004, glycerine sales offset the expenses related to investments, utilities, labour and maintenance and some of the methanol, i.e. about 10% of the total price of biodiesel, thereby having a strong influence on the economic profitability of the sector. Mainly due to a significant drop in the price of glycerol, in 2007, sales of this product in 2007 in the United States represented no more than about 2% of the final cost of biodiesel.

If trituration of oilseeds and manufacture of VOME are carried out on the same site, the labour and maintenance costs, for example, are considerably reduced.

With soya oil costing \$660/t, the cost of producing biodiesel in the United States, per unit energy, is in the region of \$22/GJ.

In mid-2009, the cost of producing biodiesel in Germany was about €800/m^3 [29].

All countries producing and consuming this biofuel are trying to make this price more competitive with that of Diesel, through the use of financial incentives.

In the United States, the cost of producing biodiesel synthesised from soya oil amounted to \$2.85/gallon, whereas during the same period, from mid-2006 to early 2007, Diesel was cheaper, varying between \$1.69 and \$2.72/gal. Grants of \$1/gallon were allocated on the biodiesel/Diesel blend, for example B99, manufactured from vegetable oils or animal fats, provided only that the blend is made in the United States, irrespective of the origin of the biodiesel, which can be imported from Brazil, Argentina or Malaysia, etc. Because of these grants, B99 was extremely cheap and could then be massively exported to Europe, creating an imbalance in the European market, a practice known as "splash and dash", where biodiesel is shipped through the United States to collect the tax credit. The \$1/gal grant has been stopped at the end of 2009.

To limit the impacts of these imports in Europe, where production figures dropped considerably, the European Commission imposed penalties in July 2009. They consist of anti-dumping and anti-subsidy taxes, ranging from a total of €213 to €409/t, and apply to biodiesel imports from the United States over a five-year period.

Still in the United States, the small producers benefit from a credit of \$0.1/gal and some states, with biodiesel production sites, grant credits of \$0.05 to \$0.3/gal.

In Europe, each country is free to make its own economic choices, to promote biofuels.

Germany, for example, set up a system of mandatory quotas, stipulating that each litre of Diesel must contain at least 4.4% biodiesel. A penalty of €0.5/l is applied to companies failing to respect this content.

In France, the biofuel support scheme is based on a double mechanism which consists in reducing the domestic tax on petroleum products (*Taxe Intérieure de Consommation*, TIC) and penalising fuel distributors failing to comply with the biofuel incorporation targets set at 5.75% in 2008, 7% in 2010 and 10% in 2015. This sanction consists in payment of a General Tax on Polluting Activities (TGAP) which amounts to a highly dissuasive €780/m^3 for Diesel not blended with biodiesel.

The domestic tax on petroleum products (TIC) was cut to compensate for the higher costs of producing biofuels compared with fossil origin fuels. This tax cut of €0.15/l in 2009 is expected to drop to €0.08/l in 2011. This loss of income for the State is nevertheless offset by the financial returns related to creating and/or maintaining jobs, environmental expenses avoided and lower imports of Diesel, cakes, soya and glycerine.

The other European countries also apply tax cuts, to various degrees.

REFERENCES

[1] Agritrade (2008). Oléagineux : Note de synthèse, http://agritrade.cta.int/fr, décembre 2008.

[2] Evrard J, de La Taille G et Joyeux J. Cahier Technique. L'huile de colza.

[3] Subramaniam V *et al.* (2008) Environmental Performance of the Milling Process of Malaysian Palm Oil Using the Life Cycle Assessment Approach, American Journal of Environmental Sciences, **4**, 4, pp 310-315.

[4] Publication de l'ONIGC (Office National Interprofessionnel des Grandes Cultures). Biocarburants 2010: Quelles utilisations des terres en France ?, octobre 2007.

[5] Chauvel A, Lefebvre G et Castex L (1985) Procédés de pétrochimie. Caractéristiques techniques et économiques, Tome 1 – Éditions Technip, Paris, pp 90-104.

[6] Loupy A *et al.* (1993) The Synthesis of Esters under Microwave Irradiation Using Dry-media Conditions. Canadian Journal of Chemistry, **71**, 1, pp 90-95.

[7] Hillion G and Proriol D (2003) Glycerolyse of Free Fatty Acids, 25th World Congress and Exhibition of the ISF, Bordeaux.

[8] Stern R, Hillion G, Rouxel JJ et Leporq S (1996) Procédé de fabrication d'esters à partir d'huiles végétales ou animales et d'alcools. Brevet français 2 752 242, US Patent 5 908 946 (1999).

[9] http://www.canren.gc.ca/tech

[10] Bloch M (2009) First Generation Bio-Diesel and Bio-jet Fuels Production. IFP Energies nouvelles Sessions, Rueil-Malmaison, 9 décembre 2009.

[11] Demirbas A (2003) Biodiesel Fuels from Vegetable Oils via Catalytic and Non-catalytic Supercritical Alcool Transesterification and Other Methods: a Survey. In Energy and Management, 44, pp 2093-2109.

[12] Clevenger MD, Bagby MO, Goering CE, Schwab AW and Savage LD (1988) Developing an Accelerated Test of Coking Tendencies of Alternative Fuels, ASAE, 31, 4, pp 1054-1058.

[13] Hawkins CS, Fuls J and Hugo FJC, Engine Durability Tests with Sunflower Oil in an Indirect Injection Diesel Engine, SAE 831357.

[14] Vellguth G, Performance of Vegetable Oils and their Monoesters as Fuels for Diesel Engines, SAE 831358.

[15] Suda KJ, Vegetable Oil Or Diesel Fuel – A Flexible Option, SAE 840004.

[16] Murayama T, Oh YT, Miyamoto N and Chikahisa T, Low Carbon Flower Buildup, Low Smoke, and Efficient Diesel Operation with Vegetable Oils by Conversion to Mono-Esters and Blending with Diesel Oil or Alcohols, SAE 841161.

[17] Hemmerlein N, Korte V and Richter H, Performance, Exhaust Emissions and Durability of Modern Diesel Engines Running on Rapeseed Oil, SAE 910848.

[18] Lance DL and Andersson JD (2004) Emissions Performance of Pure Vegetable Oil in Two European Light Duty Vehicles, SAE Paper 2004-01-1881.

[19] Ziejewski M and Kaufman KR (1982) Endurance Test of a Sunflower Oil/Diesel Fuel Blend, SAE 820257.

[20] Ziejewski M (1983) Vegetable Oils as a Potential Alternate Fuel in Direct Injection Diesel Engines, SAE 831359.

[21] Ziejewski M and Goettler H (1986) Influence of Vegetable Oil Based Alternate Fuels on Residue Deposits and Components Wear in a Diesel Engine, SAE 860302.

[22] Fort EF and Blumberg PN (1982) Performance and Durability of a Turbocharged Diesel Fueled with Cottonseed Oil Blends, Vegetable Oil Fuels Proceedings of the International Conference on Plant and Vegetable Oils as Fuel, ASAE.

[23] Bechtold RL, Alternative Fuels Guidebook. Properties, Storage, Dispensing and Vehicle Facility Modifications. SAE Edition.

[24] Grabbski MS and McCornick RL (1997) Combustion of Fat and Vegetable Oil Derived Fuels in Diesel Engines. *In* Progress in Energy and Combustion Science.

[25] Hillion G, Montagne X et Marchand P (1999). Les esters méthyliques d'huiles végétales: additif ou biocarburant ? OCL **6**, 5, Septembre/octobre, pp 435-438.

[26] ADEME (2010) Analyses de cycle de vie appliquées aux biocarburants de première génération consommés en France. Synthèse de février 2010.

[27] Well-to-Wheels Analysis of Future Automotive Fuels and Powertrains in the European Context, Concawe, Eucar, JRC, Well-to-Wheels Report, March 2007, http://ies.jrc.ec.europa.eu/WTW

[28] Bray R (2007) Biodiesel Overview, SRI Consulting, Biofuels Mission to France, Paris, 4 avril 2007.

[29] F.O Lichts, 7, p 20, June 21, 2009.

4

Lignocellulosic biomass resources

D. Ballerini, D. Lorne, A. Prieur

The total resources of lignocellulosic biomass represent a considerable source of renewable energy, equal to almost twice the world primary energy consumption in 2004. This chapter focuses on the origin and composition of these resources. The stakes related to their production and mobilisation are then detailed before concluding with an update on the long-term potentials.

4.1 LIGNOCELLULOSIC BIOMASS: SOURCE AND CURRENT USES

Lignocellulosic biomass differs from biomass in general since it belongs exclusively to the plant world and in particular to species formed from a complex substrate composed of three main fractions: cellulose, hemicelluloses and lignin. Biomass of animal origin, urban and industrial waste, algal biomass and other microorganisms, etc. are not included.

4.1.1 Two major families of lignocellulosic resources

Lignocellulosic biomass is obtained from agricultural and forest residues or wood conversion by-products, as well as dedicated crops, whether woody plants or herbaceous plants. We can nevertheless identify two major families, based on the main function of their production:

- "co-product" resources, often a plant fraction co-produced when obtaining a resource of higher added value,
- "dedicated" resources, produced specifically for energy or chemical use.

4.1.1.1 The co-product resource: agricultural and forest residues or surpluses

Currently, the co-product lignocellulosic resource consists mainly of waste left over after harvesting agricultural crops (straw, stems, tops, etc.) or forest crops (felling residues, dead wood, pruned branches, scrap wood, etc.). Surplus resources such as non-marketable harvests and growing stock, not exploited due to lack of outlets, can also be added. These resources, generally with no identified upgrading circuit and not incorporated in an existing market, are therefore potentially available for energy purposes.

4.1.1.2 The dedicated resource: lignocellulosic crops dedicated to energy

With the emergence of various bioenergy production pathways and other biomaterials, a future and a potentially high demand for biomass is taking shape, which may sometimes need to meet specific technical and economical requirements (see paragraphs 4.2 and 4.3). Due to the limited source of inevitable resources available, investigations are being conducted on new high-productivity plant species, currently with no outlets and widely known. Some annual species have already been included in the current crop systems (triticale, sorghum), while others, less well known, are studied in experimental plots and in open field by various farmers, for example miscanthus, *switchgrass,* etc.

A. Annual crops

The whole plant of several annual, non-perennial crops can be used to produce energy, especially in the context of land dedicated to “ industrial set-aside”. The presence of the grain in addition to the straw improves the energy content of the product harvested, compared with the case where only the straw is used to produce energy. Triticale, for example, can produce between 8 and 16 t d.m./ha, for an energy content between 15 and 16.5 MJ/kg (Table 4.1).

B. Herbaceous perennial crops harvested annually

Some perennial species can be harvested annually, with variable yields and applications. Miscanthus and *switchgrass*, whose productivity can reach 20 t d.m./(ha.year), are good examples. Some of these species offer niche outlets, especially for the production of paper, fuels and for the manufacture of biomaterials, bedding, etc. Moreover, some species are not cultivated but harvested wild, for example phragmite (very tall grass which can reach a height of almost 5 metres, growing in marshes and roadsides).

C. Woody perennial crops

The forestry methods used to grow Short Rotation Coppices (SRCs) or Very Short Rotation Coppices (VSRCs) are very similar to agricultural methods. Broad-leaved trees capable of producing stump shoots after clear cutting are mainly concerned. The main difference between SRCs and VSRCs is the period of rotation between each harvest: on average every 7 years for SRCs, and 2 or 3 years for VSRCs. The plantations have a lifetime of about 20 years, i.e. about 3 harvest in the first case, and 7 to 10 in the second. The trees are cut down to the ground.

Short rotation coppices are high-density crops of fast growing trees. Poplar and eucalyptus SRCs were first used as a source of raw materials for the trituration industries, but can of course form an energy source. The main species studied in France include the poplar, willow, eucalyptus and more recently the false acacia as well as fast-growing coniferous species (in particular Douglas pine and maritime pine).

The main factor limiting the development of poplar plantations is the water supply required throughout the year, either via the water table or provided by rainfall. The poplar

Table 4.1 Characteristics of various types of lignocellulosic biomass [5].

	Yield t d.m./ha	Humidity wt%	Ash wt%	NCV MJ/t d.m.
Crop residues				
Oleaginous flax straw	0.5 to 3.5	12	5	17 900
Sunflower straw	3		11.2	17 410
Jerusalem artichoke stalks	3 to 7	20		15 700
Rapeseed straw	< 2	70 to 75	9.7	16 480
Cereal straw	4 to 6	12 to 15	5.9	16 510
Annual species				
Sorghums	15	70 to 80	5	16 830
Whole crop wheat	8 to 15	12 to 15	2.8	15 000
Fibre flax	7 to 9	10 to 15	1 to 2	
Whole crop barley	5 to 10	12 to 15	2.8	15 000
Kenaf	8 to 10	20 to 30	0.6 to 1.2	
Whole crop triticale	8 to 16	12 to 15	2.8	15 000
Hemp	13 to 17		6.9	16 460
Perennial species				
Poplar SRC	10	50 to 55		
Eucalyptus SRC	16 to 22	43 to 48		
Willow VSRC	12	55		17 610
Tall fescue	7			17 170
Dactyl	5 to 6			17 170
Lucerne	10		10.9	16 830
Giant cane	25	50 to 55	2 to 6	16 740
Miscanthus sinensis	15 to 20		2.8	17 900
Reed canary grass	10	10		
Phragmite	4 to 20	20	3	10 450
Cynara	10 to 20		9.2	16 500 to 17 800

adapts to virtually all soil types [1]. The conditions in terms of water demand nevertheless limit the potential areas for development of poplar SRCs. In France, these plantations are limited to 16 departments [2], mostly in plains. Eucalyptus trees prefer hot, damp climates. The location of eucalyptus plantations is limited by the risk of freezing. Growth is favoured by regular rainfall and high average temperatures. Consequently, in France, the eucalyptus is grown mainly in 9 departments in the south west [2]. Like the poplar, the willow requires a good supply of water [3]. A number of projects worldwide are based on this type of plantation, sometimes combining several objectives, for example [4] the production of energy biomass and wastewater treatment.

Table 4.1 lists the main characteristics (yield, humidity and ash contents, calorific values of the agricultural residues, annual crops and perennial crops harvested annually or multi-annually.

4.1.2 Main figures of current sources

This paragraph deals with the raw source of biomass that can be potentially mobilised in the short term, being currently present and available, with no agronomic, economic or environmental considerations. These will be mentioned at the end of the chapter before concluding on more realistic source scenarios.

4.1.2.1 Distribution of main types of biomass

When speaking of biomass, we must distinguish between the production of biomass, expressed as an annual flow (e.g. in tonnes of dry matter per year: t d.m./year) and the stock of biomass itself or "reserves", by analogy with oil (expressed for example in tonnes of dry matter, t d.m.). Biomass developed sustainably for energy production corresponds mainly to the fraction of biomass obtained annually and not used for any other purpose. It some cases, it may also be a fraction of biomass stock resulting from an under-exploited source, like some private European forests in particular. In terms of production only, it is worth mentioning that the recoverable quantities produced are not proportional to the various producing areas. Oceans, for instance, cover 71% of the Earth's surface (Table 4.2), but the quantity of biomass produced every year from water bodies represents only 43% of the annual production, out of which only a tiny proportion can be recovered directly (mainly as algae). This reflects the large variations in energy yield of the various areas and their exploitable nature.

Table 4.2 World production and distribution of plant biomass [6].

	Area	**Annual production in Gt d.m./year**		**Recoverable production in Gt d.m./year**		**Recovered production in Gt d.m./year**	
Continents	**13 010 Mha**	**120** ***(51.6)*** [a]		**44** ***(18.9)*** [a]		**21** ***(9)*** [a]	
Crops	12%		10%		22%		46%
Pastures	26%		20%		27%		17%
Forests	30%		60%		38%		37%
Other	32%		10%		13%		0%
Water bodies	**36 500 Mha**	**90**		**~ 0**		**~ 0**	
Oceans	99%		99%		–		–
Freshwater	1%		1%		–		–

[a] *Figures between parentheses are expressed in Gtoe.*

The total recoverable production represents a theoretical maximum value of the potential biomass resources, in the current state of distribution between the areas. The fraction actually recovered is mainly intended for the production of food and energy (firewood), as well as the wood industry (building, furniture, etc.). The lignocellulosic share of this recoverable production consists almost exclusively of wood production and harvesting waste. Lignocellulosic crops, apart from wood, for example miscanthus, represent just 0.07% of the plant biomass production currently harvested worldwide.

Taking an average value of the dry matter energy content equal to 0.43 toe/t d.m. (NCV), the recoverable production represents 19 Gtoe/year, i.e. almost double the world primary energy consumption (11.7 Gtoe/year in 2006, including fuel wood). From the recovered production (21 Gt d.m./year), the share actually used currently amounts to 12.8 Gt d.m./year, i.e. 6.1 Gt d.m./year for food applications, 4.4 Gt d.m./year for energy supply and 2.3 Gt d.m./year for industrial applications. We can therefore estimate to a first approximation another outlet for biomass to energy production, including the unrecovered recoverable share and the biomass conversion waste. This potential amounts to 31.4 Gt d.m./year, i.e. 13.5 Gtoe/year (NCV). It consists mainly of forest resources and agricultural residues, or comes from dedicated crops. For forest products, special attention is often paid to the production of round wood, which represents about $1/12^{th}$ of the annual biomass increase.

4.1.2.2 The forest resource

Fuel wood currently makes up a significant share (over 1.2 Gtoe/year) in the world energy balance, far outweighing that of nuclear energy (0.73 Gtoe) and hydroelectricity (0.22 Gtoe).

A. Balance of wood removals

Although deforestation continues at a fast pace at world scale (7.3 Mha/year between 2000 and 2005), afforestation and natural expansion of the forests have, at the same time, considerably reduced the net loss of forest cover. Between 2000 and 2005, 57 countries reported an increase in their forest area, in Europe and Asia-Pacific in particular, while 83 other countries indicated a decrease. Not all of these forest areas can be exploited (Figure 4.1), either because they are designated as protected forest, or because felling is unprofitable due to their remote location or the uneven terrain.

In CIS, the tree stands are less productive and less intensively exploited than those in Europe and North America. Temperate and boreal forests are characterised in particular by the fact that in all countries, apart from Cyprus and Armenia, felling is less, sometimes considerably less, than the net annual increase in the forests available for the supply of wood. This leads to a steady increase in standing forests in almost all countries, with about **1.2 Gt of dry wood** produced in surplus and not used. This wood potential not exploited in vast areas will be discussed in paragraph 4.4, since not considered as readily available like co-products and residues.

Concerning the share of harvested wood, more than half the gross global production of round wood is intended for energy purposes (firewood, charcoal). The rest includes industrial uses of wood, in particular sawing and veneering (Figure 4.2). The conversion of wood, with low material yield, produces large volumes of waste, partially valorised as energy or raw material for trituration. In 2006, the global annual flow of round wood harvested is nearly 3.6 Gm^3/year, for a marketable volume of standing timber of about 202 Gm^3.

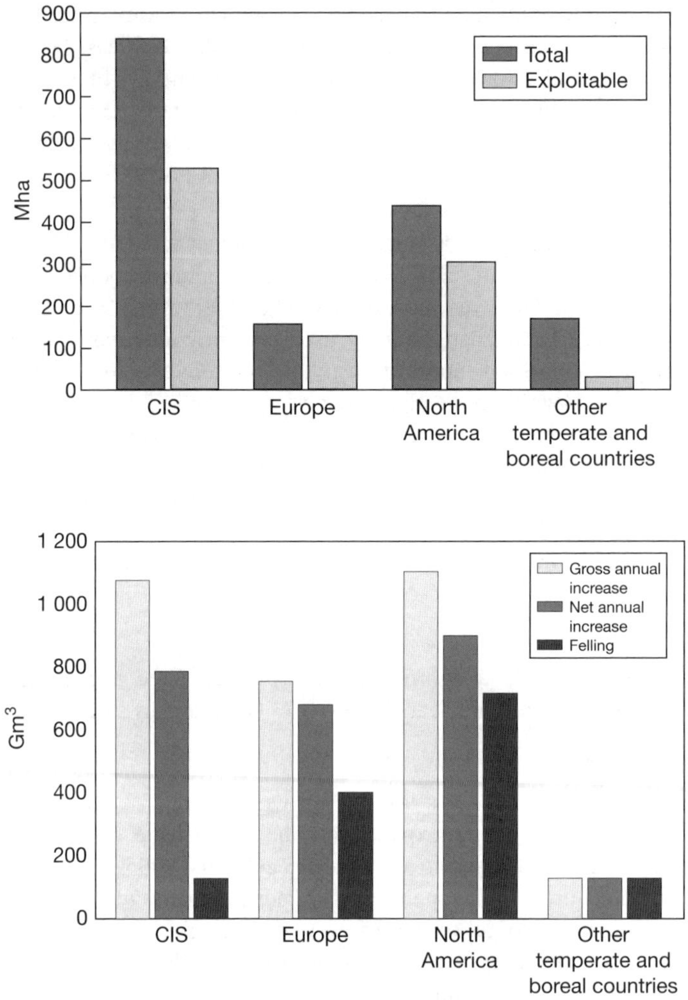

Figure 4.1

Wood availability and felling [7].

B. Forest exploitation waste

Forest exploitation waste consists of branches and twigs produced when lopping felled trees, as well as leaves and damaged or unwanted trunks. Most of this waste is left on site after felling, being of insufficient quality for use as lumber or industrial wood. The situation may vary widely depending on the regions. The proportion of residues left on site may vary depending on local conditions, but the ratio, often taken as a reference in the literature, is 1 m^3 of round wood exported from the forest for 1 m^3 of waste left on site. In certain cases, some of the waste is converted into charcoal. Assuming that the average energy content for

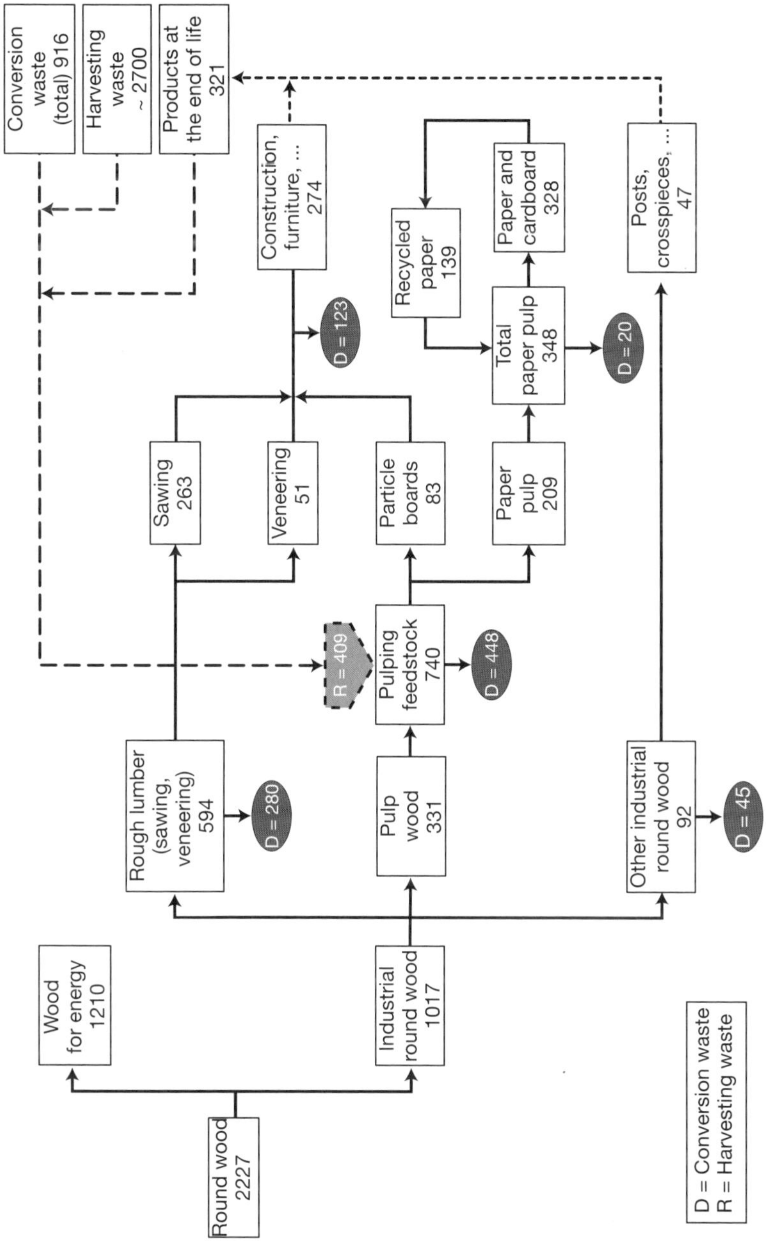

Figure 4.2

Production of waste resulting from the conversion of wood throughout the world in 2003, expressed in Mt d.m./year. According to FAOSTAT data [8] for product flows and [9] for conversion yields. D = Conversion waste; R = Harvesting waste.

these residues is 15 840 MJ/t d.m., the energy source for this waste amounts to about **1 Gtoe/year**. This worldwide potential is obviously unevenly distributed between the various regions, depending in particular on the degree of forest exploitation.

For the **European Union**, a Finnish study [10] estimates the volume of felling residues to be about 140 Mm^3, i.e. about **30 Mtoe/year**. Sweden, Finland, France and Germany are the 4 leading European producers of felling residues that can be converted into energy. The volume in France would represent about 15 Mm^3 of felling residues (excluding stump wood), i.e. almost 11% of the European stock. In the **United States,** the USDA [11] estimates a volume of felling residues similar to that of Europe, about **25 Mtoe**. However, we must also include a very large proportion of wood coming from operations designed to reduce the risk of forest fires, with a potential of 60 Mt of wood per year, i.e. **an extra 26 Mtoe**.

C. Wood industry waste

Figure 4.2 provides an estimation of the amount of waste produced during wood conversion at world scale (in Mt d.m./year). This figure attempts to provide, for 2003, a summary of the uses of round wood harvested throughout the world. The production values of each type of product (in volume round wood equivalent) are obtained from the FAO statistical database [8].

The evaluation of the weight of dry matter and of waste production at each step is based on average densities of dry matter in each product category as well as on average conversion material yields. The quantity of pulpwood produced annually, included in the FAO statistics, is not sufficient to supply the board and paper pulp factories. It has therefore been assumed that the missing material comes from conversion or harvesting waste. Lastly, the conversion wastes have been grouped together, even though the shapes and energy contents of these residues vary widely. A large proportion of this waste is already used in various industries, whether to produce energy (heat or electricity) or as raw material, for example for trituration (manufacture of paper or particle board), as was assumed. Obviously, the extent to which this waste is used depends on demand, hence the difficulty of accurately estimating the additional potential source represented by wood industry waste for energy production. Clearly, not all of these types of waste can be recovered, especially amongst the products at end of life. Assuming an average energy content of 0.43 toe/t d.m. for these various types of residues, and deducting from the total the proportion potentially used for trituration, the potential energy source of these residues amounts to **0.3 Gtoe/year**, without making any assumptions of what is used elsewhere to produce energy (heat or electricity).

In Europe, virtually all primary processing industry wood residues are already used. 40% of industry residues are recycled in the manufacture of other industrial products (boards, etc.) and 60% is used in the production of heat and electricity in the emitting plants or other thermal power plants. Currently therefore, there is zero additional potential. Scrap wood, consisting mainly of demolition and packaging wood, represents about 6 Mtoe according to EEA estimations [12].

In the **United States**, the USDA [11] and the NREL [13] estimate that between 16 and 39% of wood industry residues produced, i.e. between **1 and 2.5 Mtoe/year**, are not utilised. In this case, urban scrap wood (packaging, pallets, wood from residue reception centres, etc.) corresponds mainly to construction wood residues whose potential is estimated at 8.6 Mt d.m./year, i.e. about **3.5 Mtoe/year.**

4.1.2.3 Agricultural residues

A large quantity of lignocellulosic biomass may also come from agricultural dry residues, whose quantity and composition vary significantly depending on the species and pedoclimatic conditions.

Agricultural dry residues are defined as being the non-edible part of plants, left on the ground after harvesting, for example wheat and barley straw, corn stalks, bagasse, rapeseed or oleaginous flax straw, sunflower stalks as well as beet or potato tops. The quantity of these residues varies significantly depending on the crops. In order of importance, sugar cane residues represent 26% of the total, followed by those of wheat (18.7%), rice (18.3%) and corn (14%), i.e. for the four crops mentioned more than $^3/_4$ of all recoverable agricultural waste. Assuming that the average energy content for these wastes is 18 650 MJ/t d.m. [14], the potential energy source represented by these residues amounts to slightly more than **1.8 Gtoe/year**.

In **Europe,** most of these harvesting residues consist of cereal straw, and in particular winter and spring wheat and barley straw. These are the main cereals produced every year on the European agricultural land. It is also these straws that are the most valorised, with well-organised logistics for their use. Most of the other straws, stalks and tops are left on the field after harvesting. After deducting the usual applications for breeding (bedding, supplementary feed) from the cereal straw source, the available wheat and barley straw source in the European Union amounts to 48 Mt d.m., or **19 Mtoe**. Furthermore, a significant proportion of grain corn stalks and residues from permanent crops (pruning of vines, olive trees, etc.) could be added to this source of agricultural residues, amounting to 3.5 Mt to 4 Mt d.m. of biomass per year respectively.

American studies [13], [15], consider a wide range of crops to estimate their harvesting residue potential, in particular corn, sorghum, barley, oats, soft wheat and hard wheat. For the 2002 campaign, total production amounts to about 450 Mt d.m. of residues. After deducting the current applications of these straws, the available source would be 180 Mt d.m./year or about **75 Mtoe**.

Harvesting residue production volumes are very large in **Asia**, with in particular 755 Mt d.m. of residues in China (rice, wheat, corn, soya, other oleaginous crops, cotton, sugar cane, tubers) and over 530 Mt in India (rice, wheat, sugar cane, cotton). These countries already have large energy production systems set up for combustion on domestic and industrial scales (216 Mt/year). A study conducted by the USAID [16], assessing the potential of second-generation biofuels for 7 countries in Eastern Asia, estimates the maximum quantity of recoverable straws to be 92 Mt d.m./year, i.e. about **40 Mtoe**.

The potential volume currently not exported cannot represent the entire resource available, however, since harvesting residues are subject to numerous exploitation constraints. The other export constraints and more accurate availability levels are described in paragraph 4.3.

4.1.2.4 Dedicated crops

The dedicated herbaceous lignocellulosic crop most widely cultivated for commercial use in France and Europe is currently miscanthus. Nearly 40 ha of miscanthus were planted in France in 2005, 80 ha in 2006 and 1 300 ha were contractualised over the 2007-2008 campaign for fuel use. Other niche commercial applications can be mentioned, such as animal bedding, particle boards, etc. In the United States, *switchgrass* is the most widely cultivated species.

Major programmes to plant SRCs are also in progress, with in France 400 ha of paper poplar SRCs (compared with 6 000 ha in Italy), 300 ha of willow SRCs (compared with 20 000 ha in Sweden, 3000 ha in the United Kingdom, 2 500 ha in Rumania and 2000 ha in Poland), then 1 500 ha of eucalyptus SRCs (out of a total of 1 Mha in Europe, in particular in Spain and Portugal).

France also includes several networks to perform experiments on various lignocellulosic crops. Tests are conducted on miscanthus, *switchgrass*, fibre sorghum, giant cane, reed canary grass, etc. to study the potentials of the crops, reduce their production costs and assess the environmental impacts depending on technical routes and pedoclimatic conditions.

Since development of this type of resource is still limited, the development of dedicated lignocellulosic crops is currently assessed on the basis of scenarios in which the crops are grown on areas considered as available and suitable.

4.2 RAW MATERIAL/CONVERSION PROCESS ADEQUACY

4.2.1 Composition of the raw material

4.2.1.1 Residues

The nature and composition of the biomass contained in waste generally vary. The resource represented by this waste is therefore highly heterogeneous, which could cause difficulties during the conversion process due to lack of constant product quality. The most variable parameters are the water content and the mineral element content.

A. Agricultural waste

The humidity of crop residues is the most variable parameter, ranging from 12% for cereal straw to 75% for rapeseed straw (Table 4.1). The ash and silica contents may vary significantly between several species of the same plant and from one plant to another (Table 4.3); the same applies for the sulphur, nitrogen and chlorine compositions (Table 4.4).

Table 4.3 Compositions of various types of straw [5].

Types of straw	Mineral substances (ash) wt%	Silica wt%
Oleaginous flax	4 to 5	
Whcat	4 to 9	3 to 7
Barley	5 to 7	3 to 6
Oats	6 to 8	4 to 7
Rye	2 to 5	0.5 to 4

Table 4.4 Atomic compositions of various types of straw [5].

Types of straw	C wt%	H wt%	S wt%	N wt%	Cl wt%	Ash wt%
Wheat	46.5	5.3	0.1	1.0	0.3	5.9
Rapeseed	35 to 40	5.1	0.7	0.3	1.0	9.7
Sunflower	44.6	5.1	1.1	0.1	0.7	11.2
Oleaginous flax	47.8	5.4	0.7	0.1	0.3	4.1

B. Forest and wood industry waste

Two types of forestry waste can be identified: forest residues, defined as being all above-ground biomass left in place after harvesting or thinnings (stumps, branches of diameter less than 7 cm, etc.) and plots abandoned due to lack of demand. The characteristics of various wood species are listed in Table 4.5.

Table 4.5 Characteristics of various wood species [17].

Wood species	Weighted bark content wt%	Lignin wt%	Pentosans wt%	Cellulose wt%
Oak	10.1	21.9	21.6	–
Beech	7.1	21.9	24.9	48.7
Chestnut	7.9	20.3	18.3	46.7
Spruce	7.4	28.4	13.9	–
Maritime pine	10.4	27.0	–	44.8

Wood industry waste varies both in terms of geometry and composition. For instance, barks have a relatively high granulometry and high content in inorganic compounds while sawdust has a fine granulometry and lower content in inorganic compounds. Equally, the humidity in wood waste is quite variable, reaching high values in some cases (Table 4.6). The ash content is also highly variable and may be high, as in bark, where it can reach 20%.

Table 4.6 Characteristics of wood and its co-products [18].

Product type	NCV kWh/t	Average humidity wt%	Density kg/STCM [a]	Ash content wt%
Wood				
Broad-leaved trees	4 800-5 200	anhydrous wood	330-500	
Conifers	5 100-5 300	anhydrous wood	260-390	
Barks	5 400	variable	= that of wood	5-20
Damp sawdust	1 600-2 800	30-60	230-520	1-2
Chips and dry sawdust	4 400	10	80-110	0.5-1
Chips	2 200-3 900	20-25	180-400	0.5-1.5
Wood shreds and scrap wood	3 300-3 900	20-35	170-270	0.5-2
Pellets and briquettes	4 600	8-10	700-750	0.5

[a] *STCM: stacked cubic metre*

4.2.1.2 Dedicated crops

A. Annual crops

The composition of the various annual species of lignocellulosic biomass varies considerably. Their humidity may range from 10% to 80% depending on the plant considered. The moisture content of cereals (wheat, barley or whole crop triticale) is lower than that of species such as sorghum or kenaf. The ash content, about 2.8% for cereals, may reach up to 6.9% for hemp.

B. Perennial crops harvested annually

Since the types of perennial crops harvested annually are highly diversified, their compositions also vary widely, whether in terms of moisture or the content in inorganic compounds. The moisture content of species such as reed canary grass is low (10%) while the water content of other species such as the giant cane is high (50% to 55%). The content in inorganic compounds is rarely less than 2% and can reach almost 11% in case of lucerne.

C. Perennial crops harvested multi-annually

Tree plantations are perennial crops harvested multi-annually. In addition to carbon, oxygen and hydrogen, wood contains a certain quantity of mineral elements which depends on the soil composition and fertility, the age of the stand and the season [19]. The concentration in mineral elements will therefore be higher in young stands than in mature stands, in broad-leaved trees than in conifers, and in tropical regions than in temperate regions. Ash production also varies depending on the species (Table 4.6).

4.2.2 Characteristics of the raw material likely to affect energy conversion

4.2.2.1 Thermochemical conversion

The content in inorganic compounds of the biomass is a key parameter in case of thermochemical conversion. The lower the content in inorganic compounds, the more the biomass is adapted to thermochemical conversion. During gasification (entrained flow at high temperature), the ash melting point governs the reaction temperature and therefore the conversion yield. For a process implementing a circulating fluidised bed, the operating conditions are not determined by the ash melting point, although it may nevertheless disturb the heat exchanges. Lastly, the nitrogen and sulphur contents of the biomass may pollute the syngas with emission of H_2S, HCN, etc. which must then be purified.

The granulometry and water content also represent important characteristics of the raw material. These two variables govern the heat transfers within the reactor. High granulometry will be more suitable to slow pyrolysis since allowing better circulation of the gases, while low granulometry will be more suitable to flash pyrolysis, in which the heat transfer rate is of prime importance.

In all cases, humidity must be lowered since it reduces the energy yields and slows down the heat transfers. In addition, the water content is an important economic parameter since it increases the weight to be transported.

Lastly, the lignin content is important, although it does not affect the operating conditions. It is in fact responsible for the formation of tars, which currently represent a technological barrier with this conversion pathway.

4.2.2.2 Biochemical conversion

The lignin content is the most important characteristic of biomass as regards chemical conversion: the lower the lignin content, the more suitable the resource will be for biochemical conversion, since lignin cannot be converted biologically into biofuel. It also acts as "cement" and guarantees the stability of the lignocellulosic structure, representing a "brake" for the enzymatic hydrolysis process. After separation, however, it will be used to supply the energy required for the conversion process.

Furthermore, since the biological conversion of pentoses into ethanol is not currently efficient enough, lignocellulosic materials containing only small proportions of hemicelluloses, like some species of wood, will be the most suitable for the biological conversion process.

4.3 LIGNOCELLULOSIC BIOMASS PRODUCTION AND MOBILISATION CONDITIONS

While the mobilisable source of lignocellulosic biomass is huge, there are numerous obstacles to its utilisation. Today, wood pellets represent the only efficient biomass to energy sector in Europe and North America. As they are showing their limits, other sectors are under development. This chapter reviews the constraints which must be overcome in order to allow industrial operation of the new biomass conversion sectors.

4.3.1 Resource access constraints

Generally speaking, by-product type resources are often geographically dispersed and, in case of forest resources, even difficult to access due to topographic constraints (slopes) or lack of infrastructures (paths, roads, tracks, depots, machinery, etc.). According to a study published in a report by the Finnish Forest Research Institute [10], European wood harvesting residues would be technically available for supply on 75% of clear cuts and 45% of thinnings. On all these sites, the recovery rate was estimated to be 65% in mechanised cutting and 50% in manual cutting. Out of the 173 Mm^3 of felling residues produced annually, 62.6 Mm^3 would be technically harvestable, i.e. about **14 Mtoe.** For the United States, the USDA report [11] considers that the technical constraints to recover forest biomass reduce the potential by 40% and estimates this potential at 41 Mt d.m., or **18 Mtoe.** According to the Cemagref [20], the reduction due solely to the natural constraints hindering access to the resource in France decreases the fuel wood potential by 11%, i.e. a medium term potential of about 8 million tonnes of dry matter.

The recovery of harvesting residues is faced with a similar problem. Use of cereal straw to produce energy can only be considered if the stocks are not required for breeding activities or in cereal plains. Moreover, scattered resources may induce an economic constraint, thereby limiting the radius of supply of an energy production plant. The study conducted by the European Commission's JRC (Joint Research Centre) [21] suggests a technico-economic potential for the mobilisation of this straw based on the maximum possible collection radius to supply a power plant under good economic conditions. The price paid by a company for the resource depends partly on the transport distance. For a power plant burning 200 000 tonnes of straw/year (38 MWel), the maximum radius is 50 km. A suitability map for this type of power plant was drawn up, based on this calculation. It turns out that 38% of the total straw resource could logistically be made available, i.e. about **5.5 Mtoe** to supply a maximum of 67 reference power plants. In France, in view of the geographical distribution of the resource and the current straw outlets, the stock of 22 Mt d.m. of straw produced is reduced down to 4 Mt d.m. available per year, or **1.6 Mtoe** for energy. The USDA report [11] indicates that, based on the total amount of residues currently available in the United States (500 Mt), two constraints, one logistical and the other environmental, reduce this potential respectively by 41% and 55% (see paragraph 4.3.4), to obtain usable resources of just 113 Mt d.m./year, or **45 Mtoe**.

Another characteristic of the wood sector, in France and other European countries, is the extent of private forest land which means that the management of forest resources is not systematic and depends on the consent of owners. In addition, farmers and forestry operators remain reluctant to grow perennial crops, which implies dedicating plots for energy purposes for more than 10 years. Despite the difficulty in costing the impact of this constraint on the availability of the resources, it must nevertheless be evaluated in order to consider possible implementation of incentive measures.

4.3.2 Technical routes and conditioning

Concerning straw, the need to harvest as near the ground as possible, collect with a high-density press and protect the biomass against humidity are essential requirements which must not be neglected when evaluating the energy recovery processes.

The technical routes for dedicated crops are currently being defined. Lignocellulosic (in particular perennial) crops generally require few inputs. Currently, no phytosanitary substances have yet been approved for the new energy crop varieties. Some crops such as miscanthus may be sensitive to competition from weeds during the establishment season. In this case, mechanical and/or chemical weed control is required during the first year. While most herbaceous species can be sown with a traditional seeder, planting of miscanthus rhizomes requires special equipment and is more labour-intensive. SRCs require no inputs after planting. Water requirements vary from one species to another. Irrigation must nevertheless be ruled out to preserve a positive energy and environmental balance.

Concerning the harvesting steps, all herbaceous and VSRC species require traditional agricultural equipment, but the conditioning (silage, straw, pellet, etc.) must still be defined depending on the biomass conversion type (wet milling, dry milling) and the available storage means. SRCs must be harvested using forestry equipment. SRCs are harvested either as logs when intended for the trituration industry or as wood chips for energy purposes. Wood chip harvesting may consist either of whole-tree chipping on terrain or at roadside after transport, or in forming bundles with the felling residues or use them directly on forest plot. Chips are used to fuel automatic wood-burning energy plants. There are several types of wood chippers, ranging from carried or towed machines (agricultural system) to self-propelled or stationary machines for large capacities. Yields vary depending on the equipment used, in view of the power supply, from 1 to 100 STCM (stacked cubic meter of wood chips) per hour. The wood chips are transported from the production site to the storage or consumption site by lorries.

4.3.3 Competing uses

A large proportion of lignocellulosic biomass, especially from co-products, is used for breeding (straw), in industry, for energy purposes, or as raw material for chemistry. Consequently, there may be competition between the various uses of this biomass. This is the case with the resource from the wood industry waste which is already widely reused, either as raw material for the manufacture of fibre board, particle board or paper pulp, or for energy production, especially to dry wood.

Eventually, uses for bioproducts (biopolymers, fibres) could also generate competition, if these sectors should develop on large scale.

There may also be a certain degree of land use pressure if the demand for energy crops increases significantly. While due to their low requirements in terms of soil quality, perennial crops can initially be grown on so-called marginal, set-aside, waste land or other land types unsuitable for food crops, the possible location of these new crops must nevertheless be planned without delay in order to prevent future land use competition.

4.3.4 Environmental constraints

When they are not used for breeding purposes, cereal straw and most other harvesting residues are buried in the ground between two crops to return the organic matter back to the soil. Depending on the local conditions, however, it would appear that a limited fraction of the straw could perform the agronomic and environmental roles of straw while the remaining fraction could be exported. Although drainage, nitrate losses, N_2O and ammonia emissions do not seem to be significantly increased if straw is exported, the losses in harvesting yields and quantity of carbon stored cannot be denied. During crop rotation, if this cereal straw is to be removed without impairing soil fertility, it is essential to bury the other residues currently found in major crop growing areas (rapeseed straw, sunflower and corn stalks). The extent of the exports in a rotation must be based on local pedoclimatic conditions.

As with harvesting residues, repeated harvesting of forest felling residues is not systematically harmless: a non-negligible quantity of mineral elements may also be exported at the same time under certain conditions. On some soil types, the fertility may decline in the long term, a situation to be avoided as part of a sustainable forestry approach without external contribution. The European Environment Agency has published a map showing the areas more or less suited to the export of felling residues outside forest soils (Figure 4.3). The various adaptability thresholds do not consider fertilisation of forest soils as being a solution for the poorest soils. Similarly, the recycling of ashes to raise the threshold on podzolic soils (in particular in Finland and Sweden) is not taken into account. In addition, the map does not show the positive effect of exporting felling residues on soils with large deposits of nitrogen (some regions of Belgium, Denmark, northern France, Germany, northern Italy, the Netherlands, Poland and southern Sweden).

Some studies set an average percentage for the natural increase required to regenerate the forest soil based on the characteristics of some evaluated countries (Table 4.7).

Cultivation of lignocellulosic species seems to involve only a limited risk of contaminating surface and underground water due to the low requirements of these crops and could, in some cases, restore the water resources to good ecological condition. Like some forests, poplar plantations may play an important role in regulating flows of pollutants. As regards the reduction of nitrate content, the efficiency of poplar plantations is comparable to that of young riparian forests [1] and much greater than that of pastures. The possibility of reducing phosphate levels in surface water is quite considerable. The poplar plantation also fixes other compounds such as triazines, heavy metals and various particles of organic matter, clay, loam, thereby improving the chemical quality of the water and its limpidity.

However, some crops, including those developing extensive root systems (miscanthus, SCR in particular), may remove large quantities of water during vegetative periods. Some plants, such as the poplar, require soils well supplied with water during the summer and should not be grown in areas subject to recurrent droughts.

1. Riparian forests: woody, bushy and herbaceous formations growing along river banks.

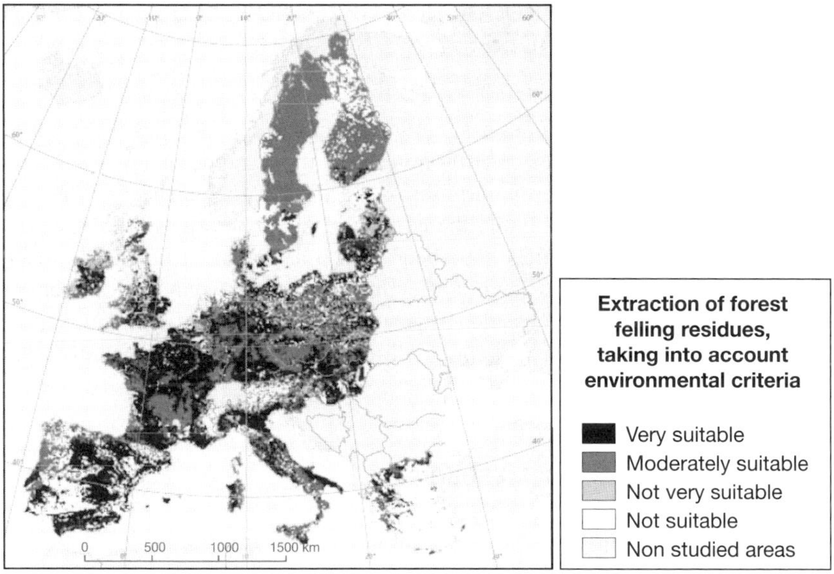

Figure 4.3

Potential for extraction of forest felling residues taking into account environmental criteria (soil erosion, compaction and fertility) [12].

Table 4.7 Proportion of biomass for soil regeneration in 5 European Union countries [9] and [10].

European countries	Increase (Mm^3)	% set	Volume to be left in the soil (Mm^3)
Austria	27.3	19	5.2
Finland	72.4	10	7.2
France	92.3	5	4.6
Portugal	12.9	10	1.3
Sweden	85.4	3.5 to 7.5	4.7

Lastly, due to the vegetative appearance and perennial nature of the lignocellulosic crops studied, introduction of this type of species in existing crops and landscapes must also be considered. Potential changes to local ecosystems with the disappearance or introduction of new flora or fauna must also be taken into account.

In this context, the country planning inherent to the development of these new agricultural, forest and energy sectors is extremely important and must include numerous technical, economical, societal and environmental factors in order to produce sustainable production systems in line with the expected bioenergy sectors.

4.4 POTENTIAL FOR THE PRODUCTION OF LIGNOCELLULOSIC BIOMASS, NEW MEANS, NEW SPACES

Large scale development programmes assuming a change in the mobilisable resource in the short, medium and long terms are currently investigating the numerous ways of improving the availability of biomass for energy purposes. Nevertheless, the perspectives and development means remain uncertain and imply the need to make assumptions often specific to the geopolitical context. Various models have been set up at global, regional and local scale. They take into account numerous factors affecting access to resources, for example food demand, climatic factors, regulations, etc.. This sub-chapter puts forward a number of points to be taken into account so that it will eventually be possible to assess the development of the lignocellulosic biomass sectors.

4.4.1 Availability of additional resources

4.4.1.1 Management of under-exploited forests

Paragraph 4.1.2.2 demonstrates that some of the world's forests have a positive balance, in other words the annual production of forest biomass (or net annual increase) remains greater than the removals. This is the case in particular for the European forests and, to a lesser extent, in North America.

The net annual growth of wood in European forests is greater than the felling volume, resulting in capitalisation of a significant proportion of growing stock, estimated at 260 Mm^3. Exploitation of these new resources will imply that more lumber or industrial wood is available and the additional production of 67 Mm^3 extra felling residues, excluding leaves and roots. However, a study mentioned in a report published by the Finnish Forest Research Institute [10] indicates that only 25% of the volume of standing biomass would actually be accessible, i.e. 17 Mm^3 of additional felling residues, i.e. nearly **4 Mtoe.** In addition to the availability of felling residues, obtained from the current exploitation of wood, the European forest potential would therefore amount to about **18 Mtoe** of wood.

In an approach aimed at a medium-long term estimation, all or some of the technical constraints could be minimised. This is the approach adopted in the study conducted by the European Environment Agency [12], which provides an estimation of the forest resources available for energy by 2010, 2020 and 2030, based on a selection of environmental criteria as well as possible competition for the wood resource, eventually favouring use for energy purposes in a context of increasing fossil energy prices. This study therefore indicates that wood energy could provide about **55 Mtoe/year**.

In **France**, forest covers over 15 Mha and its biological growth is estimated at 84 Mm^3/year [22]. In 2004, respectively [1] 12.0 Mm^3 and 22.5 Mm^3 have been used as industrial

1. For the conversion, we can consider an average density of 0.7 t d.m./m^3 for wood and an average energy content of 0.43 toe/t d.m.

wood (paper, particle board, etc.) and as lumber wood. The consumption of energy wood is estimated at 40 Mm^3/year, including about 30 Mm^3/year for domestic consumption and the rest for energy consumption by industry and collective and tertiary buildings. The French forest therefore achieves an excess balance of about 10 Mm^3 or **3 Mtoe** of unused wood.

In the **United States** [11], assuming growth in national demand for forest products, improvement of exploitation systems towards more productive forests and an improvement in the techniques used to harvest felling residues, the additional forest resources would therefore reach 23 Mt d.m. of felling residues from exploited forests and 41 Mt d.m. from forests not yet exploited, eventually resulting in an energy production of **27.5 Mtoe**. Together with the wood resource from fire-fighting measures, the total potential of unconverted wood could amount to **53 Mtoe**.

4.4.1.2 Development of production means and improvement in yields

As we have seen, most energy resources are currently either co-products of crops harvested mainly for their grains or other products intended for other outlets, or relatively undeveloped dedicated crops. Since the sectors converting these crops are still in their infancy, the crop systems are not yet optimised for the development of these new mass outlets. Significant progress can therefore be expected in terms of variety selection, technical routes and therefore crop yields, as has been the case historically for various food crops.

4.4.2 New land availabilities

In addition to changes in exploitation practices on a given area, the development of new crops can also be considered on land currently under- or not exploited or on land released following improved productivities or land use rearbitration.

4.4.2.1 Under-exploited arable land

In 2000, the FAO "Land and Water Development Division" conducted a study to assess land resource potential and constraints at regional and country levels [7]. The study resulted in particular in the creation of a database estimating arable land area by country, with two availability levels (Table 4.8).

Taking the data at face value, and in relative terms, there is greatest potential for agricultural expansion in Sub-Saharan Africa, South Asia and Central America, which together form over 70% of the global potential increase. In North Asia, Asia Pacific, North America and Europe, 54% to 64% of arable land is already in use and opportunities for expansion appear to exist although on a more limited scale. The North Africa and Near East region shows actual arable land in excess of potential, since irrigated cultivation, implemented on arable land and well represented in this area, is not included when estimating the potential (Figure 4.4).

This under-exploited or non-cultivated arable land (about 1.4 Gha) has different soil types and is sometimes degraded land.

Table 4.8 Potential arable land for rainfed agriculture, by large geographical areas [7].

	Gross potential arable land	Protected land		Settlement	Net potential arable land	Arable land (1994)	% of potential arable land actually in use (1994)
	(Rainfed cultivation) (1000 ha)	% of total area	% of potential arable	% of total area	(Rainfed cultivation) (1000 ha)	(1000 ha)	(Rainfed cultivation)%
Sub-Saharan Africa	1 119 492	8.6	4.3	1.9	1 050 083	157 608	15
North Africa and Near East	50 017	8.1	4.0	6.4	44 815	71 580	60
North Asia, east of Urals	286 800	3.0	1.5	(2.3)	275 902	175 540	64
Asia Pacific	812 551	9.4	4.7	3.9	742 672	477 706	64
South and Central America	1 048 071	10.6	5.3	1.2	979 946	143 352	15
North America	463 966	9.9	4.9	(2.1)	431 488	233 276	54
Europe	363 120	10.1	5.0	(5.8)	323 903	204 322	63
World	4 144 017	8.9	4.4	2.6	3 848 809	1 463 384	38

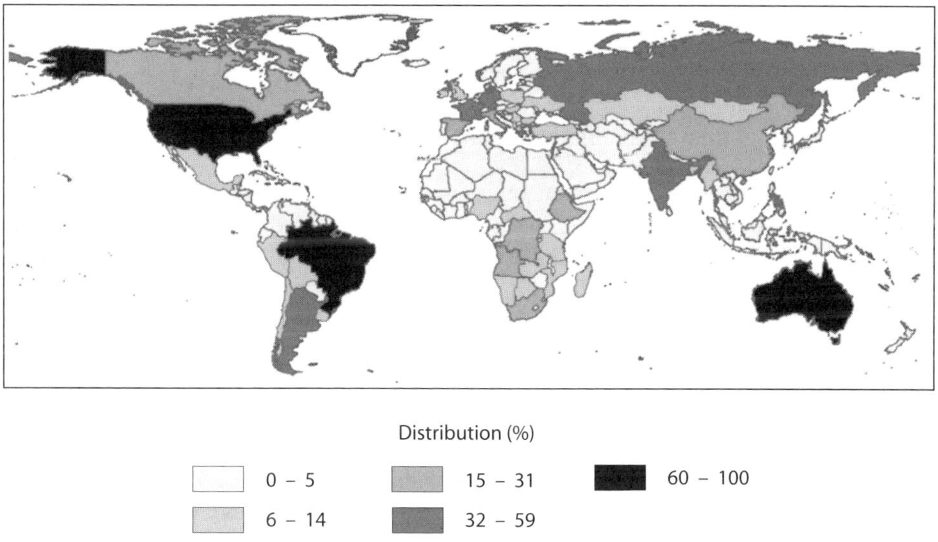

Figure 4.4

World distribution of biomass production potential on abandoned land [23].
Source: Campbell JE, Lobell DB, Geneva RC and Field CB (2008). The global potential of bioenergy on abandoned agricultural lands. Environmental Science and Technology, DOI: 10.1021/es800052w.

Uncultivated agricultural land corresponds, according to the Teruti-Lucas nomenclature, to areas that were once cultivated, which have been abandoned for financial, political or environmental reasons, but which could be re-cultivated with commonly available farming methods. This category includes non-productive and ungrazed moor land, waste land, uncultivated land, abandoned farms and permanent cultivations, improvised field paths and areas not converted into forest or urban zone. Waste land is an area of land that is excluded from the crop rotation beyond one crop season. At global scale, this type of land would extend over 386 Mha [24] with uncertainty concerning the potential use of this land for grazing. In France, the area is about 2.5 Mha every year. Note the distinction between this land and fallow land, which corresponds to arable land that is intentionally left uncultivated so that the land can rest, usually for a whole crop season.

Degraded land is characterised by a loss of ecosystem function and services. This leads to a reduction of soil fertility, then to a reduction of plant growth which in turn causes loss of protective cover and increased soil vulnerability. The total area of degraded land at world scale would extend over 500 Mha, including100 Mha in Asia, 100 Mha in South America and 300 Mha in Africa [25], [26]. In Brazil, the useful agricultural area could be doubled, compared with 2004, by reclaiming the degraded land, giving an extra 60 Mha.

The trend in non-food crops suggests a major change in terms of land rehabilitation and economic development. Some of these areas, however, may be located in sensitive regions and be used by local populations, for subsistence farming in particular. There may also be conflicts regarding the exploitation rights. Abandoned or degraded land areas must be assessed in terms of their long-term use.

4.4.2.2 Release of new areas: improvement in food crop productivities

The productivity of any crop in a given environment is defined by the main resources available in terms of light, water, inputs, the duration of the growth season and the climatic conditions. Depending on the pedoclimatic conditions (soil quality, rainfall, etc.) and the technical agricultural means, however, productivity variations are observed from one geographical area to another. In addition, for a given location, it is possible to estimate a productivity that can be achieved in the more or less long term according to a set of assumptions regarding the inputs and the agricultural practices considered. A report published by the ADAS [1] [27] evaluates various potential improvements in land productivity for a number of major crops in the main areas of the world. The most significant productivity improvements concern cereals for **Eastern Europe**, with a potential increase by 2020 of +71% for wheat and +73% for barley. For a given production, these improvements would result in savings of respectively **16 and 7.5 Mha** for these two crops alone. In view of the extent of land dedicated to monoculture, in **Asia**, a 34% increase in rice productivity, as considered in the report, would release **34 Mha**. **8 Mha** could be released on the **African** continent with a 42% increase in corn productivity alone.

A study conducted as part of the European REFUEL project [28], [29] also considers land release through improved animal feed. In a scenario taking into account these improvements, as well as reallocation of grassland to the production of energy crops, 19 Mha of grassland would be released in Europe.

Various yield improvement scenarios are considered depending on the assumptions made. This may lead to significant differences between the levels of improvement and therefore significant uncertainty concerning the area of land that can be released.

4.4.2.3 Protected or regulated buffer strips

Buffer strips are included in the requirements defined by the European Member States on the basis of the community guidelines concerning the "Good Agricultural and Environmental Conditions" (GAEC). Sensibly located buffer strips protect soils from the risks of erosion and limit the risks of diffuse pollution in soil and water. To comply with this requirement, farmers must dedicate an area of their land to buffer strips, equivalent to at least 3% of the aided area cultivated with cereals, oilseed and protein crops, fibre flax, hemp and set-aside land, and planted in priority as grass strips along watercourses. Grass strips were defined in France by decree and enforced in 2007. Buffer strips are located on plots of permanent grassland, temporary grassland of more than 5 years, set-aside land (environmental, wild fauna, flowered fallow) or on plots not dedicated to production (waste land, mountain

1. ADAS: Independent British provider of environmental and rural solutions for private and public organisations in the United Kingdom and abroad, http://www.adas.co.uk/

pastures, moorland, grazing land, wooded areas less than 5 m wide). Use of pesticides, fertilizers and organic supplements are prohibited on these areas throughout the year. In France, generalisation of this practice to all farmers could release nearly 380 000 ha of buffer strips, which could be used to grow pasture crops or SRCs for energy purposes. At European scale, this area could represent around 2 Mha.

In the United States, the study conducted by the NREL [13] considers growing dedicated lignocellulosic crops on farmland not suitable for food crops. This includes fallow land as well as Conservation Reserve Program (CRP) land. This a voluntary program set up by the USDA which offers financial and technical incentives to private landowners and/or farmers wanting to preserve soil and water resources as well as other natural resources on their land. The 2002 Farm Bill allows an area of about 15 Mha of CRP for 2007. The current effective area amounts to **11 Mha**.

Although there seem to be large areas of unused and available agricultural land, exploitation for non-food purposes is not without risks, both social and environmental. Durable exploitation of this type of land involves good knowledge of the stakes and in particular controlled environmental impacts, especially those related to modifications in the greenhouse gas balances induced by the land use changes, when developing renewable energy production systems.

4.4.3 New land exploitation balance

This chapter describes several axes for the development of new routes and new spaces potentially available for the production of lignocellulosic biomass. They include the development of forest exploitation concerning regions where the activity remains uncompetitive and/or where supply exceeds demand. In regions such as Europe and the United States, therefore, the forest wood stock potentially available for energy could reach 20 to 50 Mtoe. Agricultural areas which are under-exploited, either due to economic or environmental policies in the industrialised countries, or due to lack of production means, as in Eastern Europe and in the regions with difficult pedoclimatic conditions, are also taken into account. Development of production means to improve agricultural yields would release several tens of millions of hectares, while the area of abandoned land with agricultural potential is estimated at several million hectares. A certain degree of uncertainty nevertheless remains regarding its availability for the production of lignocellulosic biomass. In addition, numerous socio-economic and environmental obstacles are still to be removed.

4.5 LIGNOCELLULOSIC BIOMASS FOR ENERGY PURPOSES, CONCLUSION AND PERSPECTIVES

The bioenergy sectors using modern biomass conversion technologies are currently under development (heat, cogeneration, biofuels) and mainly use forest, and to a lesser extent agricultural, biomass residues. With the emergence of new sectors driven by the need for energy independence, awareness of the climatic stakes and the implementation of ambitious production objectives, biomass will play a central role. However, biomass obtained only from residues cannot meet the requirements of this demand, in a context of sustainable management of resources (dispersion, other agronomic uses, etc.). Recourse to lignocellulosic biomass dedicated to energy therefore seems inevitable. Tests are currently being conducted on new species but the level of development and commercialisation of these species, as well as the associated energy demand, is still limited. However, we can expect to see the emergence of major objectives for the development of biomass sectors in the medium term in Europe, North America and several countries in Latin America and Asia. This type of high-productivity resource (10 to 25 t dry matter/ha) raises new concerns including in particular control of the technical routes, social acceptance and the exploitation of new land, a point which is extremely important when assessing the production of this type of resource. Several species are currently being considered for cultivation on areas not affecting current market equilibria (food, industrial) or the sinks of natural resources such as carbon, biodiversity (forest and other protected areas). While it is estimated that these areas cover several hundred million hectares at world scale, local evaluations are nevertheless necessary to understand the various stakes involved in exploiting these areas and create sustainable biomass supply sectors. The European Commission and Parliament are currently in the process of defining sustainability criteria concerning the production and use of energy biomass sources and governing the eligibility of the sectors to help meet the production objectives. One strategy aims in particular to guarantee a minimum GHG emission reduction threshold under acceptable socio-economic conditions.

Prospective approaches can already be considered by defining more or less long term biomass mobilisation scenarios. The productions of forest and agricultural residues can be estimated to have the same order of magnitude depending on the sources. In contrast, the scenarios put forward concerning the mobilisation of land for energy crops may lead to widely differing results depending on the assumptions chosen. In Europe, forest resources are estimated at about **20 Mtoe** of primary resources and about **6 Mtoe for straw**. The RENEW [30] and REFUEL [28] European programmes define various land mobilisation scenarios for the cultivation of dedicated crops. The areas considered would cover 20 Mha for scenarios only based on limited changes in food crop yields, fallow areas and some of the areas for export outside the European Union. Adding the Ukraine potential and the ambitions for higher yield of the EU-12 countries amounting to 80% of those of the EU-15 countries, they may reach 50 Mha. These scenarios then lead to a dedicated biomass potential of **100 to 290 Mtoe** in Europe, including 1/3 in Ukraine alone.

In North America, the potential production of biomass from residues seems greater, with **50 to 60 Mtoe** of forest resources and **up to 120 Mtoe** of harvesting residues [11], [13]). In addition, the study conducted by USDA-DOE [11] puts forward several scenarios for the cultivation of dedicated perennial crops, with various types of land mobilised and several productivity levels expected. Between 16 Mha and 24 Mha of dedicated crops (*switchgrass,* willow and poplar SRCs) could be mobilised, including about 4 Mha of CRP [1] land, 10 Mha of temporary and artificial grassland and 2 to 10 Mha of arable land. With an average crop

1. CRP: Conservation Reserve Program, see previous chapter.

yield of 12 to 19 t d.m./ha/year and a 10% loss during harvesting, the total production of dedicated lignocellulosic crops would lie between **70 et 170 Mtoe/year**.

Note, however, that while these approaches can be used to assess resources in a given region, local feasibility studies are nevertheless required for all concrete projects. Implementation of this type of project involves thinking at the scale of the supply basin, or production basin, i.e. at local, not regional, scale. This constraint is imposed by the availability of the resource and the need for a homogeneous and continuous supply.

REFERENCES

[1] Kauter D, Lewandowski I and Claupein W (2003) Quantity and Quality of Harvestable Biomass from Populus Short Rotation Coppice for Solid Fuel Use – A Review of the Physiological Basis and Management Influences. Biomass & Bioenerg 24, pp 411-427.

[2] SOLAGRO, ADEME, IFN (2004) Évaluation du potentiel bois énergie en France, résultats disponibles sur le site www.boisenergie.ifn.fr.

[3] Tharakan PJ, Volk TA, Abrahamson LP and White EH (2003) Energy Feedstock Characteristics of Willow and Hybrid Poplar Clones at Harvest Age. Biomass & Bioenergy 25, pp 571-580.

[4] Börjesson P et Berndes G (2002) Reduced CO_2 Mitigation Costs by Multi-functional Biomass Production. Publication présentée à la 6e Conférence mondiale sur les Technologies de Contrôle des Émissions de Gaz à Effet de Serre, Kyoto, Japon, 2002.

[5] AGRICE (1998) Les cultures lignocellulosiques et herbacées pour la production de biomasse à usage non alimentaire. Dossier technique réalisé dans le cadre d'une étude AGRICE.

[6] Bonnet JF (1998) Les apports solaires implicites dans les activités humaines. Thèse de Doctorat de l'École Nationale Supérieure d'Arts et Métiers, Talence.

[7] Food and Agriculture Organization of the United Nations, Land and Water Development Division, Rome (2000) World Soil Resources Reports, Land Resource Potential and Constraints at Regional and Country Levels.

[8] FAOSTAT (2005) Banque de données statistiques de la FAO, www.fao.org, accès mai 2005.

[9] CLIP, Cahier du CLIP n° 8, (janvier 1998) Le bois énergie en Europe, Évaluation du potentiel mobilisable à l'horizon 2020, impacts sur l'environnement global et conditions socio-économiques de sa mobilisation.

[10] METLA, Working Paper of the Finnish Forest Research Institute (2004) Estimation of Energy Wood Potential in Europe.

[11] Perlack RD, Wright LL, Turhollow AF, Graham RL, Stokes BJ and Erbach DC, USDA-DOE-ORNL (April 2005) Biomass as Feedstock for a Bioenergy and Bioproducts Industry: the Technical Feasability of a Billion-ton Annual Supply.

[12] EEA (2006) How Much Bioenergy Can Europe Produce without Harming the Environment? EEA Report No 7/2006.

[13] Milbrandt A (2005) A Geographic Perspective on the Current Biomass Resource Availability in the United States. Technical Report NREL/TP-560-39181, December.

[14] Lal R (2005) World Crop Residues Production and Implications of its Use as a Biofuel. Environment International 31, pp 575-584.

[15] Kline KL, Oladosu GA, Wolfe AK, Perlack RD and Dale VH; ORNL (Oak Ridge National Laboratory) and McMahon M; Appalachian State University (February 2008) Biofuel Feedstock Assessment For Selected Countries. ORNL/TM-2007/224.

[16] Tharakan P, Bake B, Montecalvo R, Gadde B and Weingart J, United States Agency for International Development (USAID) (2009) Biofuels in Asia – An Analysis of Sustainability Options. Environmental Cooperation-Asia Clean Development and Climate Program.

[17] AFOCEL, Fiches d'identité papetière.

[18] ADEME (1999) Bois-énergie: chaufferies à alimentation automatique.

[19] Richardson J, Björheden R, Hakkila P, Lowe AT and Smith CT (2002) Bioenergy from Sustainable Forestry, Guiding Principles and Practice, Kluwer Academic Publishers.

[20] Ginisty C, Vallet P, Chabé-Ferret S, Levesque C et Chauvin C (Octobre 2007). Disponibilités en biomasse forestière pour des usages énergétiques et industriels en France. Convention DGFAR/ Cemagref E 19/06 du 20 octobre 2006.

[21] Edwards RAH, Šúri M, Huld TA and Dallemand JF, JRC (2005) Gis-Based Assessment Of Cereal Straw Energy Resource in the European Union. 14th European Biomass Conference, 17-21 October 2005, Paris, France.

[22] IFN (2005) Site de l'Inventaire Forestier National, www.ifn.fr.

[23] Berndes G, Azar C, Kåberger T and Abrahamson D (2001) The Feasibility of Large-scale Lignocellulose-based Bioenergy Production. Biomass & Bioenergy 20, pp 371-383.

[24] Field CB, Campbell JE and Lobell DB (2008) Biomass Energy: the Scale of the Potential Resource. Trends in Ecology & Evolution **23**, 2, February 2008, pp 65-72.

[25] Houghton RA *et al.* (1991) Current Land Use in the Tropics and its Potential for Sequestering Carbon. In: C. Howlett and C. Sargent, Editors, Technical Workshop to Explore Options for Global Forestry Management, International Institute for Environment and Development, p. 127.

[26] Houghton RA (1990) The Future-role of Tropical Forests in Affecting the Carbon-dioxide Concentration of the Atmosphere, Ambio, **19**, pp 204-210.

[27] Kindred *et al.* (2008) Anticipated and Potential Improvements in Land Productivity and Increased Agricultural Inputs with Intensification. ADAS.

[28] Fischer G Hizsnyik E, Prieler S and Van Velthuizen H, IIASA (2007) Assessment of Biomass Potentials for Biofuel Feedstock Production in Europe: Methodology and Results. Refuel Project, WP2. Biomass Potentials for Biofuels: Source, Magnitudes, Land Use Impacts.

[29] Fischer G. *et al.* (2009) Biofuel Production Potentials in Europe: Sustainable Use of Cultivated Land and Pastures, Part II: Land Use Scenarios, Biomass and Bioenergy, doi: 10.1016/j.biombioe.2009.07.009.

[30] Pisarek M, Ganko E, Kunikowski G, Rutkowska M, Szklarek M, Rogulska A, Ericsson K, Lantz M, Witt J, Panoutsou C, Nikolaou A, Faist Emmenegger M, McDonell K, Kennedy A, Buttle D and Blackmore J, RENEW Renewable Fuels for Advances Power Trains (2006) The Review – Biomass Resources And Potentials Assessment – Regional Studies And Experiences, Deliverable Wp5.1/D 5.1.1.

5

Conversion of lignocellulosic biomass by thermochemical pathway

D. Ballerini, L. Bournay, F. Broust, P. Rousset,
S. His, E. Marty, X. Montagne, J.M. Seiler

Thermochemical conversion of lignocellulosic biomass is a highly promising opportunity in terms of biofuel and bioenergy production.

This chapter describes the main thermochemical biomass conversion pathways, i.e. pyrolysis, torrefaction, gasification and hydrothermal conversion, the products obtained, as well as the synthesis of liquid fuels and their main characteristics.

5.1 PYROLYSIS

5.1.1 Definition – Principle

Etymologically, the word is derived from the Greek *pyr* "fire" and *lysis* "separating". Pyrolysis is the primary decomposition process of organic material at high temperature in the absence of oxygen. Increasing the temperature produces complex chemical reactions which decompose the main components of the biomass (cellulose, hemicelluloses and lignin) into three main products: a solid (biochar), condensable vapours (bio-oils and water) and combustible permanent gases (CO, H_2, CH_4 and other light hydrocarbons C_2-C_3) and non-combustible gases (CO_2). By controlling the operating parameters, these reactions can be biased towards preferential production of one of these three phases to maximise their yields (Table 5.1).

Broadly speaking, we can identify three types of pyrolysis, depending on the heat and mass transfer conditions (Figure 5.1):

- "slow" pyrolysis or carbonisation producing mostly char,
- "fast" or "flash" pyrolysis giving mostly liquid products (pyroliquefaction),
- fast pyrolysis with cracking, giving mostly gases (pyrogasification), the most rarely used pathway, which will not be detailed in this document.

Table 5.1 Operating conditions, weight yield (on dry weight of biomass) and energy yield (ratio of energy contents) obtained by wood pyrolysis [1].

	Slow pyrolysis Carbonisation	Fast pyrolysis	
		Pyroliquefaction	Pyrogasification
Typical operating conditions	Heating rate < 50°C/min Final temperature < 600°C	High heat flux density (> 100 kW/m^2) Gas temperature < 500°C Vapour residence time < 2 s	High heat flux density (> 100 kW/m^2) Gas temperature > 800°C Vapour residence time > 5 s
Char yield (%)	**30-35**	10-15	10
Liquid yield (%)	30	**70-75**	5
Gas yield (%)	30-35	10-15	**85**
Energy yield	Char/biomass 60%	Liquids/biomass 70%	Gas/biomass 80%

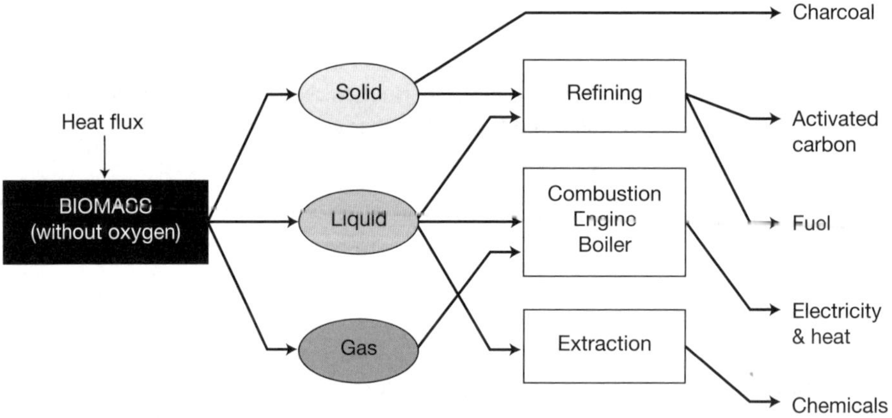

Figure 5.1

Major products obtained by biomass pyrolysis according to the temperature and heating rate conditions and their uses [2].

5.1.1.1 Slow pyrolysis

Slow pyrolysis, or carbonisation, is a very old technique used to produce biochar. It is characterised by low biomass heating rate (< 50°C/min) up to reaction temperatures of less than 600°C and residence times for the solid of a few hours.

The char yields depend on the pyrolysis conditions and do not generally exceed 35% by weight (as a ratio of the totally anhydrous wood) at atmospheric pressure, the remaining products being composed of about 30% of condensable products (tars, pyroligneous liquids) and 35% of non-condensable gases (mostly CO_2, CO, H_2, hydrocarbons) [2-4]. In this case, we obtain a maximum energy yield of 60%. Higher yields can be reached if pyrolysis is carried out under pressure (several MPa), with a theoretical limit in the region of 55% by weight [5-9] and energy yields of 75%. The condensable gases and products can also be recycled and burnt to make the processes energetically autonomous. These results are encouraging for the development of new carbonisation processes.

The density of biochar lies between 100 kg/m^3 and 190 kg/m^3 [10]. It has a net calorific value of between 28 and 33 MJ/kg. Currently, charcoal is mainly used for domestic heating in DCs, the barbecue market and production of activated charcoals in Europe as well as metallurgy (especially in Brazil). In view of the good conversion energy yields, this option is also being considered as an intermediate biomass conditioning step.

5.1.1.2 Fast pyrolysis

Developed more recently (in the 1980s), fast biomass pyrolysis is characterised by much higher heat flux densities and/or higher temperatures. Under these conditions, pyrolysis produces little char, but mainly volatile products liquid or gaseous whose relative proportions depend on the operating conditions.

In the 1970s, fast pyrolysis gained interest mainly for the production of fuel gases (pyrogasification) rich in CO and H_2 as well as light hydrocarbons (methane, acetylene). Finally, this pathway never led to an industrial application, being replaced by the gasification process (see Chapter 5.3).

Later, in the early 80s, studies on ablative pyrolysis (concept described elsewhere) demonstrated that, under specific conditions, excellent liquid yields could be obtained [11, 12]. These studies were the precursors of fast pyrolysis processes (pyroliquefaction) based on a new concept, designed to produce liquids: pyrolysis oils (or bio-oils). The prospect of obtaining liquids from biomass which could be used as biofuels with high weight and energy yields has triggered numerous research studies focusing on the development of suitable processes and on the characterisation of these highly complex liquids as well as the possible conversion pathways [2, 13].

5.1.1.3 Interest of pyrolysis

One of the main benefits of pyrolysis is that it potentially represents a biomass homogenisation and therefore preconditioning step. From a feedstock of heterogeneous lignocellulosic materials difficult to use in their current state, pyrolysis can produce solid or liquid energy intermediates which are more homogeneous and transportable, such as charcoal and in particular pyrolysis oils. These oils, of higher energy density, can then be transported for use as fuels or raw materials to industrial plants for the potential production of chemical compounds and/or biofuels. In the special case of second-generation biofuels, a sector which require extensive resources, this preconditioning step is extremely important [14]. Torrefaction, which can be defined as slow controlled pyrolysis, is another possible preconditioning process which offers the advantage of guaranteeing better energy yields. This pathway is discussed in Chapter 5.2.

In order to produce biofuels, preconditioning in liquid state by flash or fast pyrolysis offers a number of advantages in terms of storage, transportation and injection. A broader range of applications can therefore be considered and the existing infrastructures used for the liquid fuels. The remainder of this chapter focuses on fast pyrolysis. It describes the physico-chemical phenomena governing pyrolysis before reviewing the main processes considered and the application perspectives for pyrolysis oils.

5.1.2 Basic pyrolysis mechanisms

When a solid biomass grain is heated, its temperature increases more or less quickly and uniformly, depending on the heat transfer conditions (external and internal). Apart from drying, the various biomass components will then undergo chemical conversions which depend on the local temperature of the solid. The reaction mechanisms are extremely complex and lead to a very wide variety of products (several hundred). Studies have been conducted on the main isolated constituents (cellulose [15-20], hemicelluloses [18-21] and lignin [11,22-24]) and on the biomass itself.

5.1.2.1 Pyrolysis of cellulose

Numerous fundamental studies can be found in the literature on the pyrolysis of cellulose, the main constituent of biomass (up to 50% by weight) and the easiest to isolate in its original form. Several more or less detailed reaction diagrams have been proposed in the literature, the most concise being that of Diebold [25, 26].

The primary reactions of cellulose pyrolysis involve two competing pathways. Above 150°C, dehydration, decarboxylation and decarbonylation reactions reduce the degree of polymerisation, forming mainly water, CO_2 and biochar [27], whose yield is improved by the presence of minerals [28]. The second pathway, favoured at higher temperatures (> 300°C), consists of depolymerisation of cellulose by transglycosylation, leading to the formation of oligosaccharides [29]. These oligomers are in liquid state during pyrolysis [11, 30], Active Cellulose (AC) or Intermediate Liquid Compound (ILC) [31] and can either continue to be depolymerised down to levoglucosane and other sugars, or form char [32]. In addition, competing fragmentation reactions lead to the formation of hydroxyacetaldehyde, other aldehydes, ketones and furfural derivatives.

These primary compounds, in vapour state, may then undergo numerous secondary reactions in gas phase, producing simpler secondary vapours, non-condensable gases (CO, CO_2, H_2, light hydrocarbons) and, in contact with the primary char, a secondary char.

5.1.2.2 Pyrolysis of hemicelluloses

Fewer studies have been conducted on the hemicelluloses, whose structures, while similar to that of cellulose, are more variable (xylan, mannan, etc.). The reaction mechanisms are comparable to those involved with cellulose, but of higher reactivity, with degradation from 150°C and complete decomposition at about 350°C. The main products, including furan type compounds, [18] and acetic acid, are water soluble.

5.1.2.3 Pyrolysis of lignins

Lignin, a polymer with aromatic backbone, decomposes according to mechanisms that are still poorly defined. Its highly complex structure varies depending on its origin and the separation methods used. Lignin decomposes over a very wide range of temperatures: slight degradation is observed at 250°C, but pyrolysis is only complete at higher temperatures (> 500°C) [15].

Further to complex cyclisation, condensation, polymerisation and cracking reactions with dehydrogenation, the products obtained are mainly monomers (coumaryl, coniferyl and sinapyl alcohols) formed at low temperatures, substituted aromatic compounds (e.g. isoeugenol, vanillin, vinylguaiacol, etc.), phenolic compounds such as catechol, gaiacol, methyl gaiacol, as well as methane at higher temperatures [17]. Other aromatic, or even polyaromatic, compounds are formed at higher temperatures in gas phase (550-800°C) [15, 17, 23, 33]. Lastly, the vapours from lignin react strongly with the primary char, forming secondary char, CO and CO_2 [34], which is why, generally speaking, biomasses with high lignin content form the most char.

5.1.2.4 Pyrolysis of lignocellulosic biomass

Numerous authors have studied biomass pyrolysis [15, 19, 21], based on investigations conducted on its majority constituents. Some studies have demonstrated that the behaviour of biomass could, under certain conditions, be represented by a linear combination of the behaviour of these constituents [35-37]. It is now accepted, however, that this approach is limited [38-39] due to the interaction with several minerals present in biomass (especially K^+ and Zn^{2+}), which catalyse in particular the formation of chars and/or gases, always at the expense of liquid yields.

Currently, no models can provide a chemical explanation regarding the formation of all the products formed, due to the diversity of the biomass compositions and the number of reactions involved. Normal practice is therefore to define just three product classes (chars, liquids and gases) and to distinguish between:

- the **primary reactions** occurring in solid phase, which consist in depolymerisation of macromolecules into highly unstable intermediate products which form biochar and volatile products,
- the **secondary reactions** of homogeneous cracking (in gas phase) and heterogeneous cracking (in contact with char) of vapours into permanent gases or even into other solid products (coke) depending on the conditions.

Despite these simplifications, there is still considerable controversy over the kinetic models to be adopted for lignocellulosic biomass pyrolysis. Global reaction diagrams of biomass pyrolysis have nevertheless been proposed, one of the most complete being illustrated in Figure 5.2. The notion of "pyrolysis severity" must be correlated with the biomass heating rate, the temperature in gas phase and the residence time of the primary vapours. The existence of tertiary reactions has been suggested to account for the formation of highly refractory soot and volatile organic compounds at very high temperatures.

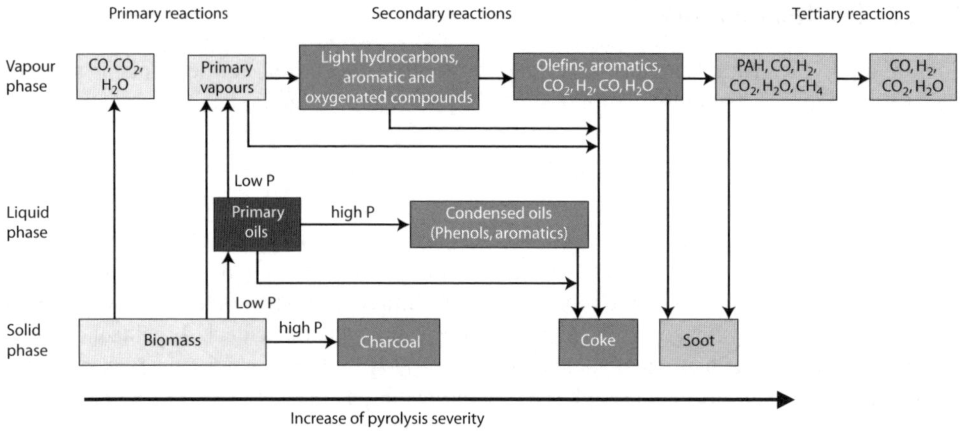

Figure 5.2

Global reaction diagram of wood pyrolysis [17].

5.1.2.5 Heat and mass transfers

Pyrolysis involves raising the temperature of the solid by external heating (conduction, convection and/or radiation). Within the solid grain, this causes temperature gradients which induce an internal heat flux mainly by conduction. These gradients may be very high and limiting due to the poor thermal conductivity of biomass. Inside the particle therefore, coupling or even competition between the heat transfers and the chemical reaction which is slightly endothermic takes place. The heating conditions, in the particular external heat transfer modes, and the granulometry of the particles are determining parameters in these competitions since they lead to more or less fast biomass heating rates and bias pyrolysis towards different product formation selectivities.

The most important mass transfer concerns the volatile species produced by pyrolysis within the solid and transferred towards the surface, mainly by convection (pressure gradient). Since the biomass and in particular the char produced are sufficiently porous, transfer of these volatile products does not represent a significant step limiting the overall pyrolysis process. However, this mass flow may be important when the residence times of the vapours are large enough to generate homogeneous or heterogeneous cracking reactions in the char layer.

5.1.3 Main pyrolysis reaction parameters

Several parameters have a significant impact on the composition and quality of the reaction products.

5.1.3.1 Heat fluxes imposed on the solid

Given that the biomass is being converted by thermochemical pathway, the reaction temperature is a determining parameter in the pyrolysis processes. Nevertheless, due to the heat transfer limitations inside the particle, this reaction temperature will depend not only on the heat fluxes imposed to the particle but also on its granulometry. These parameters will determine the rate of temperature increase inside the solid and therefore the proportion of char and vapours formed.

Low heat fluxes increase the temperature of the solid slowly, thereby favouring the formation of char. Inversely, high heat fluxes raise the biomass temperature quickly and therefore favour the formation of vapours, but this condition is not sufficient. For all the biomass to be heated quickly, the granulometry and humidity of the particle are determining parameters.

5.1.3.2 Granulometry and humidity of the raw material

Granulometry has a significant impact on the quality of the products manufactured. Apart from the special case of ablative pyrolysis, described later, the production of char is favoured by large granulometries. This is mainly due to the internal heat transfer, mentioned earlier, which limits the particle heating rate. The other reason is related to the high granulometry which forms a diffusion barrier for the primary vapours, held in the solid biomass matrix at high temperature and which react to form secondary gaseous and solid products.

Consequently, while slow pyrolysis can cope with pieces of biomass a few centimetres long, fast pyrolysis requires:

- either fine grinding of the raw material (particle diameters less than 5 mm),
- or fast removal by abrasion of the outer layer of char formed (ablative pyrolysis).

The biomass may have a high water or moisture content (up to 50% by weight when untreated) which may vary considerably over time. Humidity presents a number of disadvantages for pyrolysis:

- it has a negative impact on the energy yield, as in all thermochemical conversions (latent heat of vaporisation),
- it limits the heat transfer phenomena and therefore increases the reaction times, sometimes in considerable proportions,
- it affects the product quality, especially that of the liquids obtained by flash pyrolysis, reducing their net calorific value and favouring phase separation.

It is therefore essential for the raw material to be relatively dry. In case of fast pyrolysis, a water content of less than 10-15% by weight is recommended, which involves a prior drying step.

5.1.3.3 Temperature and residence time in gas phase

The residence time of the pyrolysis vapours can be set independently from that of the solid depending on the reactor hydrodynamics. For a given gas phase temperature, it determines the progress of the secondary cracking reactions and therefore the proportion of the liquid and gaseous fractions as well as the type of condensable compounds. A moderate temperature (< 450°C) together with a short residence time for the pyrolysis primary vapours (< 2 s) maximises the production of liquids. Conversely, a longer residence time for the primary vapours in the reactor, at high temperature, favours the secondary, and even tertiary, recombination and cracking reactions leading to the formation of gas and solids (coke and soot).

5.1.3.4 Pressure

Although less studied than the previous parameters, pressure nevertheless has a non-negligible influence on pyrolysis selectivity.

High pressures tend to favour the formation of char [6-9] at the expense of liquids and gases, since they limit diffusion transfer of vapours within the solid matrix and favour the secondary cracking reactions with the char being formed. Under slow pyrolysis conditions, solid yields of up to 40% by weight can be obtained, at a pressure of 10 bars, with similar char quality. Conversely, pyrolysis carried out under low pressure conditions (0.15 bar) favours the production of oils at the expense of gas and char [40].

5.1.4 Fast pyrolysis processes

5.1.4.1 Operating conditions required

The following conditions are required to increase the liquid yields [2,13]:

- use biomass with low moisture content,
- impose high heat flux density on the particle (> 10^5 W/m^2 if possible), generally via a heat exchange medium or a hot wall,
- reduce the particle size (< 5 mm) or rapidly eliminate the primary products on the surface (ablative pyrolysis),
- keep the gas phase temperature between 400°C and 450°C,
- limit the residence time of the pyrolysis vapours (less than 2 s) at these temperatures in the reactor, to minimise the secondary reactions, then cool the vapours quickly.

5.1.4.2 General principle of the process

Figure 5.3 shows the main steps of a pyrolysis process. After transportation from the collection site, the biomass is pre-dried to reach a 10% moisture content then ground more or less finely depending on the process, which includes following steps:

- fast pyrolysis with controlled residence time and temperature,
- separation of the solid phase and the hot vapours,
- quenching and condensation of the vapours and aerosols and recovery of the pyrolysis oils.

The by-products (char, gas) are generally used in the integrated processes, as energy supply to heat the reactor, and sometimes for the drying. In some cases, the char can be remixed with the pyrolysis oils to form a slurry which is then the final product.

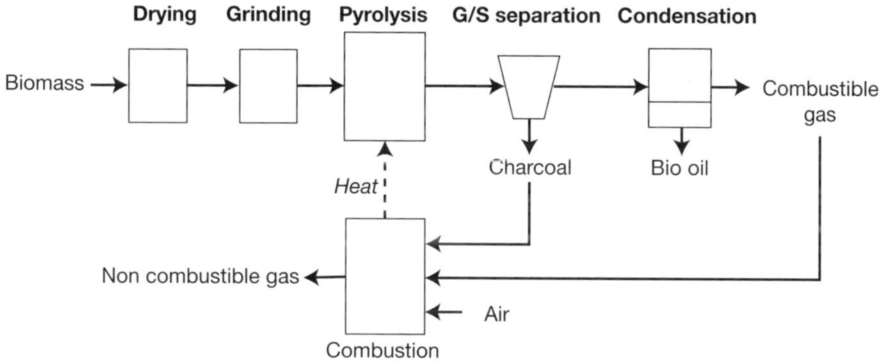

Figure 5.3

Schematic diagram of the fast pyrolysis processes.

5.1.4.3 Main reactor technologies

Numerous types of reactors have been developed or adapted to meet the specific constraints of flash pyrolysis, the main one concerning the high heat flux densities required (500 kW/m^2). Few techniques are capable of providing these high fluxes at industrial scale. Conduction and convection are the two dominant heat transfer modes, implemented in fast pyrolysis. Each one can be optimised depending on the reactor type.

Numerous bibliographic summaries list the flash pyrolysis technologies [13], the main ones still used today being detailed below. They can be classified depending on whether, or not a heat exchange medium is used. Table 5.2 lists the advantages and disadvantages of the various types of reactors, indicating the main heat transfer mode used.

A. Fluidised beds

The fluidised bed consists of a hot inert fluidisation material (generally sand) kept in suspension with a rising flow of carrier gas, as inert as possible (recycling of gases at industrial scale) into which the biomass particles are introduced. We can identify several fluidisation regimes, defined by ascending order of the gas speed, as being in dense, bubbling, fluid and entrained beds.

Fluidised beds offer the advantage of allowing very good heat transfers by convection with the pre-heated transport gas and/or conduction with the sand, but also of having a homogeneous and controllable bed temperature. In addition, they are widely used in process engineering, robust, simple to implement and easy to extrapolate with no real technological barriers [13]. However, these methods present a number of disadvantages, the main ones concerning the raw material granulometric requirements (< 5 mm) and the high carrier gas flow rates resulting in significant dilution of the vapours and therefore the need for an efficient condensation system.

Numerous companies have selected bubbling fluid bed for their development potential, based on the RTI technology, whose principle is summarised on Figure 5.4. For example,

Table 5.2 Main characteristics of the various types of reactors used in fast pyrolysis.

Reactors	Modes of heat transfer to the biomass	Advantages/Disadvantages/Characteristics	Developers
Ablative	Conduction by direct contact with the hot walls	– Transport of solid particles by gaseous fluid, not always necessary. – High mechanical abrasion of char; energy input a problem; no heat transfer by gaseous route. – Compact design; high granulometry possible.	CNRS Nancy Aston University NREL Pytech
Dense fluidised bed	Conduction with hot bed predominant Convection with the preheated carrier gas	– High heat transfer; energy input directly in the bed or in the fluidisation gas; limited abrasion of the char. – Limited biomass granulometry (6 mm). – Simple reactor design.	Union Fenosa Dynamotive RTI Wellman Fortum
Circulating fluid bed	Conduction with the hot bed with a larger contribution from convection with the preheated carrier gas	– High heat transfer; thermal cracking of vapours possible; catalytic activity possible with the char. – High solid particle contents in the oils; char and sand to be separated; limited biomass granulometry (maximum 6 mm). – Upscaling possible.	Ensyn CRES Fortum
Rotating cone	Conduction with hot sand bed predominant	– Limitation of carrier gas requirements; good energy integration by recycling and combustion of the char entrained with the sand. – Complex technology; moving parts at high temperature; difficult upscaling. – Limited control of hot vapour residence time.	BTG
Twin screw	Conduction with hot sand bed predominant	– Limitation of carrier gas requirements, good energy yields – Uncertain oil quality due to limited control of the hot vapours and relatively slow pyrolysis – Process exclusively biased towards the production of slurries; no large scale demonstration as yet.	FZK
Under vacuum	Mainly by radiation and conduction	– High granulometry; low concentrations of fines in the oils. – Low heat transfer; complicated technology; lower oil yields (< 50% by weight depending on the raw material).	Pyrovac

Union Fenosa and Wellmann which built 200 kg/h pilot units, in Spain [41] and the United Kingdom [42] respectively. The current reference, however, is Dynamotive, since it has built several units in Canada (BioTherm™), the most recent being those of West Lorne and Guelph, respectively 100 and 200 t/day [43,44], which are still in operation.

The best yields are about 75% on dry base using biomass with low ash content but they depend on the type of resource. They are obtained with bed temperatures of 475-500°C and hot vapour residence times of about one second [13].

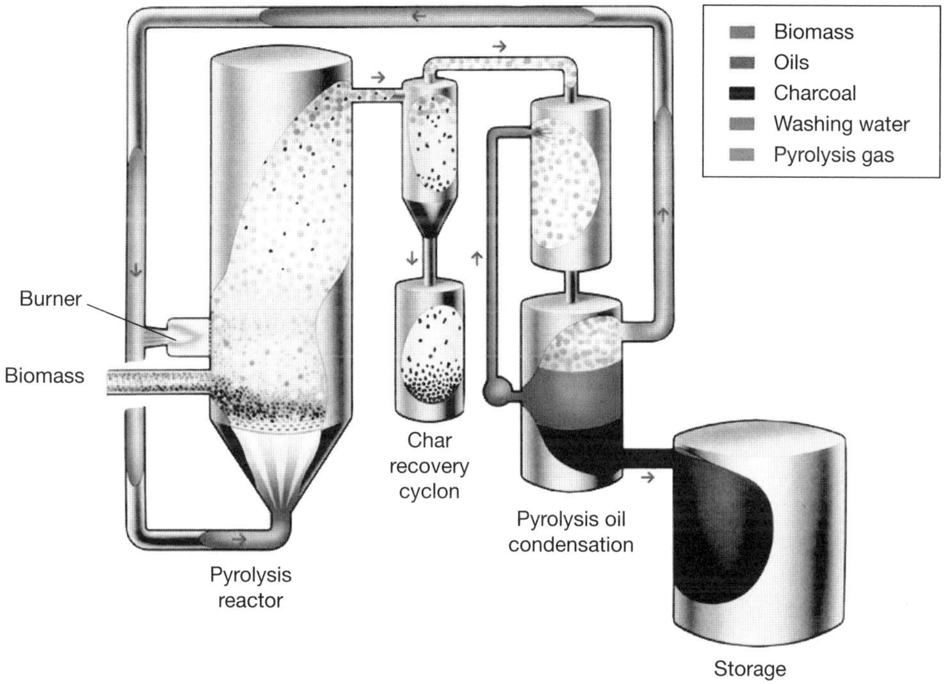

Figure 5.4

Schematic diagram of the RTI pilot process (Dynamotive/Canada) [45].

In circulating fluid bed type reactors, the higher gas speeds entrain sand and char outside the reactor. The char is burnt in a second reactor and used to heat the sand before reinjection in the bed. These processes have been developed at pilot scale and commercial scale, especially by Ensyn (Figure 5.5) which supplied a 20 kg/h unit to VTT in Finland [46] and a 650 kg/h unit to ENEL in Italy [47]. This technology is also used in several commercial units of capacities ranging from 1 to 1.5 t/h [48], by food flavour producers such as Red Arrow in the United States and Canada.

The disadvantages of this type of technology, well known in the industry, are its technological complexity, greater than the dense fluidised bed, and the high risk of wear. The high circulation speeds also cause attrition of the char particles and even of the bed. The oils therefore have a high solid micro-particle content when the condensable vapours are not correctly filtered. The only real advantage compared with the dense fluidised bed is the relatively easy extrapolation to very high capacities and the easy conversion of the carbon residue by direct combustion in the heat exchange sand.

Neste Oy has also developed a circulating fluid bed process at pilot scale [49].

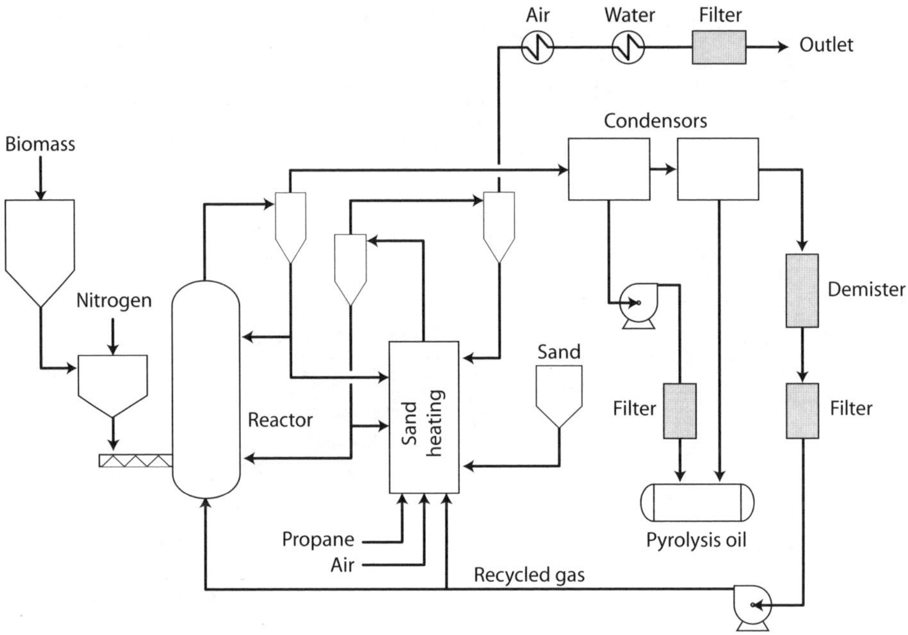

Figure 5.5

Schematic diagram of the Ensyn fluidised bed process [13].

B. Mechanically transported beds

a) Rotating cone

To overcome the problem of fluidisation gas, processes based on the principle of mechanical transport have been developed. This is the case in particular of the rotating cone developed at Twente University [50] then developed at pilot scale by BTG.

Figure 5.6 is a diagrammatic representation of the reactor. The biomass (diameter < 2 mm) and the hot sand (550°C) are introduced at the bottom of the conical reactor. Under the effect of the centrifugal force generated by rotating the cone (300 r.p.m.), the solid particles are transported upwards and rapidly mixed to trigger the pyrolysis reaction. The char produced, the ash and the cool sand are evacuated via the top of the cone and feed a combustion reactor where the char is burnt. The energy supplied by this combustion is used to heat the sand before it is reinjected in the pyrolysis cone. The pyrolysis vapours are cooled in a water tower before being stored. The oil yields are in the region of 70-75% by weight.

Two pyrolyser models have been built, the BTG 50P and the BTG 200P, of biomass capacities 50 and 200 kg/h respectively. Units of about 50 kg/h have been supplied to China and a 50 t/d demonstration unit has been built in Malaysia, operating on palm oil residues [13, 51]. Few data are available on this last facility.

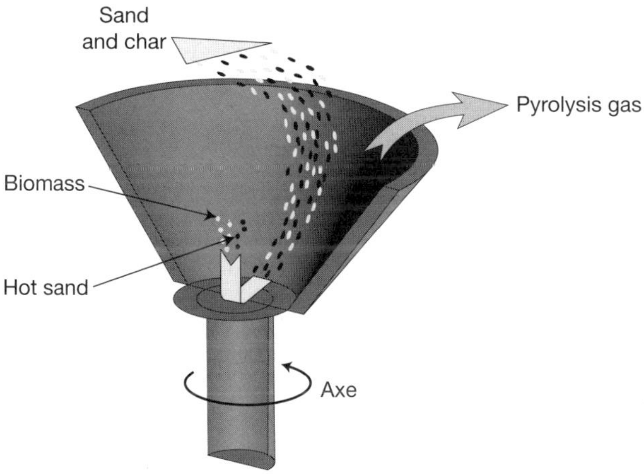

Figure 5.6

Diagrammatic representation of the rotating cone pyrolysis reactor developed by BTG [52].

b) Twin-screw reactor (Figure 5.7)

Initially developed by Lurgi-Ruhrgas (LR mixer reactor) to crack bitumens and residues and adopted by FZK (Karlsruhe University, Germany), this technology consists of a parallel twin screw rotating at between 1 and 4 r.p.m., and producing a mechanical mixture between hot sand and the Bioliq® biomass. The sand/biomass ratio can vary between 5 and 20. The sand is heated to 500-550°C by the pyrolysis gases or by an external heat source.

The residence time of the vapours is a few seconds [53-55]. The residual solid is entrained with the vapours and separated in a cyclone. The vapours are condensed in 2 steps, a dense fraction at about 100°C and an aqueous fraction at about 20°C. One of the advantages of this process is that it eliminates the neutral fluidisation gas. In addition, since the screws rotate at low speed, the mechanical forces and the risks of rapid wear are limited. The principle was successfully tested on straw in 2003. A β unit with a capacity of 500 kg/h was designed for coupling to an entrained flow gasifier, but at this stage there is still insufficient hindsight with respect to the potential of this technology at large scale.

The quality of the crude oils is poor since the residence times of the hot vapours are quite long. However, this type of process is characterised by the fact that the pyrolysis oils are remixed with the char to form a slurry, which has the advantage of increasing the overall conversion yield [54-55].

C. Ablative processes

The principle of these processes is based on heat transfer by surface conduction between the biomass and the hot walls of the reactor. The biomass is pressed against a hot spinning surface which literally melts the raw material, releasing the pyrolysis vapours, which are rapidly evacuated from the reaction zone before being condensed. The advantage of this type of reactor is that large particles can be used while retaining excellent heat transfer.

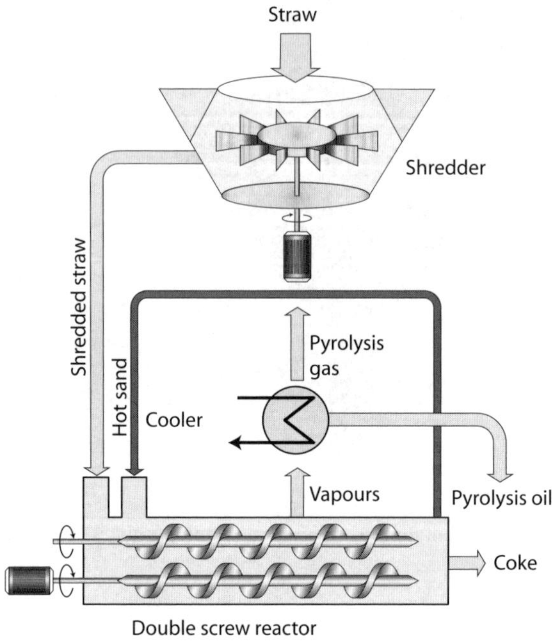

Figure 5.7

Diagrammatic representation of the twin screw pyrolysis reactor developed by FZK [53].

Exploratory studies on ablative pyrolysis have been conducted by CNRS in France on a hot spinning disc. They correlated the speed of ablation with the pressure exerted between the particle and the hot wall, the wall temperature and its relative speed with respect to the particle [12]. These studies also demonstrated that, under these conditions, flash pyrolysis of biomass was similar to fusion. The NREL (United States) later developed a vortex ablative reactor [13] where the particles are pressed against the walls at high velocity to derive high tangential pressures which provide the physical contact between the biomass and the hot reactor surface.

More recently, a 3 kg/h ablative reactor has been developed by Aston University, based on the principle of the horizontal hot spinning disc (Figure 5.8). A variant has been developed by PyTech in collaboration with Hamburg University [56, 57]. In this process, wood chips are pressed against a vertical heated abrasive disc by pistons. The vapours are rapidly tapped off and condensed. A 100 kg/h pilot was built in 2005 and is still in operation.

This variant of fast pyrolysis offers the advantage of proposing potentially more compact systems with high yield (between 60% and 75% by weight on dry feed) and avoiding the prior step of grinding into fine particles. These processes nevertheless remain fairly complex due to the presence of moving parts at high temperature (600°C) and their industrial feasibility has not been demonstrated due to the high construction costs.

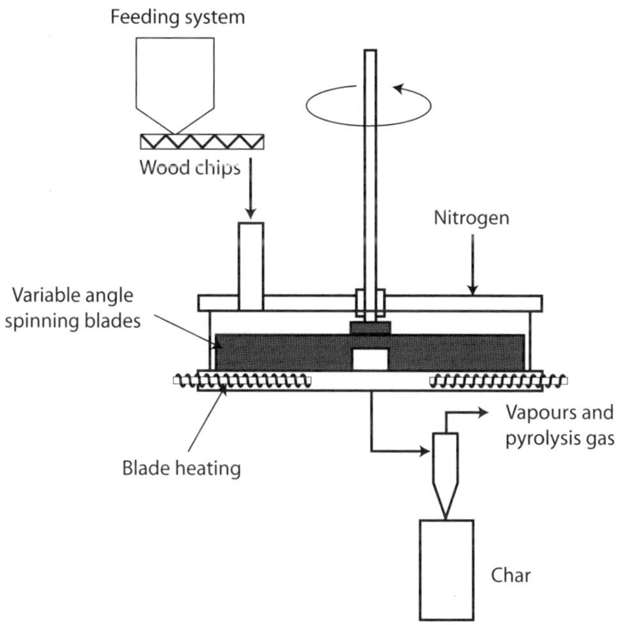

Figure 5.8

Schematic diagram of the ablative pyrolysis reactor designed by Aston University [13].

Cyclone reactors may also be classified as ablative reactors. The principle of this type of process is similar to that of the Vortex reactor. Originally developed by CNRS in France [58,59], it consists of a cyclone heated at its walls on which the biomass, transported by the carrier gas, is pyrolysed by direct contact. Its main advantage is that pyrolysis and separation of the char are performed simultaneously in a compact unit. It has only been studied at laboratory scale, however, although a variant has been developed at pilot scale by Twente University (Pyros reactor), which uses sand as heat exchanger and abrasive agent and includes a novel high-temperature filter system built into the cyclone chamber for efficient particle separation [60-61]. Although the oil yields and qualities are satisfactory, the perspectives for extrapolating this type of process to units of industrial size are limited. Use of sand impairs the initial simplicity of the technology, which in this case is similar to that of a transported bed reactor.

D. Low pressure (vacuum) reactors

The principle consists in keeping the reactor atmosphere at very low pressures, about 15 kPa [13]. Consequently, although the heating rates are relatively low and correspond to those encountered in slow pyrolysis processes, the vacuum generated in the process means that very short primary vapour residence times are possible. The vapours are therefore evacuated very quickly from the reaction zone, which avoids secondary reactions. Only one process of this type has been developed, in Canada. A 50 kg/h pilot was initially implemented for R&D

activities. A 3.5 t/d unit was then built in Quebec [62]. The facility has now been shut down without ever having really been tested. Figure 5.9 shows a schematic diagram of the process. The oil yields achieved by this process are about 50% by weight and always less than those obtained by fast pyrolysis processes.

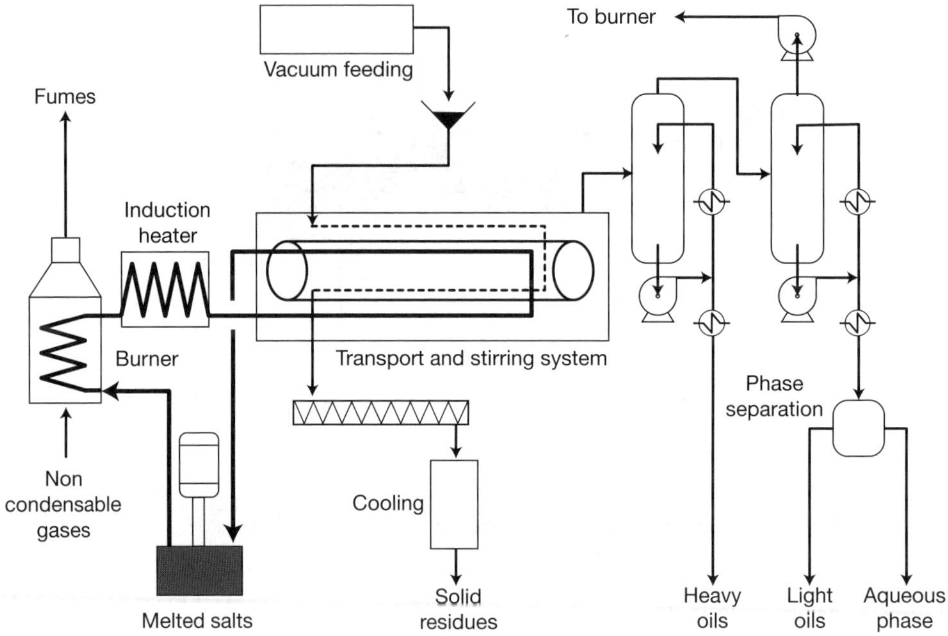

Figure 5.9

Schematic diagram of a vacuum pyrolysis process [13].

The advantages offered by this type of process are use of raw materials with large granulometry, absence of carrier gas for treatment of the gas phase and lower concentrations of solid particles in the oils. It is more difficult to make this technology viable when oil production needs to be maximised, however, due to its complexity related to limited heat transfers. Moreover, the oil quality is poor.

5.1.4.4 Separation and condensation techniques

Separation of char particles and condensation of oils are extremely important operations to guarantee good yields and in particular good bio-oil quality.

Cyclones are generally used to separate the char particles at reactor output. They are maintained at temperatures of about 400°C to avoid condensation and cracking of the vapours. However, the cut-off diameters achieved with cyclones are typically in the region of a few microns, which means that fine char particles will be entrained in the bio-oils.

These particles are likely to accelerate oil ageing and instability. Finer separation processes have been studied using the principle of hot vapour filtration, with metallic or ceramic filters. Good quality bio-oils can therefore be obtained, with a very low char content [25], but at the expense of oil yields (10-15% reduction) due to cracking of the vapours in contact with the accumulated char on the filter surface. NREL, VTT and Aston University are the main organisations which have worked or which are still working on these processes.

The condensable pyrolysis vapours must be cooled quickly to minimise secondary cracking reactions and avoid yield losses. Condensation is generally carried out by cooling the hot vapours with a quenching agent injected in a column. At industrial scale, cooled bio-oils are often recycled and used to act as quenching agent. Condensation may require large volumes when a carrier gas is used in the pyrolysis reactor, so as to limit the risks of losing the light fractions.

Condensables consist of a non-negligible fraction of submicronic aerosols which are generally not recovered in the condensers and are entrained by the gas. The most efficient and most widespread technology used to collect these aerosols is undoubtedly electrostatic filtration, installed in addition to the condenser, which also has the advantage of limiting pressure losses.

5.1.4.5 Mass and energy balances

The maximum bio-oil yields by weight reported are in the region of 70-75% on biomass with 10% moisture content. These values depend largely on the type of reactor and the initial composition of the biomass, especially its content in inorganic compounds which favour the production of gas or char at the expense of bio-oil yields. For raw materials with higher ash content such as bark and especially straw, the oil yields may drop down to 50-55%, always on dry feed basis.

From the energy standpoint, the maximum theoretical yield can be defined as being the oil/biomass energy content ratio. It is about 70% for flash pyrolysis. The overall energy efficiency for the entire process must take into account the energy required to heat the reactor and pre-dry the biomass. Although few real data are available, based on the studies conducted on the existing units energy balances nevertheless show that combustion of the by-products (gas and char) can make the process globally independent.

5.1.5 Pyrolysis oils (bio-oils)

5.1.5.1 Chemical composition

Pyrolysis oils are usually dark brown liquids with red or green tints depending on the type of raw material and the presence of microscopic carbon particles.

They contain several hundred chemical compounds in highly variable proportions, from low molecular weight compounds up to more complex, high molecular weight compounds. Their composition typically includes organic acids, sugars, alcohols, aldehydes and ketones, as well as phenols and gaiacols, furans, methoxylated compounds, mono- and polyaromatic hydrocarbons, and nitrogenated aromatic hydrocarbons. The composition in 12 chemical families of 13 oils produced by currently operational pyrolysis processes has been determined [63]. Oil compositions vary considerably, with variations ranging from 1 to 5 for a given family. Similarly, it is worthwhile pointing out that irrespective of the pyrolysis process (slow or fast) and the raw material (broad-leaf, energy crop or conifer), all chemical families are represented. Note that the highly toxic polyaromatic hydrocarbons (PAHs) are present in very low concentrations.

Bio-oils are complex matrices which generally require prior fractionations and several simultaneous analytical techniques to cover the broad range of constituents. Despite the significant progress made in analytical methods, mostly based on chromatography, a non-negligible proportion of unidentified constituents still remains. The composition of a non-water soluble organic fraction called pyrolytic lignin has not yet been fully determined, but its structure would suggest that it comes from partially converted lignin. Further investigations are therefore necessary to precisely determine their composition.

5.1.5.2 Physico-chemical characteristics and qualities of pyrolysis oils

Given the chemical complexity of the oils, it is general practice to only determine their most important physico-chemical properties, which must be known to identify the potential applications of these products. These properties represent good criteria for comparison with traditional fuels. Table 5.3 summarises the main properties of an oil produced from wood.

The composition of bio-oils is similar to that of biomass. They are characterised by high oxygen content, unlike the traditional liquid fuels such Diesel. The total oxygen content of the oils is due both to the presence of water and organic oxygenated compounds.

Table 5.3 Characteristics of pyrolysis oils compared with those of conventional fuels.

		Biomass	**Pyrolysis oils**	**Diesel**	**Heavy fuel**
Density	kg/m^3 at 15°C	550	1 220	850	960
Composition	% C	50	48.5	86.3	86.1
	% H	6	6.4	12.8	11.8
	% O	44	42.5	–	–
	% S	–	–	0.9	2.1
Viscosity	cSt at 50°C		13	2.5	350
Flash point	°C		66	70	100
Freezing point	°C		– 27	– 20	21
Ash	wt%	< 5	0.13	< 0.01	0.03
Sulphur	wt%	–	0	0.15	2.5
Water	wt%	20-60	20.5	0.1	0.1
NCV	MJ/kg	18	17.5	42.9	40.7
Acidity	pH	–	3	–	–

The calorific value is logically comparable to that of dry wood, given the similarity of their elementary compositions. It is low, however, compared to that of fossil fuels, due to the relatively low C/O ratio.

The density of the oils, about 1.2 kg/litre, is very high compared with that of the traditional liquid fuels. Combined with their calorific value, this feature confirms that the oils densify the energy content of the biomass. Note that this property also has direct consequences on the design of the equipment required to use these fuels, and in particular pumps.

Pyrolysis oils are characterised by their high water content, which may vary from 15% to 30%. In these proportions, they form a stable and homogeneous liquid phase. Above 30%, phase separation is generally observed [2, 13]. Consequently, the liquid is neither completely miscible with water nor fuels derived from oil.

The viscosity of liquid fuels is an important factor in most applications involving pumping operations. The viscosity of pyrolysis oils can vary from 25 cSt up to more than 1 000 cSt, depending on the water content, the quantity of light compounds and the oil storage time.

The content in ash or solid matter in suspension (fine particles of char or sand), found in the oils in variable proportions depending on the type of process, undeniably represents one of the most restricting constraints for their direct use, in addition to corrosion problems, in particular. These suspended particles also acts as catalysts and, depending on their types, they affect the oil stability by causing polymerisation. High-temperature ceramic filters seem to give good results in terms of particle filtration, but at the expense of oil yields.

Furthermore, once the pyrolysis oils have been collected, they can no longer be vaporised. Above 80°C, in fact, they rapidly form a solid carbon residue of about 50% by weight and a distillate containing primary and secondary products in addition to water. Due to this thermal instability, no distillation operations can be performed on the oils, which are also photo- and thermolabile.

5.1.5.3 Environment and safety

The perspective of large-scale production of bio-oils implies that they must meet applicable environment and health standards, especially as regards their handling, transportation and storage. It is therefore important to determine the impact of oils on flora and fauna, their toxicity and lifetime.

Various toxicology tests were conducted during the European BIOTOX project: eco-toxicity, toxicity and biodegradability [63-64]. The bio-oil test conclusions, irrespective of the production process and raw material, are as follows:

- non-toxic when orally ingested (tests on animals),
- low polycyclic aromatic hydrocarbon concentration (PAHs are known carcinogens), below 10 ppm,
- no eco-toxicity for fauna and flora; at low concentrations, bio-oil may have a small fertilizing effect on flora, while at higher concentrations growth is slightly inhibited although it was impossible to determine whether this was due to a toxic effect of the oils,

- biodegradability better than heavy hydrocarbons (index between 32% and 50%, as compared with 24%),
- mutagenic: for each oil, at least one out of the five strains of bacteria tested is positive to the test according to international regulations.

The study demonstrated that the oils are suitable for transportation, and exhibit average environmental behaviour in case of accident, especially in terms of their biodegradability. The safe handling and transportation conditions were defined after this study. In particular, Material Safety Data Sheets (MSDS) were produced. Although the oils are not explosive, they must be classified as hazardous product for transportation, due to the presence, even in small quantities, of volatile products. A standardisation procedure has therefore been initiated to reference bio-oils as a specific product and increase their quality control means.

5.1.6 Potential applications for pyrolysis oils

Over the last 20 years, several laboratories and companies throughout the world have studied various potential applications for pyrolysis oils. Most of these studies have been conducted at pilot scale for three main applications: as fuel in boilers and gas turbines, as fuel in Diesel engines and as biofuel production intermediate. The chemical conversion pathway has also led to extensive research work at demonstration, and even commercial scales. The latter pathway will not be discussed in this chapter, however.

5.1.6.1 Combustion in boilers

Low-power boilers are widely used to produce heat and electricity for domestic and industrial purposes. Although less efficient than motors and turbines, they can burn a large range of solid and liquid fuels, especially domestic and heavy fuel oil.

Tests conducted on combustion of pyrolysis oils in boilers have identified a number of critical points, generally concerning the viscosity, acidity and instability of the oils. This leads to a number of difficulties regarding handling, design of atomisers, behaviour during ignition, flame stability as well as the presence of pollutants in the combustion flue gases. Unlike domestic fuel oil, therefore, with pyrolysis oils the supply line must be preheated right up to the storage tank to reduce their viscosity. The preheating temperature lies between 40°C and 90°C depending on the burners used and remains limited to prevent degradation of the oils by polymerisation.

Clogging of the lines and atomisation systems has been eliminated by technical innovations such as implementation of particle filters, increasing the diameter of the atomisation holes, addition of solvents such as ethanol, use of comburent air/vapour mixture [65] or use of atomisers operating under pressure (between 5 and 10 bar) [66]. The latter solution seems to be the most promising due to its simplicity of use and low cost.

Lastly, due to the acidity of the oils, some components (pumps, atomisers, etc.) must be made from stainless steel to avoid corrosion.

The boiler operating modes require no modifications in steady state regime compared with boilers fired on conventional fuels. Some precautions are required when starting up and shutting down the boilers, such as initial preheating of the installation to allow auto-ignition of the pyrolysis oils in the combustion chamber and draining the oil supply circuits when the boiler is not in use (generally by injecting ethanol) to avoid the possibility of coking which would clog the atomisers and generate solid carbon particulates in the smoke.

Lastly, concerning polluting emissions, NO_X emissions caused by the combustion of pyrolysis oils lie between 140 and 300 ppm and are generally less than those obtained with conventional fuels. The lower flame temperature and the presence of oxygen in pyrolysis oils could explain this reduction in NO_X contents. The CO contents, between 30 and 50 ppm, although generally greater than those obtained with conventional fuels, remain compatible with applicable regulations.

5.1.6.2 Direct use in Diesel and Stirling engines

Diesel engines represent a more interesting pathway for the production of electricity than that based on the steam cycle, since the production yields are higher (up to 45%). The specifications concerning fuel quality are more restricting, however, and vary depending on the engine constraints.

Use of pyrolysis oils for these applications is technically feasible, but involves a number of problems inherent to the specific nature of this fuel [67, 68]. For example, corrosion of the systems in contact with the oils (e.g. pumps and injectors) was rapidly identified as being a problem. At the same time, coking of the oils in the supply circuits periodically blocks the injectors, a problem which has not yet been resolved. More specifically, engines running at high speed require high auto-ignition rates not compatible with pyrolysis oils. Addition of expensive additives to improve the cetane number has proved necessary for this type of application.

In more recent studies [69] where the same corrosion problems occurred, the engine developed an average thermal efficiency of 35%, comparable for all fuels tested. Auto-ignition was improved by preheating the air to a temperature of at least 55°C. This temperature must be increased to 130°C in order to obtain the same auto-ignition characteristics as Diesel.

Lastly, in terms of polluting emissions, the analyses carried out on the exhaust gases indicate particulate, total hydrocarbon, CO and NO_X contents always greater than the values obtained with conventional Diesel. The combustion characteristics can nevertheless be improved by high-temperature filtration of the oils (reduction of ignition delay and coking) by reducing the molecular weights [69].

More recent tests conducted in Germany on Stirling engines in a 25 kW CHP unit demonstrated that the emissions were compatible with the standards applicable at that time. The electrical and thermal yields still need to be improved (50-60%) [70].

5.1.6.3 Combustion in gas turbines

Gas turbines are used for a very wide range of applications, including in particular industrial generation of electricity and aeronautics. Combustion of pyrolysis oils in gas turbines has been studied at various scales on 40 kWe and 2.5 MWe turbines normally supplied with kerosene and Diesel respectively [71].

In both cases, the formation of deposits on the surface of the combustion chambers and high particulate contents in the exhaust gases indicate the difficulty of using pyrolysis oils in a gas turbine. The turbines must be modified in order to limit these effects as well as the acid or alkali corrosion problems.

Orenda Aerospace Corporation (Canada) has the most extensive experience on high-power turbines, capable of running on poor quality fuels. This company set up a partnership with Dynamotive to fuel a 2.5 MWe unit on oils produced at West Lorne [72].

5.1.6.4 Use as raw material in the biofuel sector

The opportunity offered by fast pyrolysis, to convert heterogeneous solids into a homogeneous and easily transportable liquid, with a fairly good energy yield, clearly represented the most determining factor which led to the studies conducted in the 1980s on this technology. It offered the perspective of an interesting technical solution to replace petroleum products by lignocellulosic biomass in the transport sector.

There are two possible pathways for the conversion of pyrolysis oils into fuels that can be used in the transport sector: hydrocracking or gasification followed by Fischer-Tropsch or methanol synthesis.

A. Catalytic upgrading through cracking and hydrotreatment processes

These processes are designed to remove the oxygen contained in pyrolysis oils, the cause of numerous undesirable characteristics, starting with chemical instability. Highly significant results are obtained, ranging from stabilisation of the oils to production of high-quality fuels [73]. The chemical and catalytic processes for complete deoxygenation of pyrolysis oils to produce fuels that can be used in the transport sector have been studied since the 1980s.

Hydrotreatment consists of a catalytic operation at high temperature and high hydrogen pressure. It leads to complete elimination of the oxygen contained in the oils and hydrogenation/hydrocracking of the heaviest molecules. The operating conditions and catalysts used are generally the same as those employed when refining petroleum cuts (sulphided Co/Mo or Ni/Mo on alumina support). Due to the risks of polymerisation, however, moderate hydrotreatment is carried out beforehand at low temperature (250°C) before the actual hydrotreatment (450°C) [74].

Catalytic cracking in vapour phase leads to deoxygenation by dehydration and by decarboxylation using zeolite type catalysts. At 450°C and atmospheric pressure, hydrocarbons and gases such as CO and CO_2 and steam are the majority products [75]. In all cases, fast catalyst deactivation has an impact on the cost of processing the oils, making these technologies uncompetitive compared with the use of petroleum fuels [76].

A second approach consists in partial treatment to chemically stabilise the oils, thereby allowing transportation and use in internal combustion engines. Hydrotreatment is therefore carried out under less severe operating conditions since only the carbon double bonds are hydrogenated. The oxygen can be partially removed from the oils during the heat treatment. The coke deposits can be selectively reduced through the use of activated carbon-based catalysts [77].

Good performance has also been obtained with palladium catalysts under mild operating conditions (low hydrogen pressure, less than 0.3 MPa) [75]. Cracking in vapour phase has also been studied under these conditions with, in particular, use of basic zeolite-type catalysts, instead of acidic, and injection of steam. The results obtained have demonstrated partial deoxygenation of the oils and better hydrocarbon yields [78].

Other pathways are currently being investigated to produce hydrocarbon fuels from pyrolysis oils, such as catalytic treatment in fixed bed reactor filled with K_2SO_4 on alumina for the production of Methyl Acrylic Ethers (MAE). These products could be used in fuels to improve the octane number [79].

B. Gasification of oils (or even slurries)

This is the other possible pathway to produce a syngas composed mainly of hydrogen and carbon monoxide (see 5.3), which can act as reactants for the catalytic synthesis of methanol or biofuels similar to Diesel, via the Fischer-Tropsch synthesis (see 5.4). This pathway is derived from the BTL pathway by direct gasification of biomass.

The need for a liquid intermediate to implement this biofuel sector on industrial scale is justified mainly by the quantities of resources (about 2 to 3 Mt of biomass per year) required for the Fischer-Tropsch pathway to become profitable. A highly diversified supply of biomass therefore seems inevitable, possibly requiring preconditioning and homogenisation before conversion.

In addition to the advantages already mentioned regarding supply and energy densification, flash pyrolysis could remove the technological difficulties inherent to direct gasification (see 5.3): the gasifiers must be supplied with a feedstock that is easy to pulverise, in a pressurised chamber, and consisting of a product that is more standardised and easier to handle.

Conclusion

Most flash pyrolysis units currently in operation are R&D units, with some others at demonstration, pre-commercial stage. Little or no information is available on these plants or their true performance.

Fast pyrolysis offers two main advantages compared with the other biomass thermochemical conversion pathways. The first, which is the most important, is the possible decoupling between production and use. It is in fact possible to set up an oil production unit in an area with abundant resources in order to minimise the raw material transportation costs and benefit from the scale effect on the investments related to a unit converting oils into biofuels.

The second advantage concerns the versatility of the technology which, depending on the operating conditions, can be used to bias the reaction to liquid, solid or gas and thereby separate or concentrate undesirable fractions (mineral materials) in one of the phases.

Technologically, the processes proposed by a few suppliers have reached a satisfactory level of maturity and are very similar to each other in terms of yield and oil quality. Although very few plants are in operation throughout the world, especially as regards industrial facilities, the lack of outlets for the oils, which are highly heterogeneous, limits the number of demonstration units.

In terms of applications, the first tests conducted on engines for power generation have shown that, given minor modifications, the engines could run satisfactorily on pyrolysis oils. This is undoubtedly the most promising application for pyrolysis. The possible use of these oils in gas turbines is more limited in view of current technologies.

Direct application of pyrolysis oils is still faced with numerous problems, mostly related to their physico-chemical characteristics and the variability of their composition depending on the types of process and raw materials employed. In our opinion, there is very little hope of controlling the implementation precisely enough to bias the reaction towards a given composition. Similarly, the solid fraction contained in the oils (fluidisation sand, ash and char fines) still represents a problem for all energy applications.

Chemical co-conversion, often put forward as being the way of making plants profitable, in view of the high added value to be expected on some molecules, must be kept in perspective. Firstly, the markets concerned are niche markets and a single plant would be capable of saturating the world market, at least at attractive prices. Secondly, this market is already partially satisfied by more conventional processes. Lastly, since the oils are thermolabile and therefore unsuitable for distillation, unless they are intended for immediate use, in this case pyrolysis loses its major advantages.

The most promising option seems to be gasification of the oils and/or biochar for the production of a syngas which will then be converted into methanol or hydrocarbons by Fischer-Tropsch synthesis. The quality as defined until now (stable and homogeneous chemical composition, reproducibility, etc.) becomes of secondary importance compared with the conversion yields. The main objective therefore consists in supplying a centralised gasification unit with raw material of clearly defined physical properties from a diversified source of various types of biomass, including waste.

5.2 TORREFACTION

5.2.1 Torrefaction stakes

Most promising technologies for the thermochemical conversion of biomass for advanced applications such as the production of biofuels or hydrogen generally involve a supply of finely divided products, which poses no special difficulties with coal (80 μm) or sprayable liquids. With biomass, the fibrous and elastic structure of the material makes the product more difficult to atomise; direct submicronic grinding is possible but consumes a considerable amount of energy. In addition, the particles obtained exhibit low sphericity (globally elongated shape), making fluidisation difficult and complicated when injecting the biomass by dry milling in an entrained flow gasifier.

Optimised solutions must be found to prepare the biomass, simplifying its division to obtain a fine, homogeneous and easily fluidisable powder and increasing its energy concentration.

A number of preparation techniques have been considered, including torrefaction, and although few studies have been conducted on this process since the 1970s, it offers interesting characteristics which could make it the reference solution, since:

- it is a mild thermal process (between 200 and 300°C) under inert atmosphere (nitrogen, CO_2, steam, torrefaction gas) requiring no complex technological developments or high energy inputs (deep drying), which are the disadvantages of fast pyrolysis,

- it produces almost exclusively solid matter with limited weight and energy losses (higher weight and energy yields, ranging between 78-95% and 83-97%, depending on the temperatures and residence times). Consequently, there is no need to manage gas production to limit mass losses, one of the disadvantages with slow pyrolysis,
- it produces an easily grindable solid due to the significant reduction of its mechanical properties (major obstacle to the development of its application in the field of lumber wood), with low and controlled moisture content since the product is hydrophobic,
- it partially degrades the lignocellulosic biomass fibres (mainly due to decomposition of the hemicelluloses). The powder obtained after grinding is less fibrous and has a finer granulometry than untreated wood powders, which improves its fluidisation behaviour.

The torrefaction process also includes certain disadvantages, however. It does not fundamentally modify the biomass composition. Compared with pyrolysis, it homogenises the products to a lesser extent, the heterogeneousness of the supply sources probably being partially conserved.

The energy concentration per unit mass of the resource is in the region of 10 to 20%, possibly even 30%, less than that of fast pyrolysis oils. The effect of possible decentralisation is therefore reduced. However, the low mechanical strength of the product may possibly be put to advantage to consider its densification via the production of pellets at low energy cost and therefore its transport over longer distances, regaining the advantages of decoupling.

While torrefaction is not a new process as such and has been extensively studied in the past, it is the final targeted application, i.e. a simple and energetically interesting preparation method to inject all types of biomass in the co-combustion or pressure gasification processes, which must promote its development. Most studies conducted today aim at extending the perimeter of use of lignocellulosic biomass [84] through the development of a system for the preparation of biomass by torrefaction, followed by grinding and possible densification, prior to injecting the product in a gasification reactor, for example [85].

5.2.2 Basic torrefaction mechanisms

Biomass torrefaction is a controlled pyrolysis process implemented at temperatures of between 200 and 300°C, designed to produce a fuel with better energy characteristics. This treatment is characterised by partial or total absence of oxidising agents such air and corresponds to an initial pyrolysis phase implemented to obtain a solid fuel. The difference between torrefaction and carbonisation is due to the globally endothermic nature of the reactions occurring at temperatures below 300°C. Above this temperature, the reactions become exothermic and produce char. Torrefied wood is therefore an intermediate product between anhydrous wood and char. During torrefaction, chemical changes, of the intermolecular bonds in particular (chemical bridges between the wood components, cross-linking of the lignocellulosic material), give the material better dimensional stability, at the expense of its mechanical properties, including its resilience. The weight loss as gaseous emissions (steam and volatile organic compounds) may reach 30 to 40%, representing a 10% loss of the initial energy content.

Figure 5.10 shows the main physico-chemical phenomena occurring during torrefaction.

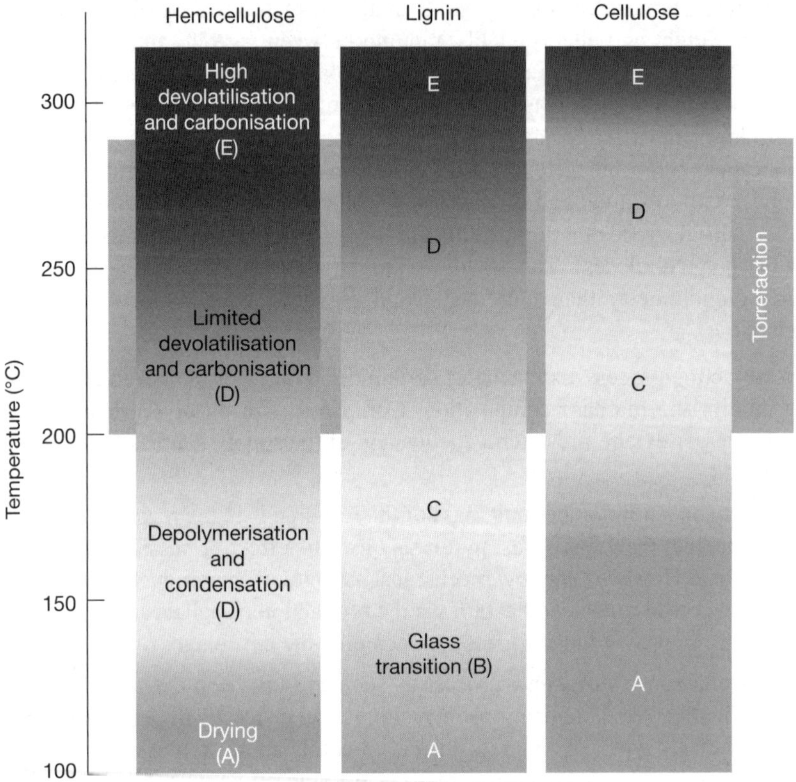

Figure 5.10

Physico-chemical phenomena occurring during torrefaction heat treatment [86].

For each compound, area (A) represents the drying phase. When the temperature is increased, the reactions are shifted towards areas (C) and (D) which represent respectively the phases of depolymerisation/recondensation and devolatilisation/carbonisation of the previously formed polymers. Area (E), which can be considered as the start of slow pyrolysis, corresponds to intense devolatilisation and carbonisation of the compounds formed in (D).

The properties of the lignins, and in particular the glass transition temperature, would be proportionally correlated to their degree of condensation, with the transition phases fluctuating between 170 and 190°C in area (B) [87]. Hemicelluloses are generally the most reactive polymers, followed by the lignins and cellulose [88].

Fewer studies have been conducted on modelling the torrefaction of wood than on pyrolysis and give inconclusive results ([89]-[92]). Generally developed to predict the overall behaviour of wood, these models do not give a precise description of the reaction mechanisms and changes to the main wood components during low temperature pyrolysis, apart from the model of Rousset *et al.* [93].

5.2.3 Processes

Torrefied wood was first considered as a substitute for charcoal. Research on the production of lumber wood accelerated since 1985-1990, especially in France, Finland and the Netherlands, where the studies led to a new family of technologies: ThermoWood® process (VTT, Finland), Plato® process (Plato International BV, the Netherlands), OHT (Oil Heat Treatment) process (Menz Holz GmbH, Germany), retification® process (New Option Wood SA, France), Perdure® process (PCI Industries inc., Quebec), Prodéo process (Avenches, CH), etc.

Although the properties of wood treated at high temperature in small scale pilot kilns are now relatively well-known, there are some doubts concerning the true characteristics of wood treated in industrial kilns. Further to the multiplication of heat treatment projects and their industrial development, all newly created companies involved in this field were faced with numerous technical problems concerning the reproducibility and homogeneity of the treatment throughout the feedstock treated. Between the definition of optimum operating conditions and the industrial process, the technological breakthroughs have often been poorly grasped. Although several studies are available, it remains difficult to guarantee the gain on the properties sought and to compare these with the losses on the qualities to be preserved. This observation shows that fundamental studies are still required to obtain a precise understanding of the relevant treatments.

Industrial processes for the production of torrefied biomass for energy purposes are virtually non-existent or at any rate not commercially available. Since the revival of interest generated by preconditioning for combustion and/or gasification, new concepts have emerged but never moved beyond the demonstration pilot stage [94].

In the 1980s, Pechiney built and ran a demonstration plant. The plant operated on wood that could be used as a substitute for coke in the electrometallurgy sector. Technically, the tests were highly conclusive (Figure 5.11).

The financial aspects, however, were much more negative and were to lead to the closure of the two plants located in France, including Pechiney's 20 000 t/year plant at Laval de Serre in the early 1990s. The processes developed at the time (continuous for Pechiney, discontinuous for Vicarb and Fages Haberman) aimed at maximising the yield. In the electric kilns torrefied wood was used instead of a reducing mixture composed mainly of charcoal. Its physico-chemical characteristics such as good crushing strength and especially its hydrophobic nature made it a first class reducing agent.

The company Transnational Technology Stubbing developed a continuous torrefaction process, known as Airless Torrefaction, based on a patented drying process, known as Airless drying (Figure 5.12) [96]. The airless drying and airless torrefaction processes are heated by superheated steam. The steam is created exclusively from the humidity contained in the biomass. According to its designers (Airless Systems), it offers very fast drying times and operates at high energy efficiency. In 2003, this continuous process was used to torrefy 20 tonnes of wood chips which were delivered to a thermoelectric plant in Holland for co-pulverisation and co-combustion tests with char. The torrefied wood/char mixture proved to be efficient [97]. To our knowledge, there have been no further experiments based on this process.

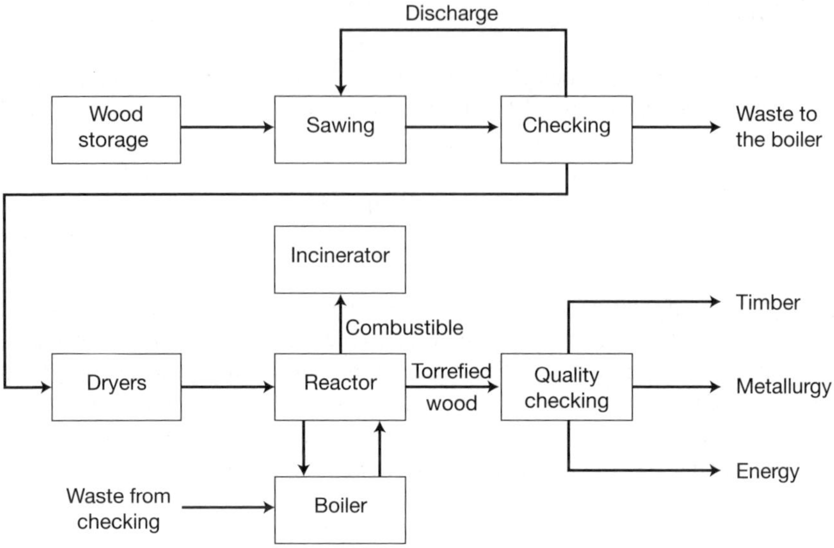

Figure 5.11

Conceptual diagram of the Pechiney process (France) [95].

More recently in Europe, ECN in Holland revived interest in the torrefaction of lignocellulosic biomass for energy applications, aimed at co-combustion of torrefied wood in power plants. A laboratory project called "*TOP Process*" is currently being developed [98]. The TOP process combines **to**rrefaction and **p**elletisation (densification) in two steps instead of one. Unlike the Pechiney process which used an exchanger, the Dutch process is based on direct contact with hot gases. The gases from the torrefaction itself are repressurised and reheated after each cycle. The energy required for predrying and torrefaction is produced by the combustion of pyrolysis gases. When there is not sufficient energy to maintain the process, other external energy sources can be used (fuel oil, gas). The reactor, reserved exclusively for torrefaction to allow better control of temperatures and pressure, operates according to the moving bed principle (Figure 5.13).

Based on laboratory experiments, modelling of this process indicates 92% energy efficiency, i.e. 4 points more than a conventional pelletisation system.

Several patents on wood torrefaction processes have been filed, the main ones being included in the references [99-108].

Other companies are currently investigating how to adapt industrial technologies on calcination or heat treatment applications to torrefaction.

CMI (Cockerill Maintenance & Ingénierie) is a Belgian company which owns the NESA multiple hearth furnace technology.

This type of furnace consists of a series of circular hearths, placed one above the other. Material is fed to the top hearth, and rabbled across it to pass through drop holes to the

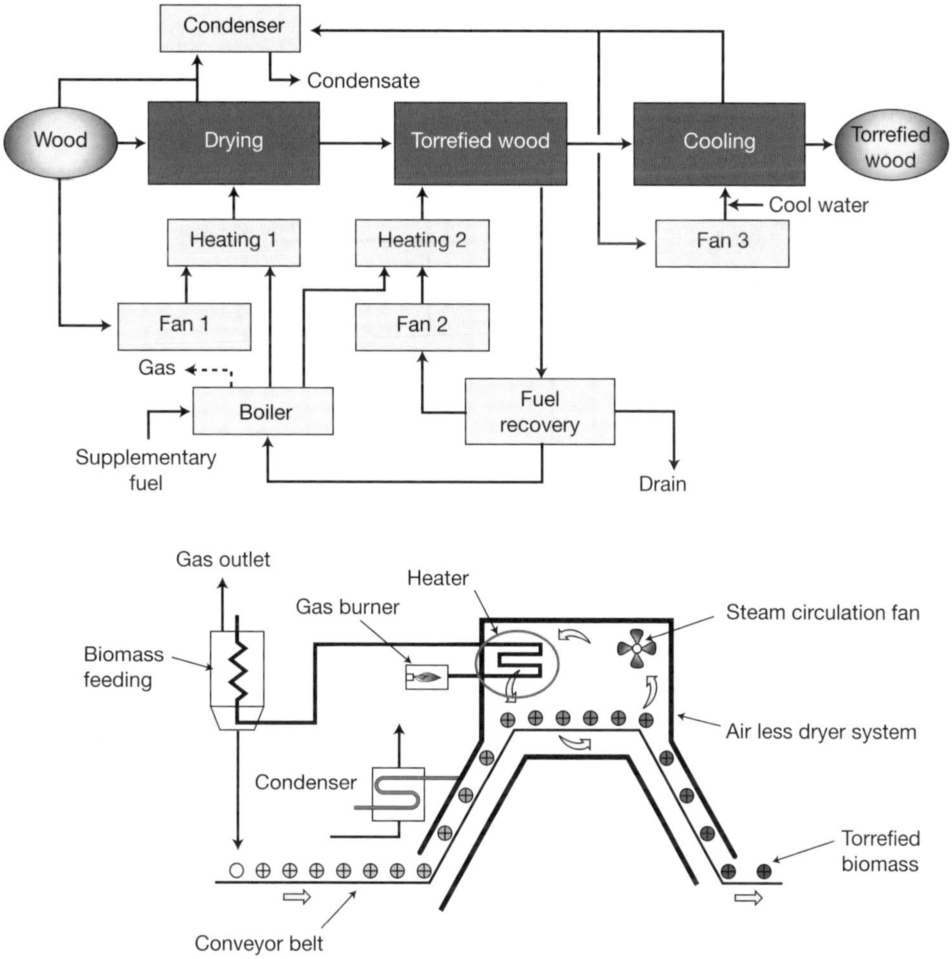

Figure 5.12

Conceptual diagram of the technology developed by Trans Technology (US) and details of the Airless dryer system.

hearth below. The furnace has a vertical rotating shaft. The temperature of each hearth is regulated by burners positioned around the outside: theoretically, a vertical thermal profile can be created in the furnace. The technology is mainly used for heat treatments between 400 and 900°C (calcination).

The American company Wyssmont is specialised in the design and manufacture of industrial dryers, especially in the chemistry, pharmacy, mineral industry and food sectors. The technology has been present on the market for 75 years.

Their Turbo-Dryer® technology consists of a stack of trays rotating around a vertical axis (Figure 5.14).

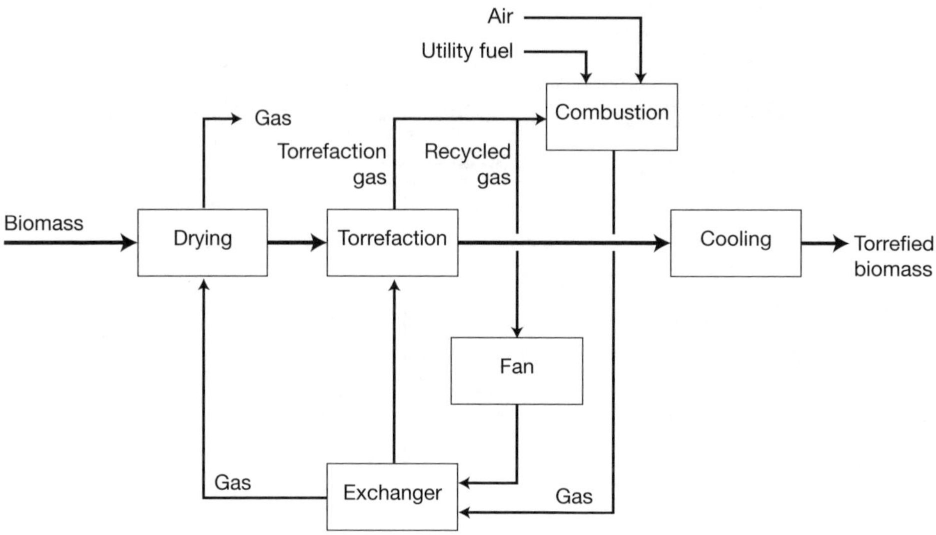

Figure 5.13

Torrefaction diagram. *TOP Process* technology developed by ECN [98].

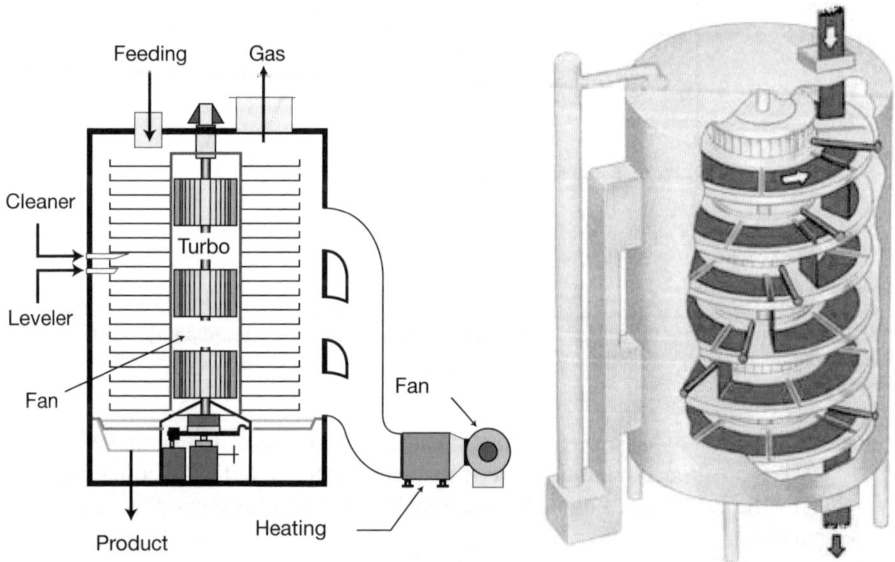

Figure 5.14

Turbo-Dryer technology.

The material is fed onto the top tray. After each revolution, the material is wiped onto the next lower tray. The bed is mixed while falling then levelled by a fixed arm system. The dryer is heated by hot air or gas circulated by internal fans in the dryer axis. Drying and torrefaction take place in the same reactor. The reactor atmosphere is kept neutral by the release of steam during the drying phase, with steam or nitrogen added if necessary.

The Torftech group licences and provides the Torbed® cyclone reactor technology (registered trademark of the Mortimer Technology Holdings Ltd group), used in numerous industries: food, catalyst production, ore and production of energy from biomass, waste and other solid fuels.

Developed by Polow Energy Systems in the Netherlands, the Torbed® technology can be applied to drying, combustion, gasification and gas treatment.

Amongst the variants of this technology, the Expanded Bed Reactor (EBR) is used for torrefaction. Inside the EBR, the velocity of the hot gases is high enough to entrain all the bed biomass particles in a vortex spinning inside the reactor (Figure 5.15). Particles separate by centrifugal force from the inner vortex to the outer wall before returning to the base of the reactor creating a diffuse toroidal bed. The base layer of the bed is subjected to high velocities and therefore high heat and mass transfer rates.

The Torbed technology was adapted in 2007 by Polow to torrefaction conditions, to produce torrefied biomass pellets. Numerous types of biomass are concerned, in particular straw, rice and wood residues. The treatment temperature defined is between 250 and 350°C.

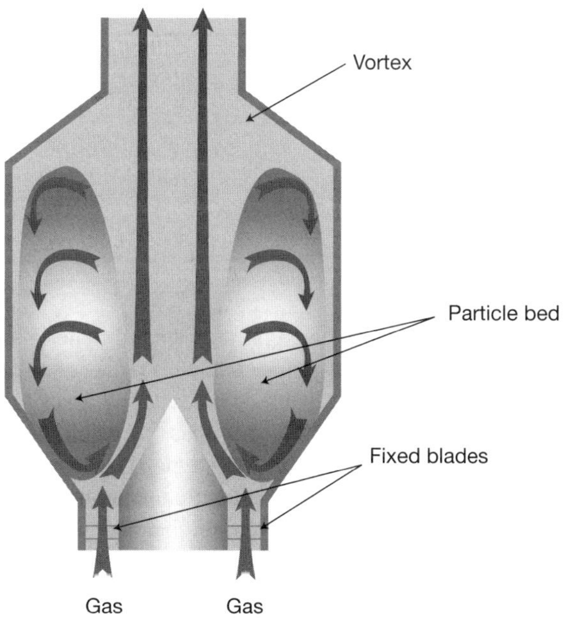

Figure 5.15

Diagram of the Torbed reactor.

The wood chips are dried before they are fed into the torrefactor which comprises 3 treatment stages (Figure 5.16). The torrefaction gases are burnt in a boiler to produce some of the heat required for drying and torrefaction.

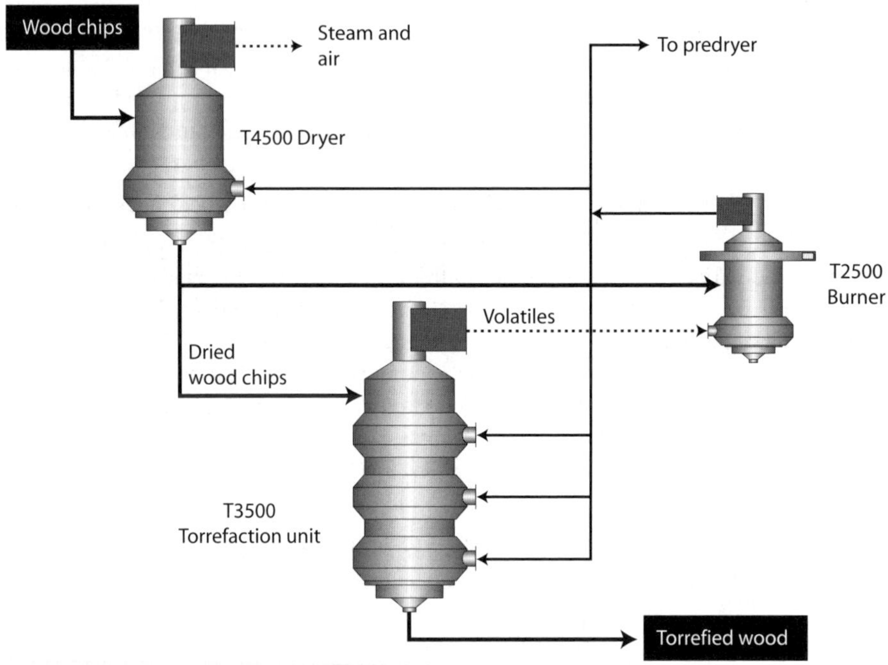

Figure 5.16

Torbed torrefaction technology.

5.3 GASIFICATION

Gasification is the thermal and chemical conversion of a solid material (char, biomass, etc.) in the presence of a gaseous reactant (oxygen, steam, etc.) unlike pyrolysis (or thermolysis) which is a thermal process conducted in the absence of reacting gases (see 5.1). Gasification converts the solid into a gaseous mixture called syngas, containing mainly H_2, CH_4, CO, CO_2 and H_2O in proportions which vary with the type of process, whose performance depends on the characteristics of the solid to be converted. Operating conditions are adapted to fit the final use of the syngas. The gaseous mixture obtained can in fact be used for various applications, in particular:

- engine combustion for automobile traction, with for example the De Lacotte wood gas generator,
- combustion for electricity and heat cogeneration. The electricity conversion efficiency is improved, compared with the simple combustion of biomass, by increasing

the temperature of the combustion gases. The efficiency can reach 40% of the biomass Lower Heating Value (LHV) [109];
- biofuel synthesis (see 5.5). In this case, the advantage lies in the fact that gasification can convert all lignocellulosic biomass components into fuels such as methanol, dimethyl ether (DME) or hydrocarbons obtained by the Fischer-Tropsch synthesis.

For combustion, syngas may contain methane (CH_4) and heavier compounds, whereas the synthesis of biofuels involves additional constraints since the aim is to obtain a syngas with as much CO and H_2 as possible. In addition, an H_2/CO molar ratio of about 2 is required, for subsequent synthesis of methanol, DME or liquid hydrocarbons via the Fischer-Tropsch reaction.

Figure 5.17 shows a general view of the gasification process followed by the Fischer-Tropsch synthesis for production of Diesel and/or kerosene.

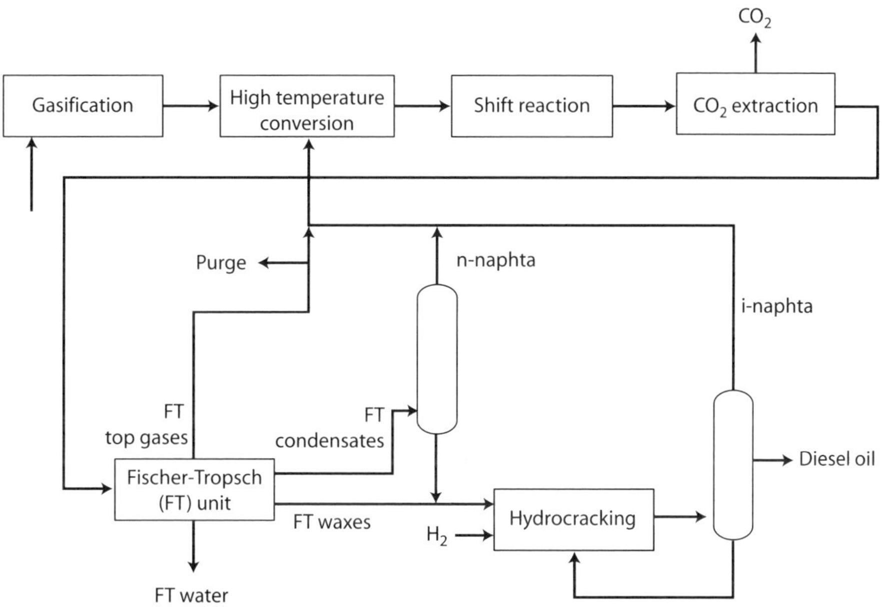

Figure 5.17

General view of the gasification process associated with the Fischer-Tropsch synthesis.

The gas obtained after gasification and high temperature conversion contains mostly CO + H_2. The H_2/CO ratio is adjusted using the water-gas shift reaction. Impurities in the mixture and CO_2 present are eliminated during the purification step before it is sent to the Fischer-Tropsch unit which can produce after different catalytic treatments mainly Diesel and kerosene and also other gaseous species (top gas) and naphtha. Naphtha can be used as raw material in the chemical industry but it can also be recycled, with the top gases to the high-temperature conversion stage or to a steam reforming unit for additional synthesis of CO and H_2 Recycling the naphtha and top gases substantially increases the overall Diesel and kerosene production yield.

5.3.1 Chemical reactions and thermodynamic constraints for the production of syngas

5.3.1.1 Main reactions

From a scientific standpoint, gasification is a heterogeneous endothermic reaction between a solid and a reactive gas. From a practical standpoint, gasification reactions only become significant when the biomass has been decomposed by pyrolysis into gas (see 5.1) and into a solid product with high carbon concentration. To a first approximation, and to understand the major reaction trends, we simply need to consider the main reactions on carbon as heterogeneous reactions. The reactions are grouped below with the reaction enthalpy ΔH, taking carbon in graphite state and water in vapour state as reference. The homogeneous reactions in gas phase then follow.

Heterogeneous reactions:

$C + O_2 \rightarrow CO_2$ $\quad \Delta H = -393$ kJ/mol

$C + {}^1/_2\, O_2 \rightarrow CO$ $\quad \Delta H = -110$ kJ/mol

$C + H_2O \Leftrightarrow CO + H_2$ $\quad \Delta H = +131$ kJ/mol

$C + CO_2 \Leftrightarrow 2\, CO$ $\quad \Delta H = +172$ kJ/mol

$C + 2\, H_2 \Leftrightarrow CH_4$ $\quad \Delta H = -75$ kJ/mol

Homogeneous reactions:

$CO + {}^1/_2\, O_2 \rightarrow CO_2$ $\quad \Delta H = -283$ kJ/mol

$H_2 + {}^1/_2\, O_2 \rightarrow H_2O$ $\quad \Delta H = -241$ kJ/mol

$CO + H_2O \Leftrightarrow CO_2 + H_2$ $\quad \Delta H = -41$ kJ/mol (water-gas shift reaction)

$CO + 3\, H_2 \Leftrightarrow CH_4 + H_2O$ $\quad \Delta H = -206$ kJ/mol

Globally, the pyrolysis gas and carbon in the presence of oxygen and steam react to form a mixture of CO, CO_2, H_2, H_2O and CH_4. These species may subsequently react together in gas phase. The production of CO and H_2 by reaction of carbon with steam of CO_2 is highly endothermic. A significant amount of heat is therefore required to obtain syngas rich in CO and H_2. The easiest way of providing the energy is to burn some of the biomass in the presence of additional oxygen or air. This will inevitably lead to partial conversion of biomass into combustion gases CO_2 and H_2O which will be no longer available for biofuel synthesis and which, in addition, must be separated from the syngas.

5.3.1.2 Biomass representation and ideal steam gasification reaction

Dry biomass can be represented by a model molecule with the average approximate composition of biomass: $C_6H_9O_4$. It is quite interesting to note that the compositions of the various types of biomass always remain relatively close to this average [110].

The heat of combustion for dry wood at constant pressure (LHV) varies from 15.5 MJ/kg to 19 MJ/kg. This dispersion takes into account the variability of the data in the literature, mainly due to the diversity of the vegetable species considered. Due to its high resin content, pine wood has a higher value, about 21 MJ/kg. The standard enthalpy of formation of biomass has been determined [111] from data in the literature: $\Delta H_f = -6.1 \pm 1.5$ MJ/kg.

Based on the approximate chemical formula of biomass, the ideal gasification conversion with an H_2O/dry biomass weight ratio of about 25%, should take place according to the following theoretical equation:

$$C_6H_9O_4 + 2\ H_2O \rightarrow 6\ CO + 6.5\ H_2$$

Note, at this stage, that the H_2/CO molar ratio obtained by this ideal reaction is about 1, therefore much lower than the H_2/CO molar ratio required to synthesise fuels. The hydrogen content of the gas is increased via the water-gas shift reaction. The endothermic enthalpy of reaction associated with the ideal reaction is + 4.8 ± 1.5 MJ/kg of dry biomass.

5.3.1.3 Steam gasification reactions at thermodynamic equilibrium

Figure 5.18 shows the results of calculations used to determine the composition of the gas produced at equilibrium against temperature, at atmospheric pressure. The calculations were performed using the CHON data basis taken from the Thermodata COACH bank (Bank 00 version 06-00) and the GEMINI Gibbs energy minimisation software [112]. The products of the ideal reaction (6 CO and 6.5 H_2) are only obtained at a temperature above about 1000°C at atmospheric pressure. Below 1000°C, the gas contains CO_2, H_2O and CH_4 in non-negligible proportions, which increase further when the reaction equilibria are taken into account. Total conversion into CO and H_2 is impossible below 1000°C at atmospheric pressure. The effect of pressure is relatively easy to predict since, according to Le Chatelier's principle, increasing the pressure at constant temperature will reduce the number of moles of gas produced and increase the production of methane. In addition, since the reaction is endothermic, increasing the temperature favours the formation of CO and H_2. Consequently, to bias the reaction towards the production of CO and H_2 in a pressurised system, the pressure increase must be compensated by a temperature increase. This effect can be evaluated using thermodynamic calculations. At 10 bar for example, a minimum temperature of 1300°C is required for complete conversion into CO and H_2.

The sensible heat associated with heating ideal syngas up to 1 000°C is in the order of + 2.5 MJ/kg of dry biomass. The total energy expenditure, calculated using the theoretical equation at 1 000°C, will therefore be in the region of 2.5 + 4.8 ± 1.5 MJ/kg, i.e. + 7.3 ± 1.5 MJ/kg of dry biomass. This energy expenditure represents about 40% of the heat of combustion (LHV) of the dry biomass.

Since the endothermic reactions producing CO and H_2 only occur significantly once the biomass has been decomposed by pyrolysis around 500°C (see 5.1), most of the energy must

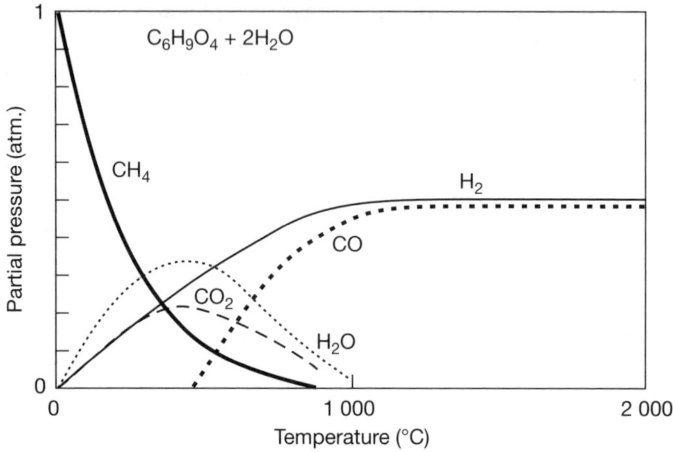

Figure 5.18

Graph of gas composition at equilibrium against temperature for the gasification reaction at atmospheric pressure.

be supplied at high temperature, above 500°C. The thermodynamic calculations show in fact that the energy associated with pyrolysis is, in comparison, relatively low, being about + 0.7 ± 1.5 MJ/kg of dry biomass, pyrolysis could even be exothermic. One important consequence concerning the process: only a small fraction of the thermal energy available at temperatures below 500°C, which comes from the gas shift reactor or from the liquid fuel synthesis reactor, installed downstream from the biomass thermochemical conversion process, can be recycled in a pyrolysis and gasification reactor.

To increase the H_2/CO molar ratio, the H_2O/biomass weight ratio can be increased at reactor input to favour the water-gas shift reaction. The thermodynamic approach nevertheless indicates that to increase an H_2/CO molar ratio from 1 to 1.5 (at 1 000°C and 1 bar), the H_2O/biomass weight ratio must be increased at input by 25 to 75%. The drawback, however, is that additional energy is required to raise the temperature of the excess steam. Consequently, performing the gas shift reaction at high temperature in the gasification reactor is not necessarily advantageous.

In addition, increasing the oxygen concentration at reactor inlet reduces the H_2/CO molar ratio and using air induces, firstly, an additional energy expense to heat and compress the inert nitrogen and, secondly, formation of nitrogen compounds which must be eliminated.

5.3.1.4 Reaction equilibrium considerations

The kinetic data available in the literature show that:

- the kinetics of homogeneous methane steam reforming, i.e. conversion of methane into CO + H_2 in the presence of steam, is very slow below 1 000°C. In practice, the homogeneous reaction only occurs at temperatures close to 1 500°C. Methane steam reforming can take place heterogeneously at lower temperatures, for example at about 800°C in the presence of a nickel catalyst. In this case, however, there is a problem of catalyst poisoning by all the inorganic species present in the biomass and partially released in the gas,

- carbon gasification kinetics are also very low below 1 000°C [113]. The rate of reaction is two orders of magnitude lower than that of the combustion reactions. Consequently, the temperatures in char gasification plants are generally 1 200°C, or even more. The rate of reaction is extremely sensitive to the effect of temperature below 1 200°C and depends on the reactivity of the carbon solid. This reactivity plays a key role in gasifiers.
- in addition to a solid and a syngas, pyrolysis and gasification produce tars, greater quantities being produced at lower conversion temperature. Between 20 and 30% of the dry biomass may be converted into tars at about 500°C. This quantity can be reduced to just a few percent at around 850°C, i.e. about 10 g/Nm^3 of gas produced. This value must be compared with the maximum 0.01 to 0.1 g/Nm^3 permissible for a cogeneration application and with the maximum 0.0001 g/Nm^3 permissible if the gas is used in a Fischer-Tropsch synthesis to form liquid biofuel. These figures show the need to implement tar elimination systems to solve this major problem in the design of gasification processes.

5.3.1.5 Autothermal and allothermal processes

The easiest way of providing the energy input required for steam gasification is to burn some of the biomass in the presence of added oxygen or air. Processes using partial combustion of some of the biomass to convert the other part into syngas are known as autothermal processes. This operation implies a loss of carbon and hydrogen, converted into CO_2 and H_2O, which will therefore not be available for conversion into fuels.

Another option would be to replace this partial combustion by external energy input. Adding about 40% of the feed biomass LHV would double the production of fuel (see 5.3.3). These processes are qualified as allothermal.

5.3.1.6 Gasification reaction in the presence of hydrogen

The HYNOL process, based on this reaction [114], owes its originality to the fact that the aim is to produce CH_4, not CO and H_2 in the gasification reactor. The advantage [115] seems to be that this gasification requires no energy input, which is always difficult to achieve with a fluidised bed technology. In this case, all the energy input would be concentrated at the second stage, gas reforming, where CH_4 is converted into H_2 and CO. In this process, natural gas is added to the syngas in the reforming stage which operates at 1 000°C and which produces excess hydrogen. The H_2 can then be recycled in the first gasification stage. The thermodynamic approach shows that carbon conversion is more difficult in the presence of hydrogen than in the presence of water. High hydrogen content, high pressure of about 30 bar and a relatively low temperature of about 500°C is required to obtain efficient conversion on thermodynamic basis. At 500°C, the kinetic effects will probably be very important and calculation at thermodynamic equilibrium can only act as a reference indicating the trend. It is also likely that a large quantity of tars will be produced, even though they will probably not be a major problem since they can be converted in the second stage, which operates at high temperature. The problem could occur when converting biomass at high pressure and low temperature in the presence of hydrogen. Despite a significant excess of hydrogen, the reactor could clog up due to the accumulation of carbon.

5.3.2 Industrial processes

5.3.2.1 Overview

The biomass gasification processes are derived from processes developed for coal gasification [113]. Most are described in Figure 5.19 [116] and are classified in the following categories:

- countercurrent or cocurrent fixed beds,
- fluidised beds,
- circulating fluid beds,
- entrained flow reactors.

This paragraph describes the main types of reactor, the generic problems related to their implementation and the main characteristics of the gases produced.

Table 5.4 lists the companies which have studied the various technologies.

Table 5.4 Summary of companies which have studied the various gasification technologies.

Gasification technology	Companies
Countercurrent fixed beds	Wellman [117], Volund [118], Bioneer [119]
Cocurrent fixed beds	DTU [120], TKE [121], Biomass Engineering [122] Fluidyne [123], De Lacotte [122]
Bubbling fluid beds	EPI [121], Carbona [124], ASCAB [121]
Circulating fluid beds	FICFB [126], Lurgi [124], Foster Wheeler [121]
Twin circulating fluid beds	Ferco Battelle [127]
Entrained flow reactors	Texaco [128], Choren [129a], Future Energy [125] Lurgi MPG, Shell, Uhde [129b]

5.3.2.2 Main types of reactor

A. Countercurrent fixed beds

In countercurrent fixed beds, the solid moves slowly downwards through a vertical reactor. The gaseous reactants are introduced at the bottom. Moving upwards through the reactor, we first have a high temperature zone (~ 1 200°C) where the combustion reactions prevail, then an area of decreasing temperature where reduction reactions occur, producing CO and H_2. The gases escape as they cross the upper pyrolysis area. The ash is evacuated at the bottom.

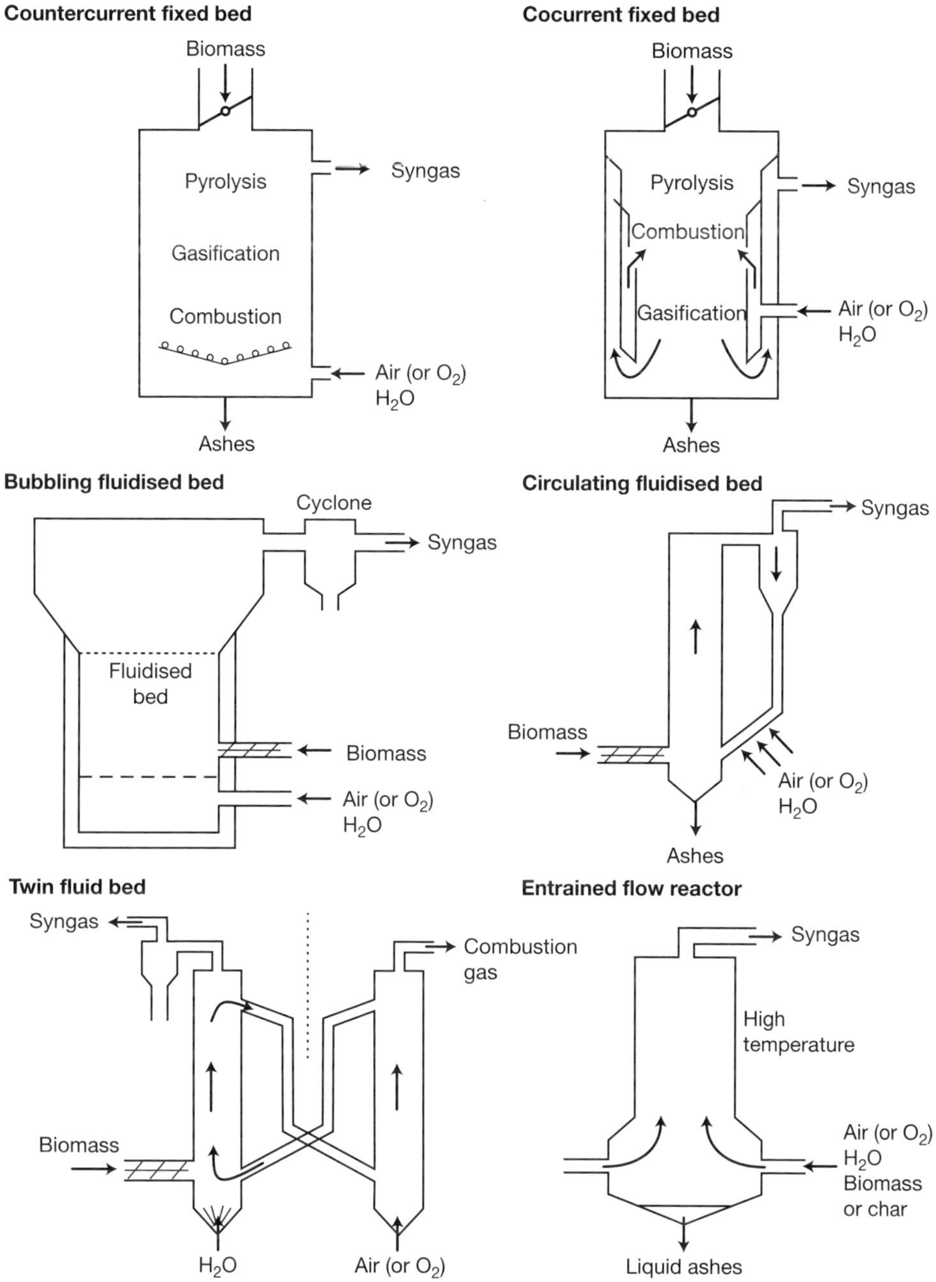

Figure 5.19

Main types of gasification reactor.

The technology is simple, reliable and proven for solid particles of uniform size with no fines. This arrangement favours the production of a dirty gas at low temperature with high tar content (about 100 g/Nm^3), which limits its use for combustion. It would appear that this type of technology is limited to dry biomass flow rates of about 4 t/h of biomass (20 MWth). The Lurgi process, adapted to char gasification, can reach a production of 100 MWth.

B. Cocurrent fixed beds

In cocurrent fixed beds, the solid moves slowly downwards through a vertical reactor in which the gaseous reactants are introduced from the top and/or from the side (the arrangements vary considerably). Most reactions take place within a limited area. The reaction gases leave the reactor from the bottom at high temperature. The technology is simple, reliable and proven for solid particles of more or less uniform size of about 5 mm. The gas produced is fairly clean with a low tar concentration of a few g/Nm^3. Due to instability problems (appearance of gas channelling) the technology seems limited to a biomass flow rate of about 500 kg/h (2.5 MWth).

C. Bubbling fluid beds

In bubbling fluid bed reactors, the pre-dried biomass is entrained in a fluid bed by reactant gases introduced at the bottom. The fluid bed consists of a fluidising material, which is either inert (sand) or catalytic (olivine, dolomite, etc.). Due to the mixture intensity, the bed temperature is uniform and easy to control. The CO and H_2 content of the gas produced increases with the bed temperature, which is nevertheless limited by technological constraints (reactor material) and by the risk of bed agglomeration due to molten ash. In practice, the temperature rarely exceeds 950°C in the current applications, and generally lies between 700°C and 850°C. The ash is extracted with the agglomerated particles, which means that the fluidisation material must be treated and replaced. The tar content in the gas is average, about 10 g/Nm^3. The technology is relatively proven and the maximum size seems to be located in the region of 10 to 15 t/h of biomass (~ 75 MWth).

D. Circulating fluid beds

The pre-dried biomass is introduced in a circulating fluid bed, in which recirculation takes place via a "down leg". The syngas is separated from the solid flow in the upper part of the reactor. This technology increases the circulation speed and therefore the reaction kinetics. Otherwise, it offers the same advantages and constraints as the bubbling fluid bed technology. Given the increased complexity, the technology is only financially advantageous for high biomass flow rates (15 t/h or more).

E. Twin circulating fluid beds

The twin circulating fluid bed technology is an extension of the circulating fluid bed technology. In the return circuit, a recirculation system mixes the fluidisation material and charcoal in a separate reactor where the residual solid is burnt. This system separates the combustion gases from the syngases. The recirculating material becomes a heat exchange material which transfers combustion energy to the gasification reactor. The temperature in the gasification reactor is in the region of 800°C to 900°C, as with the fluidised bed or circulating fluid bed technologies. The temperature in the combustion reactor rarely exceeds

950°C to 1 000°C due to technological reasons and risks of agglomeration. The small temperature difference between the gasification reactor and the combustion reactor (100°C to 150°C) means that the heat exchange material must recirculate quickly. In practice, the fluidisation material recirculation rate will be several dozen times higher than the dry biomass flow rate. This type of installation is relatively complex and would only be economically viable for high flow rates (> 10 t/h of biomass).

F. Entrained flow reactors

The material, either liquid or finely ground solid (~ 100 µm) is injected at high speed, in suspension in a mixture of pyrolysis gases and reactant gases containing steam, oxygen (or air) into the reactor. The reaction (flame) zone can be raised to high temperature (1 300°C to 2 000°C). At these temperatures the reactions are very fast, just a few seconds, favouring the formation of CO and H_2, steam reforming of methane and destruction of the tars. The ash necessarily melts. Some of the molten ash particles are recovered by sedimentation on the reactor walls, or can be more or less separated from the gas flow by inertial separation. Some of the carbon solid may be entrained by the gas.

This technology eliminates the problems caused by use of a fluidisation material and favours the production of CO and H_2, reforming of methane and destruction of the tars. To achieve sufficiently fast reaction kinetics, however, the biomass must be injected as very fine particles (100 to 200 µm). This type of reactor involves expensive pretreatment of the biomass. Several preparations can be considered:

- preparation by slow pyrolysis: the biomass can be pyrolysed beforehand in a separate reactor. The pyrolysis gas is separated from the residual char. This finely ground char is then injected at high speed, in suspension in a mixture of pyrolysis gases and reactant gases (containing steam, oxygen (or air)) into the reactor;
- preparation by torrefaction: grinding the woody biomass very finely consumes large quantities of energy. One way of reducing the energy cost of grinding is to make the particles mechanically fragile, which could be achieved by prior torrefaction. Research studies are being conducted in this area. Torrefaction nevertheless induces a loss of mass;
- preparation by fast or flash pyrolysis (see 5.1): the biomass is first subjected to pyrolysis by fast heating at intermediate temperature (~ 500°C). Pyrolysis is followed by fast quenching of the pyrolysis gases. This technique produces a liquid mixture (also called "green oil") and a char. The solid can be easily ground finely and mixed with liquid to obtain a slurry. The slurry is transportable and can be easily pumped and injected into the reactor. This process nevertheless raises a certain number of problems related to significant mass loss (between 15 and 30%) and problems checking the properties of the mixtures (risk of phase separation, acidity, high viscosity, etc.).

The entrained flow reactor technology also requires special care when designing the high-temperature reactor walls and the gas sensible heat recovery exchangers.

This technology is used for coal gasification [129] with units of ~ 500 MWth.

5.3.2.3 Syngas characteristics

Table 5.5. provides the LHVs of the gaseous mixtures obtained at gasifier output.

Table 5.6 indicates the typical gas compositions [130].

Table 5.5 Syngas LHVs

Gasifier	Reactant gas	Syngas LHV (MJ/Nm^3)
Partial oxidation	Air	~ 7
Partial oxidation	Oxygen	~ 10
Separate combustion (twin bed reactor)	Steam	~ 15

Table 5.6 Syngas compositions

Gasifier	FERCO BATTELLE	CARBONA	FICFB
Type	Twin fluid bed	Fluidised bed	Circulating fluid bed
Reactant gas	Steam	Air	Steam
Fluidisation material	Olivine	Sand	Silica sand
Biomass	Wood	Wood	Wood
Gas composition (vol.%)			
H_2	26.2	21.7	31.5
CO	38.2	23.8	22.7
CO_2	15.1	9.4	27.4
N_2	2	41.6	2.8
CH_4	14.9	0.1	11.2
C_2+	4	1	4

5.3.2.4 Generic problems related to the biomass gasification technologies

Several generic problems related to the biomass gasification technologies must be solved:

- steady supply at high flow rate of a solid into a pressurised reactor,
- how to obtain a syngas with no tars and other impurities,
- agglomeration of the fluidised bed, mainly due to the ash content in the raw material,
- corrosion generated by the presence of mineral elements in the biomass.

A. Solid biomass supply

In some cases, gasification in a pressurised reactor may be an attractive option, since the size reduction significantly reduces the investment costs. Moreover, by performing gasification under pressure, there is no need to compress the gas to perform the gas shift or fuel synthesis reactions, which represents a non-negligible energy saving. However, this raises the difficult problem of introducing solid biomass into the pressurised reactor.

Most of the existing technologies used to introduce the solid are based on screw conveying systems and/or implementation of lock hoppers. These technologies involve numerous problems related to clogging, wear, blocking of sealing parts and introduction of inert gases such as nitrogen, resulting in high maintenance costs.

One way of solving the problem could consist in first converting the biomass into a com bustible liquid by flash pyrolysis (see 5.1) [131].

B. Generation and destruction of tars

As described previously, gasification produces tars which are entrained in the gas, in variable quantities depending on the process used. For example, the fixed bed or fluidised bed processes described above produce a gas containing between 1 g/Nm^3 and 100 g/Nm^3 of tars. Since the synthesis reactions have very low tolerance to tars (~ 0.0001 g/Nm^3), they must be eliminated. This represents a key point for the biomass gasification processes.

Numerous studies have been conducted on the tars generated by gasification; a classification into primary, secondary and tertiary tars has been proposed [132], based on increasing order of refractory characteristics (see Figure 5.20), as well as typical lists of tars found in the gas for the various gasification technologies.

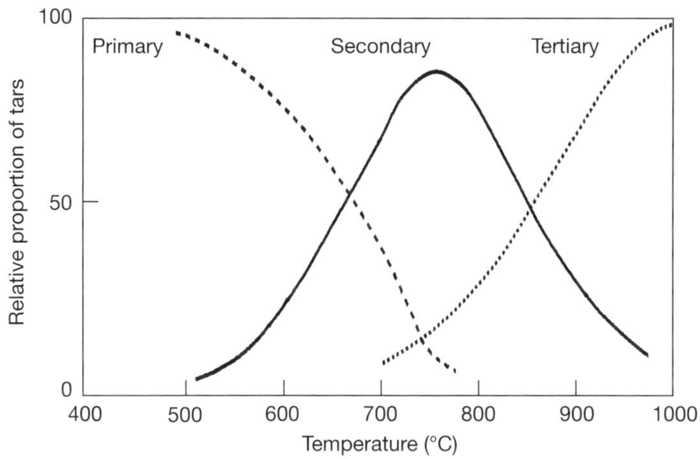

Figure 5.20

Graph of genesis of the various tar classes against temperature.

As the temperature rises (above 800°C), virtually only the tertiary tars remain, composed of monoaromatic hydrocarbons such as benzene and toluene together with polycyclic aromatic hydrocarbons (PAHs) such as naphthalene, anthracene, etc. These chemical species disappear at high temperatures (1 300°C to 1 500°C), one of the most refractory being naphthalene.

Numerous studies have been conducted on the tar destruction technologies [132] [133]. Emission of tars can be reduced at source by using a fluidisation material capable of catalysing the conversion of tars in the gasification reactor. It has been demonstrated that naturally-occurring materials such as olivine [$(Mg,Fe)_2SiO_4$] and dolomite [$CaMg(CO_3)_2$] have a catalytic effect [134]. Ni-based metal catalysts [135] have been developed. One of the problems encountered is catalyst deactivation due to interactions with the inorganic compounds such as sulphur contained in the biomass.

These catalysts may also be used in secondary beds positioned downstream of the gasifier.

Taking into account the various parameters related to the use of catalysts in gasification reactors whose outlet temperature is less than 1 000°C, the tar content could be reduced to about 0.1 g/Nm^3.

More complete elimination could be achieved after cooling the gas, by low-temperature washing in the gas treatment stage.

C. Agglomeration of fluidised beds

Agglomeration of fluidised beds is related to the physico-chemical reactions between the ash and the fluidisation material [136] [137]. Ash contains low melting point species such as:

- oxides melting between 500°C and 1 000°C: Na_2O, K_2O, PbO, MnO, B_2O_3,
- nitrates such as KNO_3 and $NaNO_3$ which melt at temperatures below 350°C,
- phosphates such as $Cu(PO_3)_2$ and KPO_3 which melt at temperatures between 700°C and 1 000°C,
- chlorides such as NaCl and KCl which form a liquid azeotrope at about 650°C, whereas KCl melts at about 770°C. Some species, such as oxides and chlorides, may also react with the fluidisation material, generally an oxide, to form eutectic mixtures or other liquid oxide or oxychloride compounds at low temperature, starting at 500°C. The liquid compounds formed cause the particles of fluidisation material to agglomerate. There would seem to be two agglomeration regimes:
  - a homogeneous regime, at a maximum temperature of about 800°C where ash decomposition products agglomerate around the particles whose size may increase from 200 μm to a few millimetres,
  - a heterogeneous regime, at higher temperature, between 800 and 1 000°C where large particles form in the fluidised bed. This regime would appear to be self-accelerated since the large particles create hot spots which in turn increase the agglomeration.

For a given ash temperature and composition, some fluidisation materials such as silica (SiO_2) will form a larger mass of liquid compounds than other more refractory and more fragile materials like MgO and CaO. This explains why some oxide compounds such as dolomite offer better resistance to agglomeration.

Note that it will not be possible to extract low melting point species from the eutectic mixtures and other compounds formed with the fluidisation material.

D. Corrosion

Problems of corrosion on metal structures occur mainly in the high temperature areas (pipes, filtration devices, exchangers, etc.). They are due to the presence, in the biomass and therefore in the syngas, of volatile inorganic species, in particular chlorides (NaCl, KCl, HCl), sulphides (H_2S) and a few metals (Na, K, Hg, etc.) [138]. The corrosion mechanisms are complex and depend on numerous parameters, in particular the pressure and local temperature. The development of special corrosion-resistant steels is an important technological stake for the operation of gasifiers.

E. Gas treatment

Biomass contains numerous elements apart from C, H and O. Although present in low concentrations (about 1% in wood), these elements will be expensive to be eliminated, since they represent up to 500 kg/h for a plant processing 50 t/h of wood. Elimination is complex due to the wide range of elements present and their significant variability depending on the type of biomass treated. Table 5.7 gives an idea of all elements present and their concentrations.

Table 5.7 Order of magnitude of the inorganic element concentrations in dry biomass.

Species and elements	Weight per kg of dry biomass	Element	Weight per kg of dry biomass	Element	Weight per kg of dry biomass
Ash ($CaCO_3$, SiO_2, etc.)	~ 10 g				
N	~ 1 g	Ba	~ 14 mg	Na	10 to 50 mg
S	~ 0.1 g	Cr	~ 5 mg	Ni	~ 5 mg
Cl	50 to 100 mg	Fe	~ 50 mg	P	~ 100 mg
F	~ 3 mg	K	~ 0.5 to 1 g	Pb	~ 1 mg
Al	~ 50 mg	Mg	~ 500 mg	Sb	~ 50 mg
As	~ 1 mg	Mn	~ 50 mg	Zn	~ 5 mg

In a gasifier, these elements will chemically recombine amongst themselves and with hydrogen, oxygen and carbon, producing various compounds such as carbonates and refractory oxides.

Some of the alkalis will form ash mixed with the fluidisation material or entrained as particulates and aerosols by the syngas, which must be eliminated by filtration. The low melting point oxides and chlorides may agglomerate with the fluidisation material. At high temperature (> 800°C), the carbonates may decompose into oxides.

The minor gaseous species present in the largest quantities in the reducing syngas are H_2S and HCl, but other species such as ZnS, NaCl, KCl may be encountered, which may condense above 400°C and lead to corrosion phenomena, like some chlorinated species at high pressure.

These solid, gaseous or condensable species will be eliminated by installing relatively complex gas cleaning devices [139].

5.3.3 Yields [1]

5.3.3.1 General approach to the problem of yields

The weight yield largely depends on the type of fuel produced. This is simply due to the fact that the energy content varies from one fuel to another. The fuel production is related directly to the energy yield, however. We may therefore be led to think that the energy yield represents a sufficient basis to compare the processes in terms of fuel production. In fact, although the fuel production weight yield is indeed proportional to the energy yield for the autothermal processes, irrespective of the type of fuel produced and due to the fact that biomass represents most of the incoming energy, the same is not true for the allothermal processes, for which we must distinguish between the weight and energy yields. The various processes are compared by taking as common base the same type of fuel produced, in this case Diesel from the Fischer-Tropsch synthesis. Chapter 5.5 provides details of this synthesis. The production of hydrogen is also discussed.

Yields do not depend on the gasification technology alone but also on the choices made in the overall framework of the integrated process, both in the upstream steps (direct use of biomass or of products derived from slow or fast pyrolysis) and in the downstream steps, especially as regards the utilisation of by-products from the Fischer-Tropsch synthesis, such as the top gases and naphthas. The top gases can be burnt to provide some of the energy required to operate the industrial unit. The naphthas can be upgraded separately. These products can be recycled and totally or partially converted into CO and H_2 (Figure 5.17).

5.3.3.2 Main points of the process used to increase the fuel production yield

The main points of the process which can be used to increase the fuel production yield correspond to the items where significant losses are observed. They include in particular:

- the autothermal gasification item where the biomass gasification energy is provided by partial combustion of biomass. This combustion consumes up to 40% of the initial carbon and hydrogen inventory,
- the H_2/CO ratio adaptation item where some of the CO in the syngas is converted into hydrogen by the water-gas shift reaction to reach an H_2/CO ratio value of about 2,

1. *Carbon yield*: proportion of carbon atoms from biomass found in the Diesel type fuel produced.
Hydrogen yield: weight of hydrogen produced divided by the maximum weight of hydrogen that could be produced from the dry biomass. The latter value represents about 17.5% of the weight of dry biomass.
Weight yield: weight of **Diesel equivalent fuel** produced per unit weight of dry biomass at process inlet. A carbon yield of 100% corresponds to a Diesel weight yield of about 58% (this figure therefore represents the maximum weight yield).
Plant energy yield: sum of the usable energies (fuel, electricity, etc.) produced divided by the sum of incoming energies (biomass, electricity, natural gas, etc.).
Primary energy yield: this yield also includes the incoming energy production yield, for example the production yield of the electricity used in the process.

necessary for fuel synthesis. In this reaction, some of the biomass carbon is lost as CO_2 and cannot be transferred into the final fuel,
- the Fischer-Tropsch synthesis item which, in addition to fuels, generates naphtha and top gases. As mentioned previously, these products can be valorised separately, but in this case the carbon they contain cannot be valorised in the fuel.

Consequently, in order to significantly increase the fuel production yield, alternative technological solutions must be found for these various items. The stake is important since the fuel production weight yield of a standard autothermal process is in the order of only 15% (i.e. 150 kg of Diesel for 1 tonne of dry biomass) whereas the maximum theoretical yield is about 58% (depending on the initial biomass composition).

A number of technological solutions can be considered:
- to replace partial combustion of biomass in the reactor, by supplying external energy. Over the last few years, several ways have been explored along these lines:
  - supply of electrical energy, for example by injecting a plasma gas at high temperature in the entrained flow reactor,
  - injection of hydrogen or methane in an entrained flow reactor. In this case, calculations show that the combustion reaction consumes mainly the hydrogen injected and not the biomass carbon which will be almost completely converted into CO at reactor outlet,
- to avoid the water-gas shift reaction, by injecting more hydrogen upstream from the fuel synthesis reactor,
- to avoid the loss of carbon in the top gases or naphthas, by reforming all or some of these products into CO and H_2.

Implementation of these solutions would multiply the weight yield of a traditional autothermal process by a factor of 3, reaching values of about 45%.

Note also that finally, these technologies would represent a way of storing both electricity and hydrogen in a fuel, without storage or distribution problems and without the need for modifications to current vehicle engines.

5.3.3.3 Yield calculations for various process options

The various yields (weight, energies, carbon, hydrogen) expected for a number of process options are discussed below in more detail.

The results shown were obtained by calculation.

Table 5.8 describes the carbon and hydrogen yields calculated for various processes. It appears that autothermal processes have low carbon yields, less than 30%. Providing external energy, for example as electricity injected in the gasifier in the high temperature gas reforming stage or as addition of hydrogen, may increase these yields significantly.

Table 5.9 lists the Diesel production weight yields and the energy yields.

The results clearly show the increase in the Diesel production weight yield obtained by adding external energy. The weight yield increases from slightly under 20% in autothermal operation to about 40% in allothermal operation. Assuming an annual availability in France

Table 5.8 Biomass gasification and liquid fuel synthesis or hydrogen production. Calculated change in the Carbon or Hydrogen yield depending on the energy sources used for the various stages in the process (A: autothermal; E: external input)

Gasification at 850°C	HT rectification (1 300°C)	Adjustment of the gas composition for fuel	Carbon yield (%)	Hydrogen yield (%)
Fluidised bed process				
A (oxygen or air)	–	A (gas shift reaction)	*11 to 18*	*28-45*
A (oxygen or air)	A (oxygen)	A (gas shift reaction)	*18 to 25*	
A (oxygen or air)	E (electricity)	A (gas shift reaction)	*25 to 31*	*51 to 68*
A (oxygen or air)	E (electricity)	E (H_2 complement)	*31 to 40*	
E (HTR heat)	E (electricity)	E (H_2 complement)	*40 to 65*	
Autothermal entrained flow process				
A (oxygen)		A (gas shift reaction)	*18 to 22*	
Allothermal entrained flow process				
E (electricity)		A (gas shift reaction)	*46 to 65*	
E (electricity)		E	*60 to 95*	*80 to 95*

Table 5.9 Biomass gasification and production of Fischer-Tropsch Diesel. Calculated change in the carbon, Diesel weight and energy yields depending on the energy sources used for the various stages in the process (A: autothermal; E: external input)

Gasification	HT rectification	Adjustment of the gas composition for fuel	Diesel weight yield (%)	Carbon yield (%)	Plant energy yield (%)	Primary energy yield (%)
Fluidised bed process						
A (oxygen or air)	–	A (gas shift reaction)	*7 to 11*	*11 to 18*	*20 to 26*	*20 to 26*
A (oxygen or air) + FT recirculation	E (electricity)	A (gas shift reaction)	*15 to 19*	*25 to 31*	*31 to 37*	*20 to 26*
Autothermal entrained flow process						
A (oxygen)		A (gas shift reaction)	*11 to 13*	*18 to 22*	*28 to 34*	*28 to 34*
A (oxygen) FT recirculation		A (gas shift reaction)	*16 to 20*	*27 to 33*	*39 to 45*	*39 to 45*
Allothermal entrained flow process						
E (electricity) without FT recirculation		A (gas shift reaction)	*16 to 20*	*27 to 33*	*31 to 37*	*19 to 25*
E (electricity) FT recirculation		A (gas shift reaction)	*34 to 38*	*56 to 63*	*48 to 54*	*25 to 30*
E (hydrogen in the reactor); FT recirculation		E (H_2 complement)	*45 to 50*	*77 to 86*	*53 to 57*	*30 to 33*

of 50 Mt of biomass, the production of liquid biofuel could reach, ignoring other uses (production of ethanol or energy as steam or electricity) between 7.5 Mtoe by autothermal pathway and 15 to 22 Mtoe by adding external energy, a significant value compared with the French annual fuel consumption.

The energy yields associated with the production of hydrogen are greater than the energy yields for production of Diesel, due to losses related to the Fischer-Tropsch synthesis.

Optimisation of the processes does not depend on the technological choices alone, but also on the economic and environmental aspects.

5.3.4 Technico-economic aspects

Figure 5.21 below shows a detailed diagram of a complete system for the production of biofuels from lignocellulosic biomass [140].

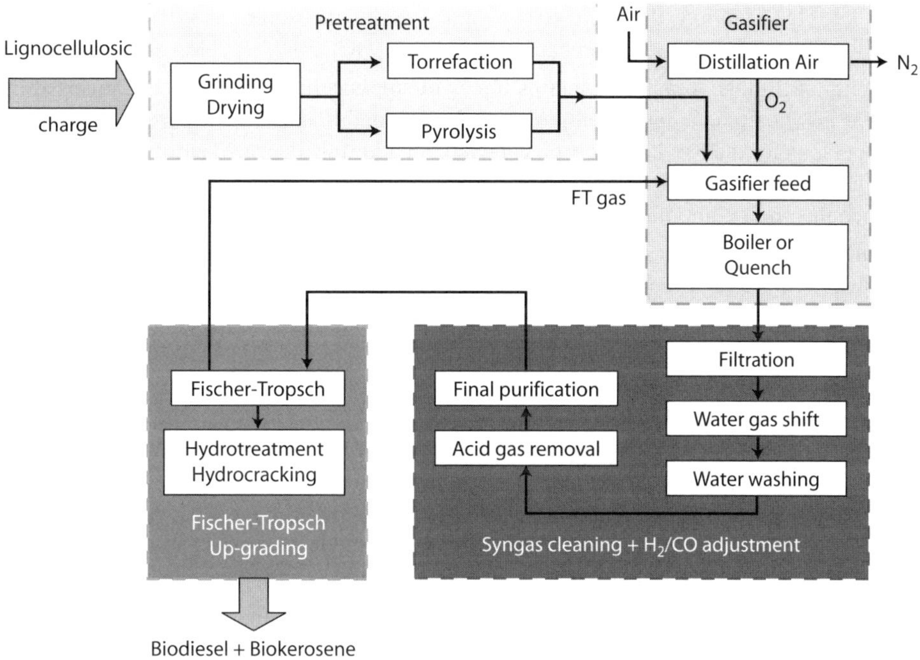

Figure 5.21

Detailed diagram of a complete system for the production of biofuels from lignocellulosic biomass.

This system includes pretreatment of the feedstock, the gasification process followed by a gas purification/conditioning step before the Fischer-Tropsch synthesis for the production of Diesel and/or kerosene.

The crude biomass is first dried before undergoing pretreatment to allow introduction in a pressurised entrained flow gasifier operating under pure oxygen. The cost of an air distillation unit is largely compensated since the presence of nitrogen in the syngas presents two major disadvantages:

- lower productivity of the catalytic steps (Fischer-Tropsch, *Water-Gas Shift*) by dilution of the reactive species,
- greater size and therefore greater cost of all downstream treatment steps.

The gas obtained after gasification and conversion at high temperature contains mostly $CO + H_2$. To reduce its temperature, it may be sent to a heat recovery boiler to produce high pressure steam or undergo quenching in water which offers the advantage of washing out some of the entrained solid particulates, also called fly ash, while lowering its temperature to the dew point (200-250°C depending on the gasification reactor pressure). In addition, this device allows the syngas to become saturated with water in proportions close to those required by the *water-gas shift* reaction.

A filtration step eliminates all solid particles which could clog the catalytic beds downstream. The H_2/CO ratio is then adjusted to about 2 during the water-gas shift reaction. Washing with water eliminates the halogenated compounds (HCl, HBr, HF, etc.) and ammonia.

The lignocellulosic biomass contains traces of sulphur, most of which is converted into H_2S after gasification. Sulphurated compounds in general, and H_2S in particular, are well-known poisons for the Fischer-Tropsch catalysts. In addition, a large quantity of CO_2 is generated during the gasification step then by the water-gas shift reaction. This is a disadvantage since this inert compound dilutes the other gases during the Fischer-Tropsch synthesis.

Acid gases such as CO_2 and H_2S must therefore be removed from the syngas by washing with solvents, for example methanol or amines. These processes produce a stream of relatively pure CO_2 for possible future sequestration. After a final purification step on absorbents, the syngas is sent to the Fischer-Tropsch unit such as for example the Gasel^{TM} process under Axens licence developed jointly with ENI and IFP Energies nouvelles which will produce mainly Diesel and kerosene, but also gases (top gases) and naphtha.

These aspects show that this type of system is complex and generates low weight and energy yields, leading to considerable investment and operating costs. In 2007, for a unit with a production capacity of 200 000 t/year of BTL, the investment costs were estimated at $670 M [140a]. Another study [140b] evaluated investment costs of $850 M for a production of 480 000 t/year and operating costs of $1.8 per gallon of fuel equivalent.

For a demonstration unit (Choren Beta), capable of treating about 70 000 tonnes of biomass per year, the investment costs amount to 50 M euros.

To be economically viable, BTL complexes must therefore take up the challenge of improving the weight and energy yields by process integrations and removal of technological barriers but also use the scale effect to reduce the cost of the end product. Increasing the capacity of this type of plant will reduce the proportion of the conversion cost related to investment amortisation.

The problems of biomass availability and mobilisation then arise. Due to its nature, lignocellulosic biomass is a dispersed resource. Mobilisation of large quantities, in the region of a million tonnes per year, creates considerable logistics problems.

In addition, the transportation cost has a significant impact on the cost of accessing to the biomass. Beyond a certain supply radius, the costs and environmental impact related to transportation of large quantities of biomass over long distances may prove prohibitive.

The flexibility of BTL complexes with respect to feedstocks (wood, straw, energy crops) helps overcome this difficulty so as to optimise the exploitation of a given supply basin. This flexibility is a technological challenge in itself since the physico-chemical characteristics of biomasses vary widely. This will induce a considerable impact on the syngas pretreatment and purification step.

Another technical solution based on co-processing biomass and fossil feedstocks could be implemented to solve all these constraints.

Introduction of fossil feedstocks in a co-gasification process offers a number of advantages:

- industrial units large enough to benefit from minimum scale effect while using quantities of biomass that are compatible with durable, economic supply having very little impact on the environment (transportation),
- problem of biomass seasonal variability (in terms of quantity and quality) overcome,
- biomass upgraded in regions where use of the resource would not be profitable.

5.3.5 Environmental aspects

The European study [140c] indicates major gains in terms of fossil energy consumption, about 95% for the BTL sector from lignocellulosic materials. The gains in greenhouse gas emissions, expressed in grams of CO_2 equivalent, should reach 87%.

5.3.6 Existing demonstration units and R&D requirements

Industrial units for the production of fuels by coal gasification (CTL) already exist (e.g. Sasol). Until now, biomass gasification reactors have mainly been developed for cogeneration applications (Varnamo, Battelle, FICFB, PRME, etc.). Some plants, however, were already considering the production of fuels, for example the ASCAB demonstration plant at Clamecy (France), which operated during the end of the 1980s. Due to the drop in oil prices at this time, these projects were either slowed down or even abandoned. Developments have resumed in recent years.

The company Choren developed an entrained flow gasifier combining feedstock preparation and gasification: biomass is sent to a first low-temperature gasification stage (NTV) and converted into a carbonaceous fraction and gas. The gas is then sent to the burner of the high-temperature entrained flow gasification reactor (Carbo V) where chemical quenching is carried out using the carbonaceous fraction (Figure 5.22).

A 1 MWth (Alpha plant) pilot unit was operated at Freiberg (Germany). A 45 MWth demonstration unit (Beta plant) was then built on the same site at Freiberg on the basis of data acquired on the pilot unit. This unit is currently in startup phase.

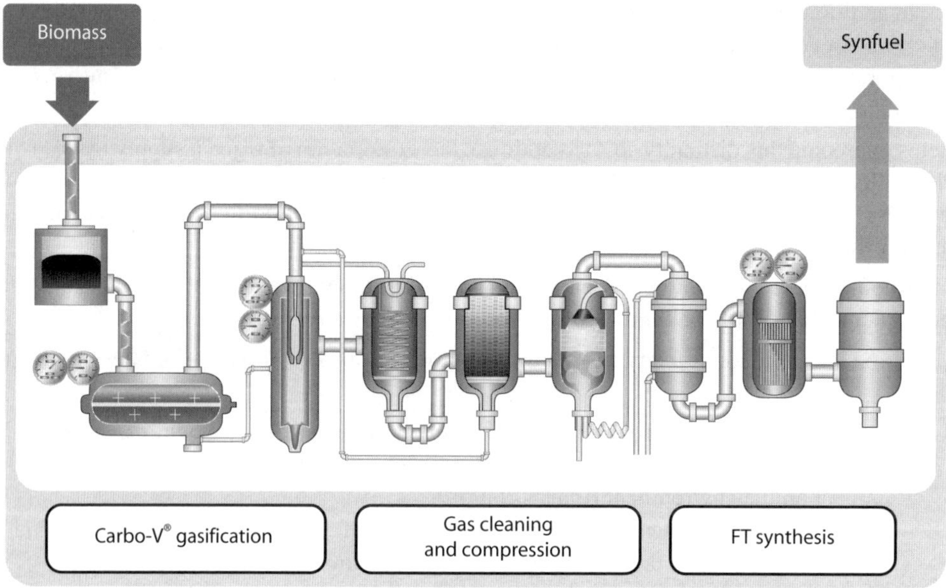

Figure 5.22

Diagram of the Choren Carbo V system.

The Swedish company Chemrec, specialised in the gasification of black liquor, a co-product of the paper industry, has developed a pressurised entrained flow gasification reactor operating at moderate temperature (1 000°C), below the ash melting point. The internal wall is protected by a refractory coating. Chemrec built a 3 MWth demonstration pilot plant in 2005 at Pitea (Sweden), on a pressurised system (30 bar), using pure oxygen. By the end of 2007, this pilot plant had been running for more than 3 600 hours (Figure 5.23).

Chemrec's objective is to build a demonstration unit of ~ 50-60 MWth in a paper pulp plant to demonstrate the feasibility of bio-DME (dimethyl ether) production.

Lurgi and FZK (ForschungsZentrum Karlsruhe) developed BIOLIQ, a concept based on a decentralised approach: collection and first conversion at local scale of biomass into oil by fast pyrolysis, then gasification and production of methanol at regional scale or gasification and Fischer-Tropsch synthesis of fuels in a centralised approach.

The BIOLIQ project involves four successive steps which will eventually be used to demonstrate the concept (Figure 5.7):

- BIOLIQ 1: decentralised fast or flash pyrolysis,
- BIOLIQ 2: gasification/synthesis,
- BIOLIQ 3: syngas purification and methanol synthesis,
- BIOLIQ 4: fuel production, either from methanol (MtSynfuels, bio-DME) or Fischer-Tropsch synthesis.

The basic technology chosen to conduct pyrolysis is based on the LR-mixer (LR = Lurgi Rührgas) reactor. After fine grinding, biomass is heated to 500°C for a few seconds using a

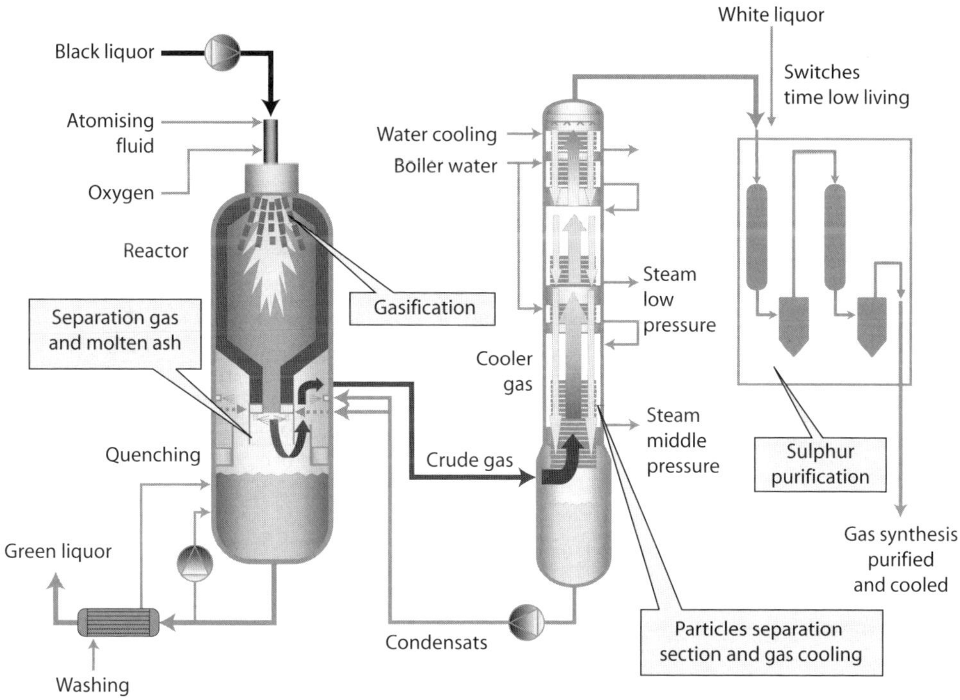

Figure 5.23

Chemrec reactor.
Source: from Chemrec.

solid heat exchange material (sand, steel balls). The liquid organic fraction and the solid phase are mixed to form a slurry. Injected under pressure, this slurry acts as feedstock for the gasification reactor (Lurgi Multipurpose Gasifier).

A gasification process in circulating fluid bed operating at low pressure and low temperature (~ 850°C) was developed by the Technical University of Vienna (TUV) and commercialised by Repotec.

An 8 MWth demonstration unit was built in 2000 at Güssing (Austria) and operated since 2002. The syngas supplies a gas engine and a boiler to generate heat and electricity for the town of Güssing.

This type of gasifier produces syngas containing a large quantity of methane (> 10% mol) (Figure 5.24).

After a basic research phase (reaction kinetics at high temperature and variable pressures: methane reforming, destruction of tars, gasification of the tar residue; behaviour of inorganic species: condensation, corrosion, aerosols, gas purification), the development and industrialisation of the biomass to biofuel gasification processes involve demonstration steps to:

– adapt existing technologies [141] to biomass feedstocks (gasification, AGR, WGS),

– develop technologies dedicated to BTL systems (torrefaction),
– validate the feasibility of the complex sequence of processes,
– validate, using experimental data obtained on tools of significant size, the process integration diagrams defined by technico-economic and environmental evaluations.

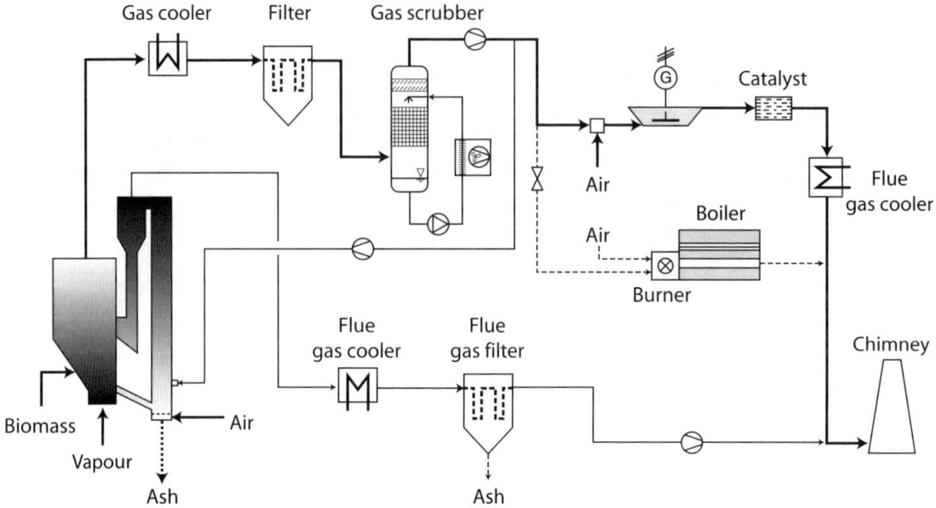

Figure 5.24

Diagram of the Güssing demonstration unit.

Source: from International Slovak Biomass Forum, Bratislava, 9-10/02/2004. Markus Bohlàr-Nordenkampf, Hermann Hofbauer, Institute of Chemical Engineering, Vienna (Austria).

5.3.7 Conclusion

Biomass gasification technologies are very similar to the industrial technologies already implemented for coal. Biomass nevertheless involves specific problems concerning its collection, preparation and treatment of minerals. For biomass, a number of technological bricks already exist for implementation of an autothermal process, but there is no industrial experience on the operation of an integrated process.

The advantage of this technology lies in the fact that it opens numerous process options which can be used to adapt the same technology to different resources, coal (CTL), heavy oils, biomass (BTL), natural gas (GTL), even organic waste, while possibly considering addition of various energies (electricity, hydrogen, natural gas) to significantly increase the synfuel production yield. Moreover, various types of fuel can be produced (dimethyl ether, Diesel, kerosene, methanol and even hydrogen).

Lastly, advanced processes would allow electricity and hydrogen to be stored in a liquid Diesel without any storage or distribution problems and without the need for modifications to current vehicle engines.

5.4 DIRECT LIQUEFACTION OR HYDROTHERMAL CONVERSION

A priori, several strategies can be used to produce liquid fuels from lignocellulosic biomass: gasification followed by the manufacture of a fuel from syngas, pyrolysis to produce usable oils after improvement of their physico-chemical characteristics, and lastly direct liquefaction. It is easy to see, in fact, that while the aim is to produce a liquid fuel, it may be simpler and therefore more economically efficient to manufacture a liquid compound directly and then improve the properties, rather than going via an intermediate. This paragraph focuses on the only direct biomass liquefaction processes, those based on hydrothermal conversion. They involve two main reactions: solvolysis (depolymerisation through physical and chemical interactions with the solvent), and hydrogenolysis (cleavage of chemical bonds by hydrogenation). These two reactions generally take place simultaneously.

5.4.1 Principle and operating conditions

The principle of direct liquefaction consists in dissolving the ground biomass in an aqueous or organic solvent, possibly a hydrogen donor solvent, and adjusting the various operating parameters (temperature, total pressure, partial pressure in reducing gas, presence of catalyst) to produce an organic liquid of reduced oxygen content and an H/C ratio similar to that of hydrocarbons.

Obviously, as with all thermochemical conversion processes, direct liquefaction of biomass leads to a mixture of reaction products whose respective proportions depend on the operating conditions: liquid organic compounds, light gases (H_2, CO, CO_2, CH_4) and solid residues (mineral residues, carbon). Numerous studies have been conducted on the effect of the operating conditions on direct liquefaction of biomass, in particular sawdust [142].

The optimum temperature lies between 250°C and 450°C. Below 250°C, the reaction rates are too slow, while above this temperature thermal degradation phenomena prevail over the solvolysis and hydrogenolysis reactions. The pressure conditions vary depending on the processes, but liquefaction is generally carried out under a total pressure of 150 to 200 bar. It may be interesting to work under reducing atmosphere (H_2) to favour, for example, hydrogenation of the cuts produced.

Water is the solvent most frequently used for direct liquefaction. Organic solvents such as petroleum aromatic cuts, oxygenated organic solvents and cresols have also been used. One essential parameter to be taken into account is the viscosity of the reaction mixture, which must remain pumpable at the reaction temperature.

Out of the numerous catalysts which have been tested in direct liquefaction, the most interesting are the alkali and alkaline-earth metals as carbonates and in particular sodium carbonate. Hydrogenation metal catalysts (nickel-, cobalt- and molybdenum-based) can also be used if deeper hydrogenolysis is required.

5.4.2 Processes

The first studies focusing on the development of direct liquefaction processes started in the 1920s, but it was mainly after the two oil crises, on the initiative of the American Bureau of Mines, that significant technical progress led to the construction of the first pre-industrial liquefaction pilot units. The oil companies developed a number of processes, in particular those using petroleum solvents to dissolve the biomass.

Mobil, for example, filed patents for the production of liquid fuels by direct liquefaction of biomass [143] and tested its concept at prototype scale. The process consists in forming a slurry by mixing sawdust with a petroleum cut distilling between 230 and 600°C and in heating this mixture for about 1 hour at a temperature of between 250 and 275°C. The fuel obtained has the viscosity of a heavy fuel oil at ambient temperature. The wood conversion may get close to 100%. This process is adapted to the production of fuel to supply thermal power plants and offers the advantage of operating at atmospheric pressure without catalyst. It operates well with solvents such as the aromatic residues from the FCC unit. The resulting fuel, however, must be filtered while hot to eliminate the solid particulates in suspension. This process has not found any industrial applications, since the outlets for this type of fuel are not economically viable.

Recent developments in the field of liquefaction have focused on the possibilities of using water as solvent. Apart from the environmental advantages, conversion of biomass, in particular that with high water content, in supercritical water environment offers numerous advantages. At temperatures and pressures greater than the critical values of water (374°C, 221 bar), the reaction of organic compounds with water is kinetically favoured. Numerous recent developments are based on this technology which, despite the reaction step which occurs in liquid environment, is more like a gasification process when the temperature exceeds 374°C. Below 374°C, conversion is biased towards the synthesis of high-molecular weight products [144].

HTU® (Hydrothermal Upgrading) developed by the TNO-MEP research centre in the Netherlands, in collaboration with other Dutch universities, is an interesting process (Figure 5.25). Following on from the developments made by Shell in the 1980s, this biomass liquefaction process consists in reacting the biomass in aqueous environment at high temperature and pressure. One advantage offered by the process would be its efficiency in reducing the biomass oxygen content from 40-50% to 10-15%, by dehydration to produce water and by decarboxylation to form carbon dioxide. The composition of the products obtained from the HTU process would correspond to that given in Table 5.10. The aqueous phase obtained contains 50 to 100% of the salts in the biomass feed as well as hydrophilic organic compounds such as acetic acid, methanol, acetone, etc. The oil phase, known as biocrude, is obtained after separation of the aqueous phase. It has a relatively high calorific value, about 30-35 MJ/kg, higher than that of the oil produced during fast pyrolysis processes. Its average elementary composition is as follows: C 75-82%, H 6-8%, O 10-18%. This biocrude can be used as fuel to produce electricity or hydrotreated to improve its characteristics and convert it partly to products like Diesel [145].

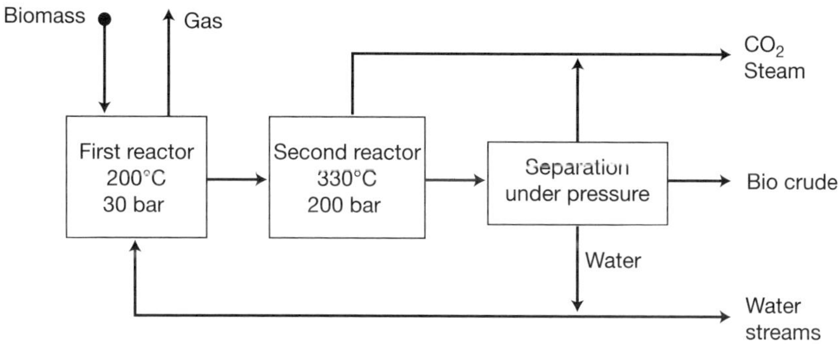

Figure 5.25

Schematic diagram of the HTU® process.

Table 5.10 Composition of effluents obtained from the HTU® process.

Products	% weight
Biocrude	
MW fraction [a] 760	7.5
MW fraction 500	10.0
MW fraction 380	17.5
MW fraction 260	12.5
Methyl-n-propylether equivalent	2.5
CO_2	25
Methanol	5
Ethanol	3.5
Water	16.5

[a] *MW: Molecular Weight.*

5.5 LIQUID FUELS AND HYDROGEN PRODUCED FROM BIOMASS SYNGAS

Syngas obtained by thermochemical conversion of biomass can be used as raw material to produce hydrogen and liquid fuels such as methanol, dimethyl ether and BTL fuel.

This paragraph successively describes the various reactions involved in the synthesis of these compounds as well as their properties as fuels.

5.5.1 Methanol synthesis [146], [147], [148], [149]

Methanol synthesis from syngas is an exothermic reaction. It occurs according to the following main equation:

$$CO + 2H_2 \Leftrightarrow CH_3OH \ \Delta H298\ K = -90.64\ kJ/mol$$

The operating conditions of this reaction are now far less severe than in the past. Pressures in the region of 250-350 bar and temperatures reaching up to 350°C have been reduced on average to 75-100 bar and 180-270°C. This has obviously resulted in a substantial reduction in the production cost since the first commercial synthesis of this alcohol in 1923 by BASF.

The main innovation which led to generalisation of less severe operating conditions concerns the reaction catalysts. The first catalysts were composed of zinc and chromium oxide (Zn/Cr_2O_3). Highly active and selective, they also displayed greater resistance to sulphur, an important element in the composition of the syngas produced, in particular from coal.

The improvement in syngas purification processes was decisive, reviving interest in the copper-based catalysts, which are more active, although more sensitive to impurities. The better activity of the copper oxide-based catalyst compared with its zinc oxide/chromium oxide-based competitor leads to better overall financial profitability thanks to less severe operating conditions. With copper-based catalysts, in fact, the equipment required is smaller and the energy consumption of the industrial unit reduced. Furthermore, a lower temperature reduces the quantity and variety of co-products, thereby simplifying final methanol purification. The last unit operating at high pressure and high temperature closed in the mid 1980s. All the catalysts currently used in the new units built are copper-based.

These improvements can be attributed in particular to the company Johnson Matthey (JM) (formerly Imperial Chemical Industries), which commercialised the first industrial unit in Taiwan in 1970.

The low-pressure process uses an adiabatic reactor into which the cold syngas is injected through the fixed catalyst beds. While the yields obtained with the successive heating and cooling phases are only average, owing to its simplicity and reliability this process is still one of the most widely used.

We must nevertheless point out that most methanol is currently produced using natural gas as initial feedstock, and the syngas obtained is cleaner than it would have been if lignocellulosic biomass had been used as raw material. In the latter case, the syngas produced may contain various impurities such as tars, ash, dust, alkaline compounds, halogens and even sulphur, which must be eliminated.

The following secondary reactions also take place:

$CO_2 + 3H_2 \Leftrightarrow CH_3OH + H_2O$ ΔH298 K = – 41 kJ/mol

$CO + H_2O \Leftrightarrow CO_2 + H_2$ ΔH298 K = – 50 kJ/mol

Due to these various reactions, there is an optimum value for the molar ratio of the gases present in syngas, defined as follows: $(nH_2\text{-}nCO_2)/(nCO + nCO_2)$. For methanol production, the fresh gas stoichiometric ratio is 2.

The main licence vendors of methanol production processes are ICI (now JM) already mentioned, Lurgi and Mitsubishi Gas Chemical (MGC). Other technology vendors obviously exist, including Acid Amine Technologie, AO Giap, Badger/Shaw, Forskningsparken A/S, Jacobs Engineering, Haldor Topsoe, Kellogg, Brown & Root, Linde AG, Methanol Casale, Polimex-Cekop Chemical Ltd, Stauffer, Uhde.

The market shares of the main players in methanol production over the period 1999-2003 are respectively 61% for JM, 27% for Lurgi, 8% for MGC, 3% for Kellogg and 1% for the others [1].

Methanol has been used as an automotive fuel since the 1920s, competing with ethanol and kerosene. In the 1940s, it was used in aviation for the higher power that could be obtained during take-off and climbing. However, it was mainly following the oil crisis of the 1970s that the possibility of blending it with gasolines was considered. Lastly, from the 1990s, with the appearance of lead-free fuels, methanol was primarily used in the form of MTBE. MTBE eliminates many of the technical drawbacks inherent to the use of pure methanol: demixing of the methanol/gasoline mixture in the presence of traces of water and higher volatility of the methanol/gasoline mixture. Use of methanol in the form of MTBE is now threatened in some countries.

Methanol has also been considered as a source of hydrogen to supply fuel cell systems, with hydrogen being produced on board the vehicle. This option, at some stage taken up by Daimler and Chrysler, is now increasingly rejected, especially since the US DOE has just decided to stop research programmes related to onboard production of hydrogen.

The main methanol producers are Methanex (Canada), SABIC (Saudi Arabia), Methanol Holding Trinidad (Trinidad) and Celanese (United States) which together hold almost 40% of the market. Methanex is leader with 20% of the market share.

Most methanol production units, generally located near the markets they supply, are now relatively small, with capacities of about 30 000 to 150 000 t/year. These new units therefore use expensive natural gas, which may represent up to 75% of the final cost of the product. The prices of methanol, which have fluctuated over the last few years between 100 and 250 euros/t in Europe, are therefore closely linked to those of natural gas.

We are currently witnessing a reversal of the trend, designed firstly to move the production sites closer to the areas where natural gas is cheap (typically between 0.5 and 1 dollar/MMBTU) and secondly to increase the size of the plants very significantly to about 750 000 tonnes/year, the capacity of the units operating in Saudi Arabia and Qatar. Some projects of over 3 Mt/year have even been studied, as announced recently in Qatar.

5.5.2 Dimethyl ether synthesis

Dimethyl ether (DME) is often mentioned as an alternative fuel. Gaseous at room temperature and pressure and currently used as a solvent or propellant in some aerosols, DME can be described as LPG for Diesel engines. It is non-toxic and non-carcinogenic. It has a half-life of just one day, before decomposing into water and carbon dioxide. In addition, its combustion is "clean" since, compared with Diesel, the NO_x, CO and particulate emissions are significantly reduced.

Most DME is currently produced by dehydration of methanol according to the following reaction (with an aluminosilicate-based catalyst):

$$2\,CH_3OH \Leftrightarrow CH_3OCH_3 + H_2O$$

1. *Source*: SRI (Stanford Research Institute).

An aluminosilicate-based catalyst is used for this reaction.

A new syngas to DME production pathway is currently being developed. The overall reaction equation is as follows:

$$3H_2 + 3CO \Leftrightarrow CH_3OCH_3 + CO_2$$

This balance equation is in fact the result of combining methanol synthesis, methanol dehydration and the water-gas shift reaction. The main reactions representative of this process are as follows:

$$4H_2 + 2CO \Leftrightarrow 2CH_3OH \quad (1)$$

$$2CH_3OH \Leftrightarrow CH_3OCH_3 + H_2O \quad (2)$$

$$CO + H_2O \Leftrightarrow CO_2 + H_2 \quad (3)$$

A fourth reaction also occurs:

$$CO_2 + 3H_2 \Leftrightarrow CH_3OH + H_2O \quad (4)$$

Reaction (3) is very slow during the process. Only methanol synthesis and methanol dehydration occur. The end products are therefore DME, CO_2 and water. In practice, however, reactions (3) and (4) occur simultaneously and, under equilibrium conditions, CO_2 is always formed.

The DME synthesis catalyst consists of a mixture of the catalysts used for methanol synthesis (copper, zinc and aluminium oxides) and methanol dehydration (aluminosilicate). Research studies are currently underway to optimise their composition.

DME is also an intermediate product in a process developed during the 1980s by Mobil Oil aimed at converting methanol into gasoline. This process is known by the acronym MTG which stands for Methanol To Gasoline.

The overall reaction is written: $nCH_3OH \Rightarrow (–CH_2–)n + nH_2O$.

This synthesis first involves dehydration of methanol into DME. The ether is then converted into a biradical**:** CH_2 on a ZSM5 zeolite-based catalyst according to the following reaction:

$$CH_3OCH_3 \Rightarrow 2{:}\ CH_2 + H_2O$$

The biradicals then generate olefins and hydrocarbon mixtures of all families.

$$n\,(:CH_2) \Rightarrow (–CH_2–)\,n$$

The hydrocarbon effluents contain 60 to 80% of liquid products (C_5+ fraction) and 20 to 40% of gas. No hydrocarbons heavier than C10 are formed. The aromatic content lies between 25 and 45%, which is a problem for certain markets. This process was implemented on industrial scale in New Zealand in 1987, but the plant owned by Methanex has since been dismantled.

5.5.3 Fuel methanol and MTBE [150], [151]

Methanol has been used as fuel in gasoline engines and even until 1940 as additive in aircraft piston engines. From 1970 to about 1990, addition of methanol to gasolines was widespread in the United States mainly as well as in some European countries, especially Germany. European Directive 85/5 36/EEC set the maximum authorised content to 3% by volume. In the United States, addition of methanol was limited to a maximum of 5% by volume.

5.5.3.1 Methanol characteristics

The characteristics of these two compounds and of other oxygenated products such as TertioButylic Alcohol (TBA) and TerAmylMethylEther (TAME), ethanol and ETBE are listed in Table 5.11.

Table 5.11 Characteristics of various oxygenated organic products.

Characteristics	Methanol	Ethanol	TBA	ETBE	MTBE	TAME
Density at 15.5°C (kg/m^3)	796	794	792	750	746	750
Boiling point (°C)	64.7	78.3	82.8	72.8	55.3	86.3
RVP at 37.8°C (bar)	0.32	0.16	0.12	0.40	0.54	0.10
Heat of vaporisation (kJ/kg)	1 100	854	510	321	337	310
Stoichiometric ratio	6.4	8.95	11.1	12.1	11.7	12.1
NCV/weight (kJ/kg)	19 937	26 805	32 560	35 880	35 200	36 500
NCV/volume (kJ/l)	15 870	21 285	25 790	26 910	26 260	27 375
Research octane number (RON)	126	120	113	117	118	114
Motor octane number (MON)	96	99	100	101	101	100
Cetane number	0 to -3	5	< 0	< 0	< 0	< 0

The low calorific value of methanol, about 50% less than that of gasoline, increases the vehicle fuel volumic consumption.

Its high octane numbers make methanol a good potential fuel for spark ignition engines. Its high heat vaporisation must not be neglected during startup and in cold weather.

While the vapour pressure of methanol is lower than that of gasolines, the measured value of the mixture is significantly increased since methanol forms azeotropes with the light compounds in gasolines. The increase in vapour pressure may reach 150 to 200 mbar after adding 2 to 3% methanol. It then stabilises and even decreases when more than 15% methanol is added.

Demixing of the gasoline-methanol mixture can be expected if the fuel contains any traces of water. This phenomenon can be avoided by adding a co-solvent such as TBA.

Although the mechanical resistance of some materials used in the distribution and supply circuits may be affected by the presence of methanol in the fuel, these problems can be solved by the use of new polymer formulations.

Other experiments have been conducted, mainly in the United States, on mixtures with 85 and 90% methanol, designated respectively as M85 and M90 [152]. After adapting, for example, the compression rate, installing methanol-resistant materials and modifying the intake circuit, the vehicles develop good performance. Mixture M85 reduces the emissions of some pollutants such as CO and HCs, but much higher formaldehyde emissions are observed.

All these attempts to use methanol have progressively been abandoned and replaced by the MTBE technology which developed from the mid 1980s, brought in as lead derivatives were progressively removed from gasolines.

5.5.3.2 MTBE characteristics

As mentioned in Table 5.11, MTBE has high octane numbers. Unlike methanol, MTBE mixed with gasoline does not increase the vapour pressure and demixing does not occur when traces of water are present. CO and HC emissions are also reduced.

Since the start of the years 2000, however, questions have been raised in the United States concerning the use of MTBE in fuel, for reasons related to protection of the environment (see Chapter 1).

5.5.4 Fuel-DME

5.5.4.1 DME characteristics

Table 5.12 shows some of the properties of DME. DME is a colourless gas at ambient temperature, highly flammable in air, with an odour similar to that of ether. Its vapour pressure is 0.6 MPa at 25°C. It can be liquefied at ambient temperature under a pressure of 5 bar. It can be transported and stored in liquid form, like LPG. DME is not corrosive to metals but is an excellent solvent of numerous elastomers, especially those used in conventional Diesel engine fuel supply circuits.

It contains oxygen and no carbon-carbon bonds, thereby seriously limiting the possibilities of forming carbon particles during combustion. It has a high cetane number, between 55 and 60, depending on the sources.

Table 5.12 Main DME characteristics.

Properties	DME	Diesel	LNG
Formula	CH_3OCH_3		CH4
Boiling point (°C)	– 25	150-380	– 162
Density at 20°C (kg/m^3)	660	800-840	420
NCV (MJ/kg)	28.4	42.5	49.4
Cetane number	55-60	40-55	
Auto-ignition temperature [a] (°C)	350 [b]	206 [c]	732

[a] *With air at 0.1 MPa.*
[b] *A value of 235°C is also found in the litterature.*
[c] *n-cetane.*
Source: IFP Energies nouvelles.

In terms of safety, DME is similar to LPG and Diesel. The aspects concerning toxicity and its potential environmental impact have been assessed by several companies and government organisations, mainly due to its use as substitute for CFCs in aerosols [153].

The conclusions of these studies reveal that the toxicity risks associated with DME are low or virtually non-existent and that its environmental impact is very limited.

5.5.4.2 Use of DME in Diesel engines

The fuel circuit and tank must be adapted to take into account the physical properties of DME (density, viscosity, volatility, etc.), quite different from those of Diesel [154].

Its lubricating power is too low for the injection systems, so lubrication additives are required. Long-chain alcohols and fatty acids of chain length between C15 and C22 could be used [155].

A. Impact on combustion

Numerous studies have been conducted to characterise the jet shape and the processes occurring during combustion. Globally, they show that DME vaporises very quickly and that no soot forms, even temporarily during combustion. Auto-ignition occurs in the lower part of the spray, while the temperature at the injector nose is low due to the latent heat of vaporisation of DME.

DME seems to be a promising fuel for HCCI engines, since auto-ignition can occur over a wide range of equivalence ratios [156].

B. Impact on emissions

Several studies agree on the fact that engines supplied with DME produce few if any particulates [157].

The results obtained concerning NO_X and HC emissions are contradictory and vary considerably depending on engine settings. It is nevertheless accepted that by delaying the injection advance, NO_X emissions are generally comparable with those generated by Diesel and that discharges of HCs, which contain few PAHs and monoaromatics, can be reduced by using an oxidation catalyst.

CO emissions are higher with DME than with Diesel [158].

C. Compatibility with materials

It is a known fact that DME damages elastomers. PTFE (Teflon) seals, for example, becomes brittle due to sudden and repeated temperature variations created by vaporisation of DME, and are no longer efficient. Metal seals would appear to be the best, provided that the metal does not produce sparks.

There is a still risk of leakage through the injectors under high pressure, after the engine has been switched off, although studies indicate that the leaks would be negligible.

5.5.5 Fischer-Tropsch synthesis [159], [160]

The Fischer-Tropsch synthesis (named after two German chemists Hans Fischer and Franz Tropsch, who discovered and studied this reaction during the 1920s) is designed to convert syngas to a Diesel type fuel composed of a mixture of hydrocarbons of various chain lengths. Until the early 1990s, this process was used mainly in countries suffering from oil shortages, for example Germany during World War II and South Africa during the oil embargo, enforced under the apartheid regime. In both cases, the synthetic fuel was produced from coal, especially by Sasol in South Africa. In the early 1990s, Shell revived interest in this sector by building a plant in Malaysia using natural gas as raw material. Numerous projects are currently being considered, but still based on the exploitation of natural gas reserves.

After the production of syngas, the hydrogen and carbon monoxide injected in the Fischer-Tropsch unit react exothermically to form alkyl radicals and water according to the following equation:

$$2H_2 + CO \Leftrightarrow -CH_2- + H_2O, \Delta H298\ K = -\ 165\ kJ/mole$$

Water molecules formed also react exothermically with the CO, according to reaction (3) already mentioned, in the presence of an Fe-based catalyst.

However, the main reaction occurring concerns the alkyl radicals which react as soon as they appear, forming, in an iron- or cobalt-based catalytic environment, olefinic and paraffinic hydrocarbon chains according to a process similar to polymerisation. The overall reaction can then be represented by the following equations:

$$n(2H_2 + CO) \Leftrightarrow CnH_2n + nH_2O \text{ (reaction catalysed by iron)}$$

$$(2n+1)H_2 + nCO \Leftrightarrow CnH_2n + 2 + nH_2O \text{ (reaction catalysed by cobalt)}$$

Iron catalysis biases the reaction to the production of olefins in particular, which can be used to make synthetic detergents. In contrast, cobalt catalysis used with an H_2/CO ratio of about 2 leads to the production of methane and mainly linear saturated paraffins ($CnH_{2n}+_2$) ranging from ethane to $C_{80}H_{162}$ with a Anderson-Schulz-Flory distribution. The aim is therefore to obtain selectivity ranging from C_5 up to heavier liquid products with most products in the C_{10}-C_{20} interval, i.e. with a distillation interval from 180 to 360°C, similar to that of a Diesel.

Since paraffinic hydrocarbon chains above C_{20} are solid waxes at ambient temperature, isomerising cracking is generally required. To obtain as much Diesel fuel as possible, a catalytic hydrocracking unit is implemented. In addition, specific catalysts can induce the isomerisation of straight hydrocarbon chains into branched chains with lower solidification temperatures and higher viscosity indices.

The Diesel obtained is of excellent quality, since it contains no sulphur, nitrogen, metals or aromatics. It has a high cetane number, about 70-75 (as compared with values of about 50 for commercial diesels). These properties make it an ideal fuel for current and future Diesel engines, especially since it considerably reduces pollutant emissions compared with a traditional Diesel (see Chapter 3).

Note that the main products to be obtained when implementing a Fischer-Tropsch process are fuels for the road transport sector, generally for Diesel engines. Currently, however, Diesel produced in Fischer-Tropsch units is always accompanied by naphtha and kerosene. Depending on the operating conditions, the product distribution at unit output is as follows: 25 to 60% Diesel, 25 to 50% kerosene and 15 to 25% naphtha. The process will be biased towards either kerosene or Diesel depending on the targeted market. Processes using coal is designated by the acronym CTL (Coal-to-Liquids), gas by GTL (Gas-to-Liquids) and biomass by BTL (Biomass-to-Liquids).

The start of the years 2000 represents a major milestone for industrial development of GTL, with the announcement of large scale projects. In 2001, for example, Sasol (South African state-owned company), associated with Chevron since June 1999, announced the first contracts in Qatar and Nigeria to build industrial GTL production plants. In 2000, Shell signed an agreement with Egypt and two years later announced the launch of a project in Qatar.

Two points are worth mentioning out of this multitude of announcements:

- after the plant in Malaysia (15 000 b/d), Shell built a new GTL unit in Qatar (140 000 b/d), planned to start in 2011-2012,
- in 2008, Sasol built a 35 000 b/d unit, once again in Qatar, whose capacity is expected to be progressively increased up to 150 000 b/d.

5.5.6 BTL fuel

With characteristics similar to those of GTL fuel, BTL fuel is perfectly adapted to the requirements of Diesel engines. The hydrocarbon cuts obtained are mainly composed of *n*-paraffins, free from aromatics and sulphurated compounds. Due to their high cetane number, above 70, these cuts are excellent fuels, especially in terms of pollutant emissions. As shown on Figures 5.26 and 5.27, compared with Diesel, they reduce CO emissions by more than 50%, HC emissions by about 50% and particulate emissions by 20 to 50%, depending on the engine technologies used. There seems to be little effect on nitrogen oxide emissions, possibly even a slight increase.

5.5.7 Hydrogen production [149]

From syngas, the quantity of hydrogen can be increased via the water-gas shift reaction. The hydrogen obtained must then be purified.

The reaction is slightly exothermic (ΔH298 K = – 41 kJ/mole) and more or less complete depending on whether it is carried out in 1 or 2 steps. The operating conditions will also depend on the type of hydrogen purification process. If virtually complete conversion of the CO is required, two shift steps will be carried out after first cooling the gaseous mixture down to 350°C. The conversion is favoured by low temperatures and excess steam; a temperature below 250°C will therefore be required to obtain virtually complete conversion of the CO. Under these conditions, however, the carbon methanation and formation reactions may also develop. Use of active and selective catalysts is therefore essential. They are generally sensitive to various poisons, in particular halogenated and sulphurated compounds.

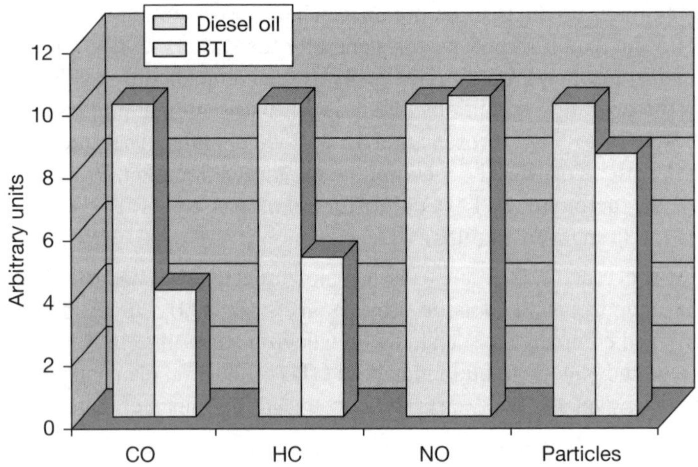

Figure 5.26

Emissions from a vehicle equipped with a common rail injection system.

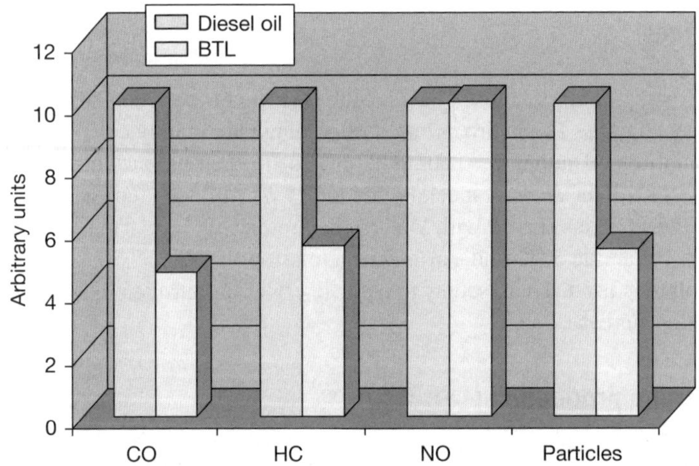

Figure 5.27

Emissions from a vehicle equipped with pump injectors.

Depending on the type of hydrogen final purification process, one or two steps will be carried out.

Low-temperature conversion takes place in the presence of copper oxide catalysts (15-30% by weight), promoted by zinc oxide deposited on alumina. The temperature is in the region of 200 to 250°C for the second step.

High-temperature conversion avoids the production of coke. An iron- and chromium oxide-based catalyst is used (90%-10% by weight). This operation is carried out at temperatures of about 350 to 450°C. The CO content at reactor outlet is 1 to 3% by vol.

The final gas contains 0.1 to 0.4% by vol. of CO of the dry gas obtained.

The final purification step consists in separating hydrogen from CO_2 then eliminating any remaining traces of impurities. This step can be carried out by selective adsorption on molecular sieves by PSA (Pressure Swing Adsorption) or by methanation. In modern plants, the CO converter is generally followed by a PSA unit to produce highly pure hydrogen. With this purification technique, the CO content at unit inlet must lie between 1 and 3%. The water-gas shift reaction can therefore be conducted at high temperature in a single step. In some cases, however, medium temperature conversion is preferable to reduce the CO content to 1% by volume. The PSA technique therefore removes the need for the traditional low-temperature conversion processes. Hydrogen purification is carried out on the gas obtained after CO conversion, which contains 65 to 70% hydrogen. The molecular sieve separation process uses activated carbon or zeolites to remove CO, CO_2 and CH_4. The separation units operate cyclically, according to adsorption/desorption cycles; the adsorbent is regenerated during the desorption cycle. The adsorption and desorption processes are carried out at different pressures. The operating conditions are as follows:

- temperature: 5 to 40°C,
- pressure: optimum at about 3.5-4 bar.

About 85 to 90% hydrogen is recovered from syngas and the hydrogen purity reaches 99.9%.

The PSA purge, which contains CH_4, CO, H_2 (10 to 15%) and CO_2, has a non-negligible calorific value and is therefore often used as fuel in the syngas reactor furnace.

Purification by methanation produces 97-98% pure hydrogen, unless it is followed by cryogenic separation of methane and nitrogen, to increase the purity. In this pathway, some of the hydrogen is consumed for deep CO conversion. To limit the consumption, high CO conversion is required, in two steps (at high temperature – residual CO content of 1 to 3% by volume – then at low temperature – residual CO content of 0.3 to 0.8% by volume).

The methanation pathway is carried out via a first decarbonation step followed by a second step for methanation of the residual CO and CO_2.

The decarbonation step consists in chemical separation of the CO_2 by liquid absorption (processes using ethanolamines or carbonates). The remaining CO_2 now represents no more than 0.005% to 0.1% by volume. The second step for methanation of CO and elimination of the residual CO_2 is carried out with a nickel catalyst, according to the following reactions:

$$CO + 3H_2 \Leftrightarrow CH_4 + H_2O \qquad \Delta H298\ K = -\ 206\ kJ/mol$$

$$CO_2 + 4H_2 \Leftrightarrow CH_4 + 2H_2O \qquad \Delta H298\ K = -\ 162\ kJ/mol$$

These two reactions are almost complete below 300°C.

The final CO and CO_2 contents are now less than 0.001%. About 98% of the hydrogen is recovered from the syngas.

5.5.8 Fuel/electricity polygeneration

Gases of low energy content (4 to 10 MJ/Nm^3) can be recovered from all the processes described above and used for example to produce electricity. Since the commercially available combustion turbines do not run properly on fuels with such low energy content, they must be adapted or used in co-combustion, for example with natural gas. In addition, if a combined cycle system is used, the steam generated can be re-employed in the process.

For the time being, co-production of fuel and electricity is rarely used at industrial scale since it incurs a non-negligible additional investment cost and requires a nearby consumer for the electricity generated.

REFERENCES

[1] Bridgwater AV (2003) Renewable Fuels and Chemicals by Thermal Processing of Biomass, Chemical Engineering Journal, **91**, 2-3, pp 87-102.

[2] Bridgwater AV, Meier D and Radlein D (1999) An Overview of Fast Pyrolysis of Biomass, Organic Geochemistry, 30, pp 1479-1493.

[3] Benabdallah B, Carré J, Khennas S, Vergnet LF and Ossoukaii P (1994) Guide Biomasse-Energie, Academia Ed. Collection Etudes et Filières, Québec.

[4] Antal Jr MJ, Croiset E, Dai X, DeAlmeida C, Shu-Lai Mok W and Norberg N (1996) High-Yield Biomass Charcoal, Energy & Fuels, 10, pp 652-658.

[5] Bridgwater AV (1996) Thermal Biomass Conversion and Utilisation – Biomass Information System, European Commission report AIR1-92-8006.

[6] Mock W S-L and Antal MJ (1983) Effects of Pressure on Biomass Pyrolysis. I Cellulose Pyrolysis Products, Thermochim. Acta., pp 155-164.

[7] Antal MJ, Varhegyi G and Jakab E (1998) Cellulose Pyrolysis Kinetics: Revisited, Industrial and Engineering Chemistry Research, 37 (2), pp 1267-1275.

[8] Richard JR and Antal MJ (1994) Thermogravimetric Studies of Charcoal Formation from Cellulose at Elevated Pressures, in Advances in: *Thermochemical Biomass Conversion*, Bridgwater AV Ed., Blanchie Academic and Professional, London, **2**.

[9] Girard P, Numazawa S, Mouras S and Napoli A (1999) Products Formed under Pressurized Pyrolysis, *in: Proceedings of the 4th Biomass Conference of the Americas, California États-Unis*, pp 1161-1167.

[10] Antal Jr MJ and Gronli M (2003) The Art, Science, and Technology of Charcoal Production, Ind. Eng. Chem. Res., 42, pp 1619-1640.

[11] Diebold JP (1980) Ablative Pyrolysis of Macroparticles of Biomass, *in: Proceedings of the Specialists Workshop on the Fast Pyrolysis of Biomass*, Copper Mountain, CO, SERI/CP-622-1096, Solar Energy Research Institute, Golden, CO, Octobre, p 237.

[12] Lédé J, Panagopoulos J, Li HZ and Villermaux J (1985) Fast Pyrolysis of Wood: Direct Measurement and Study of Ablation Rate, *in: Fuel*, 64, pp 1514-1520.

[13] Bridgwater AV and Peacocke GVC (2000) Fast Pyrolysis Processes for Biomass, *Renewable and Sustainable Energy Reviews,* 4, pp 1-73.

[14] Henrich E, Dinjus E, Koegel A, Raffelt K, Stahl R and Weirich F (2004) A Two Stage Process for Synfuel from Biomass, *in: Proceedings of the 2nd World Conference on Biomass and Energy, Industry and Climate Protection*, Rome, Italy, 10-14 May.

[15] Antal Jr MJ (1983) Biomass Pyrolysis: A Review of the Literature Part I – Carbohydrate Pyrolysis, Advances in Solar Energy, Boër KW and Duffie JA, 1, Plenum Press, p 61.

[16] Kilzer FJ and Broido A (1965) Speculations on the Nature of Cellulose Pyrolysis, Pyrodynamics, 2, p 151.

[17] Evans RJ and Milne TA (1987) Molecular Characterization of the Pyrolysis of Biomass I: Fundamentals, Energy and Fuels, 1, p 123.

[18] Evans RJ and Milne TA (1985) Applied Mechanistic Studies of Biomass Pyrolysis, *in: Proceedings of the Biomass Thermochemical Conversion*, Contractors Meeting, Minneapolis, Minnesota, 15-16/10/85, p 57.

[19] Shafizadeh F (1985) Pyrolytic Reactions and Products of Biomass, *in: Fundamentals of Thermochemical Biomass Conversion*, Overend RP, Milne TA and Mudge LK (Eds), Elsevier Applied Science Publishers, New York, pp 183-218.

[20] Soltes EJ and Elder TJ (1981) Pyrolysis, in Organic Chemicals from Biomass, Goldstein IS (Ed), CRC Press, Boca Raton, Florida, pp 96-100.

[21] Antal Jr MJ (1985) Biomass Pyrolysis: A review of the Literature, Part II – Lignocellulose Pyrolyse, Advances in Solar Energy, Boër KW and Duffie JA, 2, Plenum Press, p 175.

[22] Milne TA (1979) Pyrolysis – The Thermal Behaviour of Biomass Below 600°C, *in: A Survey of Biomass Gasification*, Vol II, SERI TR-33-239, Juillet, pp 95-132.

[23] Nunn TR, Howard JB, Longwell JP and Peters WA (1985) Product Composition and Kinetics in the Rapid Pyrolysis of Milled Wood Lignin, Ind. Eng. Chem. Prod. Res. Dev., 24, p 844.

[24] Napoli A, Van de Steene L, Rousset P and Girard P (2001) Pressured Pyrolysis of Wood for High Quality Charcoal Production, 1st International Congress on Biomass for Metal Production & Electricity Generation, Belo Horizonte Brazil, October.

[25] Diebold JP (1994) A Unified, Global Model for the Pyrolysis of Cellulose, Biomass and Bioenergy, **7**, 1-6, pp 75-85.

[26] Diebold JP, Czernik S, Scahill JW, Philips SD and Feik CJ (1994) Hot-gas Filtration to Remove Char from Pyrolysis Vapours Produced in the Vortex Reactor at NREL, Biomass Pyrolysis Oil.

[27] Shafizadeh F, Furneaux RH, Cochran TG, Scholl JP and Sakai YJ (1979) Appl. Polym. Sci., 23, pp 3525-3539.

[28] Antal Jr MJ, and Varhegyi G (1995) Cellulose Pyrolysis Kinetics: The Current State of Knowledge, Ind. Eng. Chem. Res., **34**, 3, pp 703-717.

[29] Lede J, Blanchard F *et al.* (2002) Radiant Flash Pyrolysis of Cellulose Pellets: Products and Mechanisms Involved in Transient and Steady State Conditions, Fuel, **81**, 10, pp 1269-1279.

[30] Boutin O, Ferrer M *et al.* (1998), Radiant Flash Pyrolysis of Cellulose-Evidence for the Formation of Short Life Time Intermediate Liquid Species, Journal of Analytical and Applied Pyrolysis, **47**, 1, pp 13-31.

[31] Lédé J, Diebold JP *et al.* (1997) The Nature and Properties of Intermediate and Unvaporized Biomass Pyrolysis Materials, Developments in Thermochemical Biomass Conversion, E. A. V. B. a. D. G. B. Boocock, Blackie Academic and Professional, pp 27-42.

[32] Wooten JB, Seeman JI *et al.* (2004) Observation and Characterization of Cellulose Pyrolysis Intermediates by C13 CPMAS NMR. A New Mechanistic Model, Energy & Fuels, **18**, 1, pp 1-15.

[33] Demirbas A and Demircioglu G (2000) Mechanisms Of Carbonization, Gasification, Pyrolysis, Extraction, and Liquefaction of Lignin from Biomass, Reviews in Process Chemistry and Engineering, 3, pp 175-189.

[34] Boronson ML, Howard JB *et al.* (1989) Heterogeneous Cracking of Wood Pyrolysis Tars Over Fresh Wood Char Surface, Energy & Fuels, 3, pp 735-740.

[35] Raveendran K, Ganesh A *et al.* (1996) Pyrolysis Characteristics of Biomass and Biomass Components, Fuel, **75**, 8, pp 987-998.

[36] Koufopanos CA, Maschio G *et al.* (1989) Kinetic Modelling of the Pyrolysis of Biomass and Biomass Components, Canadian Journal of Chemical Engineering, 67, pp 75-84.

[37] Svenson J, Pettersson JBC *et al.* (2004) Fast Pyrolysis of the Main Components of Birch Wood, *Combustion Science and Technology*, 176, pp 977-990.

[38] Müller-Hagedorn M, Bockborn H *et al.* (2003) A Comparative Kinetic Study On The Pyrolysis Of Three Different Wood Species, *Journal of Analytical and Applied Pyrolysis*, 68-69, pp 231-249.

[39] Piskorz JR, Scott D and Czernik S (1989) Pretreatment of Wood and Cellulose for Production of Sugars by Fast Pyrolysis, J. Anal. Appl. Pyrolysis, 16, pp 127-142.

[40] Pakdel H and Roy C (1991) Liquid Hydrocarbons from Vacuum Pyrolysis and Gasification of Wood, *Energy and Fuel*, 5, pp 427-436.

[41] Cuevas A, Reinoso C and Scott DS (1995) Pyrolysis Oil Production and its Perspectives, *in: Proc. Power Production from Biomass* II, Espoo, Mars (VTT).

[42] Mc Lellan – Wellman Process Engineering Limited (1998) PyNe Newsletter n° 7, November, pp 4-15.

[43] Dynamotive Press Release (2000) October 3.

[44] BioOil Production at Dynamotive's Guelph Plant (2007) ThermalNet Newsletter, n° 5 ; September, Aston University, UK.

[45] Thamburaj R (2000) Fast Pyrolysis of Biomass for Green Power Generation, *in: Proceedings of the 1st World Conference on Biomass for Energy and Industry*, Sevilla, Espagne, pp1519-1524.

[46] PyNE (1997) Newsletter N° 4, September, Aston University, UK.

[47] Rossi C and Graham RG (1997) Fast Pyrolysis at ENEL, *in: Kaltschmitt MK and Bridgwater AV, Editors, Biomass Gasification and Pyrolysis: State of the Art and Future Prospects, CPL Press*, pp 300-306.

[48] Underwood G (1992) Commercialization of Fast Pyrolysis Products, *in: Biomass Thermal Processing,* Eds. Hogan E, Robert J, Grassi G and Bridgwater AV, CPL Press, pp 226-228.

[49] Gust S, Niemien JP, Nyronen and Focstera TM (2003) Liquefied Wood Fuel Pilot Plant *in: Pyrolysis and Gasification of Biomass and Waste*, CPL Press, January, pp 169-174.

[50] Prins W and Wagenaar BM (1997) *in: Biomass Gasification and Pyrolysis*, Eds. Kaltschmitt MK and Bridgwater AV, CPL Press, pp 316-326.

[51] Wagenaar BM, Venderbosch RH, Carrasco J, Strenziok R and Van der Aa BJ (2001) Rotating Cone Bio-oil Production and Applications, *in: Progress in Thermochemical Biomass Conversion,* Ed. Bridgwater AV, Blackwell, Oxford, UK.

[52] http://www.btgworld.com/technologies/pyrolysis.html

[53] Henrich E, Dinjus E, Koegel A, Raffelt K, Stahl R and Weirich F (2004) A Two Stage Process for Synfuel from Biomass, 2nd World Conference on Biomass and Energy, Industry and Climate Protection, 10-14 May, Rome, Italy.

[54] Raffelt K, Henrich E, Koegel A, Stahl R, Steinhardt J and Weirich F (2004) Stable Slurries from Biomass Pyrolysis Products for Entrained Flow Gasification, 2nd World Conference on Biomass and Energy, Industry and Climate Protection, 10-14 May, Rome, Italy.

[55] Stahl R, Henrich E, Koegel A, Raffelt K, Steinhardt J, Weirich F and Dinjus E (2004) Pressurised Entrained Gasification of Slurries from Biomass, 2nd World Conference on Biomass and Energy, Industry and Climate Protection, 10-14 May, Rome, Italy.

[56] Schöll S, Klaubert H and Meier D (August 2004) Bio-oil from a New Ablative Pyrolyser, International Conference, Science in Thermal and Chemical Conversion of Biomass, Victoria, Canada.

[57] Pyne Newsletter (2001) Issue 12, September.

[58] Lédé J (2000) The Cyclone: a Multifunctionnal Reactor for the Fast Pyrolysis of Biomass, Ind. Eng. Chem. Res., **39**, 4, pp 893-903.

[59] Broust F (2003) Le cyclone : un réacteur multifonctionnel. Application à la pyrogazéification et à la pyroliquéfaction de la biomasse, Thèse de l'INPL (Nancy, France).

[60] Bramer EA and Brem G (2002) A Novel Technology for Fast Pyrolysis of Biomass: Pyros Reactor, *in: 12th European Conference and Technology Exhibition on Biomass for Energy, Industry and Climate Protection*, Amsterdam, NL.

[61] Bramer EA, Holthuis MR and Brem G (2004) Development of a Cyclonic Reactor with Internal Particle Filter for the Flash Pyrolysis of Biomass, in: 2nd *World Conference on Biomass for Energy, Industry and Climate Protection*, Rome.

[62] Yang J, Blanchette D, de Caumia B and Roy C (2001) Modelling, Scale-up and Demonstration of a Vacuum Pyrolysis Reactor, *in: Progress in Thermochemical Biomass Conversion*, Ed. Bridgwater AV, Blackwell, Oxford, UK.

[63] Girard P, Blin J, Bridgwater AV and Meier D (2004) An Assessment of Bio-oil Toxicity for Safe Handling and Transportation, *Science in Thermal and Chemical Biomass Conversion*, Vancouver Island, Canada, September.

[64] Bridgwater A V and Hogan EN (1996) Bio-oil Production and Utilization, *in: Proceedings of the 2nd EU-Canada Workshop on Thermal Biomass Processing*, CPL Press.

[65] Barbucci P, Costanzi F, Ligasacchi S, Mosti A and Rossi C (1995) Bio-fuel Oil Combustion in a 0.5 MW Furnace, *in: Proceedings of the Second Biomass Conference of the Americas,* NREL, Golden CO 80401, CP-200-8098, pp 1110-1120.

[66] Gust S (1997) Combustion Experiences of Flash Pyrolysis Fuel in Intermediate Size Boilers, *Developments in Thermochemical Biomass Conversion* (AV Bridgwater and DGB. Boocock, Eds.), Blackie Academic and Professional, London, pp 481-488.

[67] Solantausta Y, Nylund NO and Gust S (1994) Use of Pyrolysis Oil in a Test Diesel Engine to Study the Feasibility of a Diesel Power Plant Concept, Biomass and Bioenergy, 7, pp 297-306.

[68] Leech J (1997) Running a Duel Fuel Engine on Pyrolysis oil, *in: Biomass Gasification and Pyrolysis, State of the Art and Future Prospects,* Kaltschmitt M, Bridgwater A, (Eds), CPL Press, Newbury, UK.

[69] Shihadeh AL (1998) Rural Electrification from Local Resources: Biomass Pyrolysis Oil Combustion in a Direct Injection Diesel Engine, D.SC. Thesis, Department of Mechanical Engineering, MIT.

[70] Bandi A and Baumgart F (2001) Stirling Engine with Flox® Burner Fuelled with Flash Pyrolysis Liquid, *in: Progress in Thermochemical Biomass Conversion*; Bridgwater AV Ed ; Blackwell Science, Oxford, pp 1459-1467.

[71] Ardy PL, Barbucci P, Benelli G, Rossi C and Zanforlin S (1995) Development of Gas Turbine Combustor Fed with Bio-fuel Oil, *in: Proceedings: Second Biomass Conference of the Americas,* NREL, Golden CO 80401, CP-200-8098, pp 429-438.

[72] Pyne Newsletter, 18 April (2005).

[73] Maggi R and Elliott D (1997) *in: Developments in Thermochemical Biomass Conversion,* Bridgwater A and Boocock D, Eds., Blackie Ac. And Prof., pp 575-588.

[74] Laurent E and Delmon B (1994) *in: Biomass for Energy and Industry*, Hall D, Grassi G and Scheer H, Eds., Ponte Press, Bochum, pp 177-181.

[75] Chang C and Silvestri A (1997) J. of Catalysis, 47, p 29.

[76] Bridgwater A (1998) *in: Biomass for Energy and Industry*, Kopetz H, Weber T et al, Eds., CARMEN, pp 268-271.

[77] Centeno A, David O, Vanbellinghen C, Maggi R and Delmon B (1997) *In: Developments in Thermochemical Biomass Conversion*, Brigdwater AV and Boocock DGB, Eds Blackie Academic and Professional, London, pp 589-601.

[78] Williams P and Nugranad N (1998) *In: Biomass for Energy and Industry*, Kopetz H, Weber T *et al.*, Eds, Carmen, pp 1589-1592.

[79] Samolada M and Vasalos J (1998) *In: Biomass for Energy and Industry*, Kopetz H, Weber T *et al.*, Eds ; Carmen, pp 1716-1719.

[80] Arias B *et al.* (2008) Influence of Torrefaction on the Grindability and Reactivity of Woody Biomass, Fuel Processing Technology.

[81] Govin A, Guyonnet R, Repellin V and Rolland M (2008) Effect of Torrefaction on Grinding Energy Requirement: First Results on Spruce. *World Bioenergy Conference*, Jonkoping, Sweden, 27-29 May.

[82] Bergman PCA *et al.* (2005) Torrefaction for Entrained-flow Gasification of Biomass. ECN.

[83] Bergman PCA *et al.* (2005) Torrefaction for Biomass Co-firing in Existing Coal-Fired Power Stations. ECN.

[84] Bridgeman TG, Jones JM *et al.* (2007) Torrefaction of Reed Canary Grass, Wheat Straw and Willow to Enhance Solid Fuel Qualities and Combustion Properties, Fuel, **5**, 41.

[85] Prins MJP and Janssen FJJG (2006) More Efficient Biomass Gasification via Torrefaction, Energy, 31, pp 3458-3470.

[86] Bergman PCA, Boersma AR *et al.* (2005) Torrefaction for Biomass Co-firing in Existing Coal-fired Power Stations "BIOCOAL", Petten, Netherland, Energy Research Centre of the Netherlands (ECN), 45.

[87] Baumberger S, Dole P *et al.* (2002) Using Transgenic Poplars to Ellucidate the Relationship between the Structure and the Thermal Properties of Lignins, *Journal of Agricultural and Food Chemistry*, **50**, 8, pp 2450-2453.

[88] Inari G N, Petrissans M *et al.* (2007) Chemical Reactivity of Heat-treated Wood, Wood Science & Technology, 41, pp 157-168.

[89] Felfli F, Luengo C *et al.* (1998) A Numerical Model for Biomass Torrefaction, *European Conference*, 10th, Wurzburg, Germany.

[90] Bonhke I (1993) Étude expérimentale et théorique des traitements thermiques du bois, *Caractérisation physico-mécanique des bois traités*, Saint-Étienne, École Nationale Supérieure des Mines de Saint-Étienne, p 205.

[91] Rahjohnson J R, Guyonnet R *et al.* (1994) Experimental Study and Modelling of the Wood Retification Process, *Automatic Control of Food and Biological Process.*, Bimbenet JJ, Dumoulin E and Trystam G, Elsevier Science BV, pp 227-235.

[92] Younsi R, Kocaefe D *et al.* (2006) Transient Multiphase Model for the High-Temperature Thermal Treatment of Wood, AIChe. Journal, **52**, 7, pp 2340-2349.

[93] Rousset P, Perré P *et al.* (2006) Selection of a Micro-particle Wood Pyrolysis Model for Use in a Macroscopic Transport Model, Annals of Forest Science, 63, pp 213-219.

[94] Arcate J (2002) Marchés et technologies pour le bois torréfié en 2002, Bois Énergie, 6.

[95] Bergman PCA and Kiel JHA (2005) Torrefaction for Biomass Upgrading, *14th European Biomass Conference & Exhibition*, Paris, France.

[96] Lipinsky ES, Arcate JR *et al.* (2002) Enhanced Wood Fuels via Torrefaction, Fuel Chemistry Division Preprints, **47**, 1.

[97] Weststeijn A (2004) First Torrefied Wood Successfully Cofired with Coal, Biomass Pyrolysis Network, 12.

[98] Mitchell P, Kiel J *et al.* (2007) A Foresighting Study into the Business Case for Pellets from Torrefied Biomass as a New Solid Fuel, All Energy 07, Aberdeen.

Patents

[99] IFP Energies nouvelles FR2910489 WO200893012

[100] RUMPLER DE102007010696

[101] BIO 3D APPLICATIONS FR2903177 WO200800960

[102] TORR COAL TECHNOLOGY WO2007145507

[103] G & A INDUSTRIEANLAGEN GMBH DE202007013672

[104] PAOLUCCIO JOHN US2007266623

[105] BERGMAN PETER CHRISTIAAN ALBER STICHTING ENERGIE NL1030864 WO200778199

[106] JESCHAR RUDOLF WASSMANN KURT DE10317084

[107] HOCHREITER JOHANN DE19932822

[108] BERLIN CONSULT GMBH DE3721006

[109] Värnamo Demonstration Plant, the Demonstration Programme (2001) Sydkraft Publication, Berlings Skogs, Trelleborg.

[110] Robert PH Caractéristiques physico-chimiques de la biomasse ligneuse. De la forêt aux chaufferies à bois à alimentation automatique. IDF.

[111] Boissonnet G and Seiler JM (2002) Approche thermodynamique des transformations de la biomasse. Rapport CEA-R-6025.

[112] Chevalier PY, Generino G (1992) Thermodynamic Data and Calculation Code adapted to the Molten Core Concrete Interactions. *Proc. of the Second OECD (NEA) CSNI Specialist Meeting on Molten Core Concrete Interactions*, Karlsruhe, Germany, 1-3 April, NEA/CSNI/R(92)10.

[113] Loison R (1983) Gazéification du charbon. Les Techniques de l'Ingénieur J 5 410.

[114] Unnasch S (1996) Hynol Process Engineering. EPA-600/-96-006.

[115] Norbeck JN and Johnson K (2000) Evaluation of a Process to Convert Biomass to Methanol Fuel NRMRL-RTP-202.

[116] Bridgewater AV (2003) Renewable Fuels and Chemicals by Thermal Processing of Biomass. Chemical Engineering Journal 91, pp 87-102.

[117] Wellman: http://www.berr.gov.uk/files/file14930.pdf

[118] Ansaldo Volund: www.**volund**.dk/

[119] Bioneer: http://www.ieabioenergy.com/media/18_AR1999Task20colour.pdf

[120] DTU: voir Inventaire des installations.

[121] Inventaire des installations et caractéristiques: http://gi.grensy.info/installations

[122] Biomass Engineering: http://www.supergen-bioenergy.net/.../

[123] Fluidyne: http://www.fluidynenz.250x.com

[124] Carbona: www.forestprod.org/smallwood04patel.pdf

[125] Future Energy/Siemens: www.coal.org/userfiles/File/**Siemens**

[126] FICFB: Rauch *et al.* (2004) "Steam Gasification of Biomass at CHP Plant in Guessing –Status of the Demonstration Plant" 2nd World Conference and Technology Exhibition on Biomass for Energy, Industry and Climate Protection, 10-14th May, Rome, Italy.

[127] Ferco/Battelle: Preliminary Operating Results from the Battelle/Ferco Gasification Demonstration Plant in Burlington, Vermont, United States. http://www.fercoenterprises.com/downloads/seville.pdf

[128] Texaco (1995) Texaco Gasification Process, US Environmental Protection Agency, EPA 540/R-94/514a.

[129a]Choren: http://www.choren.de/

[129b]Uhde-PRENFLO^{TM} process: http://www.uhde.eu/competence/technologies/gas/techprofile.en

[130] Bain RL (2004) Biomass Gasification Overview DOE/NREL January.

[131] Bridgewater AV and Hagen EN (1996) Bio-oil Production and Utilization. Proceedings of the 2nd EU-Canada Workshop on Thermal Biomass Processing. CPL Press.

[132] Milne TA, Abatzoglou N and Evans RJ (1998) Biomass Gasifier "Tars": Their Nature, Formation, and Conversion NREL/TP-570-25357.

[133] Sutton D, Kelleher B and Ross JRH (2001) Review of Literature on Catalysts for Biomass Gasification. Fuel Processing Technology, 73, pp 155-173.

[134] Donnot A, Magne P and Deglise X (1991) Experimental Approach to the Catalyzed Cracking Reaction of Tar from Wood Pyrolysis. Journal of Analytical and Applied Physics, 21, pp 265-280.

[135] Rapagna S, Provendier H, Petit C, Kiennemann A and Foscolo PU (2002) Development of Catalysts Suitable for Hydrogen or Syngas Production from Biomass Gasification. Biomass and Bioenergy 22, pp 377-388.

[136] Olofsson G, Ye Z, Bjerle I and Anderson A (2002) Bed Agglomeration Problems in Fluidized Bed Biomass Combustion. Industrial and Engineering Chemistry Research **41**, 12.

[137] Lin W and Dam-Johansen K (1999) Agglomeration in Fluidised Bed Combustion of Biomass – Mechanisms and Co-firing with Coal. Proc. 15th International Conference of Fluidized Bed Combustion. Savannah, Georgia. May 16-19.

[138] Baker W (2004) High Temperature Corrosion in Gasifiers, Materials Research, **7**, 1, pp 53-59.

[139] Boerrigter H, Calis HP, Slort DJ and Bodenstaff H (2004) Gas Cleaning for Integrated Biomass Gasification and Fischer-Tropsch Systems 2nd World Conference on Biomass for Energy, Industry and Climate Protection, Rome, Italy, 10-14 May.

[140] Bournay L, Viguié JC, Babe C and Quignard A (2009) Biomass and Coal-to-Liquid BTL/CTL/B-XTL, IFP Energies nouvelles Sessions, December 9.

[140a]Duplan JL and Viguié JC (2007) Biomass-to-Liquids: the Future of BTL, BTL Workshop London, 1st November.

[140b]Wright MM and Brown RC (2007) Biofuels Bioprod. Bioref., 1, pp 49-56.

[140c]Well-to-Wheels Analysis of Future Automotive Fuels and Powertrains in the European Context. Concawe, Eucar, JRC, Well-to-Wheels Report, March 2007, http://ies.jrc.ec.europa.eu/WTW

[141] Coda B, Cieplik MK, de Wild PJ and Kiel J (2007) Slagging Behaviour of Wood Ash under Entrained-Flow Gasification Conditions, Energy and Fuels 21, pp 3644-3652.

[142] Barth T (2002) Optimising Reaction Conditions Relative to Product Slates in Aqueous Pressurised Pyrolysis of Biomass and Waste Samples, Pyrolysis and Gasification of Biomass and Waste, Proceedings of an Expert Meeting, Ed. A.V. Bridgwater, Strasbourg.

[143] Liquefaction of Solid Carbonaceous Material. US Patent 4 052 292.

[144] Kruse A and Gawlik A (2003) Biomass Conversion in Water at 330-410°C and 30-50 MPa. Identification of Key Compounds for Indicating Different Chemical Reaction Pathways, Ind. Eng. Chem. Res., 42, pp 267-279.

[145] Goudriaan F and Peferoen DGR (1990) Liquid Fuel from Biomass via a Hydrothermal Process, Chem. Eng. Sc. **45**, 8, pp 2729-2734.

[146] Ladebeck J (1993) Improve Methanol Synthesis. Hydrocarbon Processing International Edition **72**,3, pp 89-91.

[147] Pavone A (December 2003) Mega Methanol Plants, PEP SRI Consulting, Menlo Park, Californie.

[148] Nexant Chemsystems (December 2004), Methanol 03/04-4, Nexant Chemsystems, White Plains, New-York.

[149] Hamelinck CN and Faaij APC (September 2001) Future Prospects for Production of Methanol and Hydrogen from Biomass, Utrecht University, The Netherlands.

[150] Guibet JC (1997) Carburants et moteurs. Technologies. Énergie. Environnement **2**. Éditions Technip, Paris, pp 608-625.

[151] Dumon R, Guibet JC and Portas JY (1984) Le méthanol – Réalités et perspectives. Éditions Masson, Paris, pp126-171.

[152] Gorse RA *et al.* (1992) The Effects of Methanol/gasoline Blends on Automobile Emission SAE 920327.

[153] DuPont (2000) Robust Summary For Dimethyl Ether, DuPont Submission to the US EPA, October 11. http://www.epa.gov/opptintr/chemrtk/dimethr/c12794b.pdf

[154] McCandless JC and Li S (1997) Development of a Novel Fuel Injection System (NFIS) for Dimethyl Ether – and Other Clean Alternative Fuels. SAE 970220.

[155] Ho Teng, MacCandless JC and Schneyer JB (2002) Viscosity and Lubricity of (Liquid) Dimethyl Ether – an Alternative Fuel for Compression Ignition Engines, SAE 2002 – 01 – 06862.

[156] Shudo T, Ono Y and Takahashi T (2002) Influence of Hydrogen and Carbon Monoxide on HCCI Combustion of Dimethyl Ether, SAE 2002 – 01 – 2828.

[157] Kajitan S, Chen ZL, Konno M and Ree KT (1997) Engine Performance and Exhaust Characteristics of Direct Injection Diesel Engine Operated with DME, SAE 972973.

[158] Kapus P and Cartellieri W (1995) ULEV Potential of a DI/TCI Diesel Passenger Car Engine Operated on Dimethyl Ether, SAE 952754.

[159] Tijmensen MJA, Faaij APC, Hamelinck CN and van Hardeveld MRM (2002) Exploration of the Possibilities for Production of Fischer-Tropsch Liquids and Power via Biomass Gasification, Biomass and Bioenergy, 23, pp 129-152.

[160] Path PL and Dayton DC (2003) Preliminary Screening – Technical and Economic Assessment of Synthesis Gas to Fuels and Chemicals with Emphasis on the Potential for Biomass-Derived Syngas, NREL, Golden Colorado, United States, NREL/TP-510-34929, December.

6

Conversion of lignocellulosic biomass by biochemical pathway. Production of ethanol and ABE [1]

D. Ballerini, F. Monot

This chapter describes several pathways for the biological conversion of lignocellulosic biomass. Ethanol production is highly advanced, like acetonobutylic fermentation, whereas ethanol synthesis from syngas is still principally confined to laboratory research.

6.1 ETHANOL PRODUCTION

While ethanol production from saccharose in sugar plants and starch from amylaceous plants is a mature process implemented on industrial scale, its production from lignocellulosic raw materials is still faced with various technico-economic barriers. However, this route is undoubtedly the most interesting one from the environmental standpoint, and could be used for large-scale low-cost ethanol production. Another advantage is that alcohol-producing plant residues such as sugar cane bagasses, corn cobs or wheat straw can be used as feedstocks. Boosted by this stake, significant R&D studies are underway to improve the entire process, in particular the key steps, i.e. pretreatment, hydrolysis of cellulose into fermentable sugars and conversion of the pentoses produced from hemicelluloses into ethanol.

6.1.1 Substrate composition

Lignocellulosic biomass can be considered as being composed mainly of the plant cell wall, which is subdivided into primary wall and secondary wall. The primary wall, about 0.1 to 0.2 μm thick, is found mainly in young cells and contains a high proportion of hemicelluloses. The much thicker secondary wall is located inside the primary wall and divided into 3 layers: S1, closest to the primary wall, from 0.1 to 0.3 μm thick; S2, from 1 to 5 μm thick; S3, 0.1 μm thick which surrounds the lumen. In these 3 layers, the cellulose fibrils are

1. ABE: Acetone-Butanol-Ethanol.

oriented in different directions, and lignins and hemicelluloses are closely linked to each other and to the cellulose fibrils. The parietal cells are held together by the middle lamella of thickness 0.5 to 2 µm, formed almost exclusively of lignins.

Cellulose is a linear homopolymer consisting of D-glucopyranose units linked together by β,1-4 bonds (Figure 6.1). Each glucose residue is rotated by 180° around its axis relative to an adjacent residue, such that the basic unit is in fact cellobiose, consisting of 2 glucose molecules. The Degree of Polymerisation (DP) may vary from 100 to 14 000. For wood, the average DP is about 10 000. The cellulose chains combine together to form rigid and insoluble microfibrils, 3 to 4 nm thick in the higher plants. These chains are parallel and form highly structured crystalline domains separated by amorphous regions.

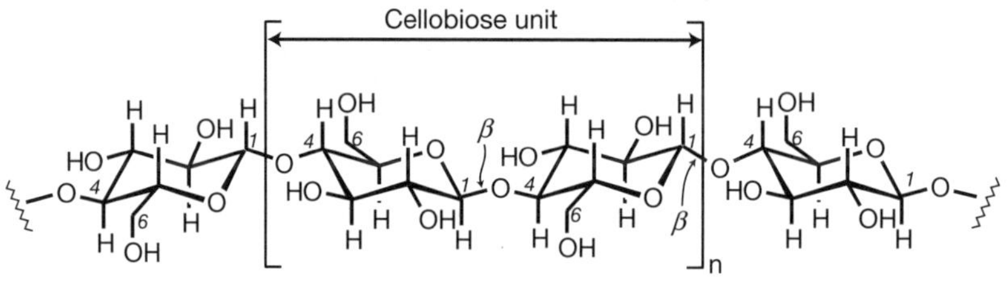

Figure 6.1

Chemical structure of cellulose.

Hemicelluloses are straight or branched complex heteropolysaccharides, of low DP (150 to 200), whose main components are pentoses (C5 sugar such as xylose and arabinose), hexoses and uronic acids (derived from sugars by oxidation of the primary alcohol group). The main chains are xylans and glucomannans. The secondary hydroxyl groups are sometimes substituted by acetyl groups.

Lignins are the main components of the intercellular space. They are highly cross-linked polymers, formed mainly by condensation of three aromatic alcohols: *p*-coumarylic alcohol, coniferylic alcohol and sinapylic alcohol (Figure 6.2). The content in corresponding phenylpropane units, respectively *p*-Hydroxyphenyls (H), Guaiacyls (G) and Syringyls (S), as well as the interunit bonds and free phenol groups are highly dependent on the types of biomass (softwoods, hardwoods and grass) whose lignin structures and organisations are therefore different [1].

The three polymers forming the plant wall are closely associated together, forming a rigid matrix. The interactions between these polymers depend on the type of plant cell and the plant maturity.

The composition of lignocellulosic materials varies (Table 6.1). Cellulose is the main constituent, accounting for 35 to 50% of the dry matter. The hemicellulose content depends largely on the plant considered and the composition of these hemicelluloses is itself highly variable. Glucomannan is the main constituent of conifer hemicelluloses which therefore contain few pentoses, unlike wheat straw whose hemicelluloses consist mainly of xylans.

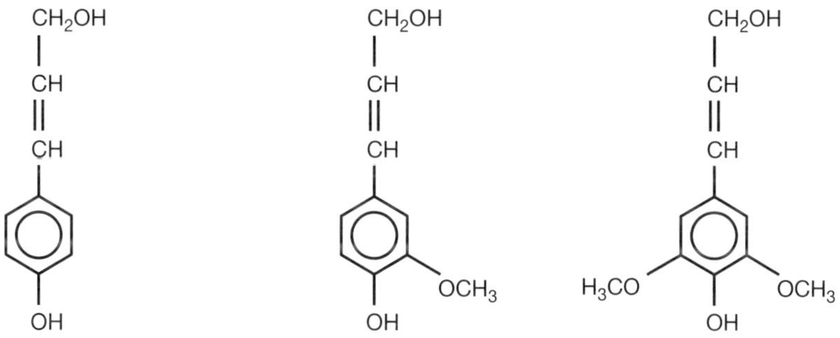

a b c

Figure 6.2

Main constituents of lignins.

Table 6.1 Composition of lignocellulosic plants (% dry matter).

Plant	Lignin	Cellulose	Hemicellulose			
			including:	xylose	arabinose	mannose
Pine	27-29	38-42	20-24	4-7	0-1	7-12
Poplar	20-24	40-49	16-27	13-18	0.5-2	2-8
Wheat straw	15-23	30-43	22-30	18-24	2-3	0-1
Corn cobs	12-17	35-41	20-27	15-22	0-3	0-1
Sugar cane bagasse	11-25	38-40	23-34	21-24	0-3	0-1

The variations observed for a given plant are due to the diversity of determination methods used, the species studied and their maturity at the time of sampling.

The variations in composition affect the entire process, the efficiency of pretreatment and hydrolysis of cellulose and hemicelluloses or the type of fermentable sugars and the inhibiting molecules affecting the fermentative processes.

6.1.2 The process

For many years restricted to R&D studies, processes designed to produce ethanol from lignocellulosic biomass are currently being extrapolated to larger scales: pilot or demonstration. The processes include several basic considerations: i) lignin cannot be fermented directly into ethanol [1], ii) the lignocellulosic matrix must be pretreated so that cellulose and hemicelluloses

1. One special case is the transition via thermochemical conversion into syngas which is then fermented into ethanol. This option is discussed in Section 6.2.

can be hydrolysed, iii) the cellulosic and hemicellulosic fractions are potential sources of fermentable sugars. Consequently, the generic diagram of the process includes the following main individual operations: pretreatment of the raw material, hydrolysis, ethanol fermentation and separation of ethanol from the fermentation wort (Figure 6.3). The hexose fermentation steps and ethanol recovery by distillation/rectification/dehydration are identical to those of proven processes for the production of ethanol from sugar and cereal plants, detailed in Chapter 2. The processes currently developed are variants of this diagram, which depend on the options chosen for individual operations and the flows of cellulose, hemicellulose and lignin (paragraph 6.1.2.6).

Figure 6.3

Generic diagram of the production of ethanol from LCB (lignocellulosic biomass).

6.1.2.1 Pretreatment

This section does not discuss biomass conditioning as such, which generally consists in grinding the fibrous materials, for example into chips for wood, so that it can be transported on conveyor belts, but the treatment required to allow cellulose hydrolysis. There are several ways of achieving this objective: lower the lignin and hemicellulose content of the solid substrate to be treated, make the matrix more porous, reduce the crystallinity of the cellulose or increase its specific area. Depending on the type of pretreatment, these actions may be combined and frequently lead to a modification of the lignin structure. The main technical constraints of this step are to prevent the loss or degradation of the sugars while limiting the formation of inhibiting products as well as discharges and excessive energy consumption. Numerous pretreatment techniques have been studied and proposed (Table 6.2). For clarity purposes, this table makes a distinction between chemical and physico-chemical processes, although physical and chemical phenomena are combined in all cases. The distinction relies more on the treatment terminology. Despite abundant literature on this subject, it is difficult to classify the efficiency of the pretreatments, firstly for methodological reasons, since efficiency measurement is often based on an enzymatic hydrolysis rate which itself depends on the implementation conditions (substrate and enzyme concentrations, action time, stirring, etc.) and which ignores the presence of compounds that are toxic for the fermentative microorganisms, and secondly due to the high dependency on the raw material considered.

A. *Physical pretreatments*

a) Mechanical pretreatments

These treatments are mainly designed to increase the accessible areas, thereby favouring subsequent hydrolysis. They may also reduce the DP of cellulose and lignins and the cellulose crystallinity. They consist in intense mechanical grinding to obtain particles of size less than 2 mm. The energy consumption depends on the raw material, especially its degree of dryness, but remains too high for this type of pretreatment to be used at large scale [2].

Table 6.2 Main pretreatment processes

Physical processes	Mechanical pretreatments Pyrolysis
Physico-chemical processes	Thermohydrolysis Steam explosion Steam explosion under acid conditions AFEX and ARP CO_2 explosion
Chemical processes	Pretreatment with dilute acid Pretreatment with concentrated phosphoric acid Alkaline pretreatment Organosolv processes Pretreatment with ionic liquids Chemical oxidation Biological oxidation

b) Pyrolysis

At temperatures below 300°C, only limited decomposition of cellulose into volatile and liquid products is observed and the solid product obtained can then be hydrolysed into sugars by mild acid hydrolysis with yields in the region of 80 to 85% [2]. However, the fermentability of the hydrolysates has never been documented.

B. Physico-chemical pretreatments

a) Thermohydrolysis

Thermohydrolysis, also known as LHW (Liquid Hot Water) pretreatment consists in exposing biomass to pressurised hot water. The process is generally discontinuous and the water is heated to a temperature of 200-230°C (see 5.3). The cooking time is generally 15 to 60 minutes. This technique may result in complete solubilisation of hemicelluloses and partial solubilisation of lignin (20 to 50%). It would also increase the pore size in the lignocellulosic matrix and reduce the cellulose crystallinity index, thereby favouring subsequent enzymatic hydrolysis. Hemicellulose solubilisation rates of between 80 and 100% and enzymatic hydrolysis rates above 90% have been obtained in the laboratory [3]. It is generally accepted that the acetic acid released during hemicellulose hydrolysis acts as catalyst during the pretreatment.

A pilot plant capable of treating 1 t/h of wheat straw and comprising 3 steps was operated continuously during the IBUS European project [4]. Straw is impregnated with water for 20 minutes at 80°C in the first reactor. During the second step, the hemicelluloses are extracted by heating at temperatures in the region of 170 to 180°C for about fifteen minutes and the fibres undergo a final treatment for 3 minutes with updraft circulation of water at 195°C. This last treatment conditions the fibres for better hydrolysis. High water/straw ratios favour extraction of the hemicelluloses and limit their degradation but have the opposite effect on the hydrolysis susceptibility of the cellulosic fraction obtained.

Thermohydrolysis offers numerous advantages:

- high pentose solubilisation rate and high enzymatic digestibility of the pretreated cellulose,
- no addition of chemical reactants,
- no inhibiting products,
- few discharges,
- high biomass concentration (> 100 g/l).

b) Steam explosion

This treatment, sometimes known as auto-hydrolysis, consists in rapidly heating the substrate (160 to 260°C) by injecting pressurized saturated steam (10 to 50 bar), keeping it at the required temperature for a time ranging from several seconds to several minutes, then in suddenly reducing the pressure down to the atmospheric one. The process has been implemented discontinuously (Iotech process) and continuously (Stake process) on large scale. The operating parameters are temperature, duration, dry matter concentration and prior conditioning of the material (chip size in case of wood, for example). This technique results in partial hydrolysis of the hemicelluloses catalysed by the acetic acid released as well as conversion of the lignin and significant destruction of the lignocellulosic matrix structure.

Hemicellulose hydrolysis is obtained by combining either high temperature and short duration (270°C – 1 min) or low temperature and longer duration (190°C – 10 min). With this process, however, hemicellulose solubilisation, obtained when hydrolysis has led to the production of monosaccharides and oligosaccharides of very low DP, is not optimised. In addition, enzymatic hydrolysis of the cellulose obtained after steam explosion generally remains below a yield of 50%. Lastly, this technique may lead to the formation of inhibiting products resulting from the degradation of sugars or lignins.

c) Steam explosion under acid conditions

Most of these disadvantages disappear when steam explosion is carried out under dilute acid conditions, for example in the presence of H_2SO_4 or SO_2, which allow the process to be conducted at lower temperatures (150 to 200°C as compared with 250°C when there is no catalyst). This allows complete hydrolysis and solubilisation of hemicelluloses, improved yield of the subsequent enzymatic cellulolysis and reduced production of inhibiting compounds. Steam explosion (16 bar, 2.5 min) of poplar wood impregnated with 0.1 N sulphuric acid enables 90% of the potential glucose and 60% of the pentoses to be recovered [5]. Dry matter losses, often due to evaporation of volatile products during decompression, do not exceed 10% (mainly furfural and acetic acid). Steam explosion in the presence of gaseous SO_2 would allow better penetration of the reactant in the lignocellulosic matrix and would be one of the rare pretreatments efficient on conifers. However, this would result in reactant losses and make its use on industrial scale difficult due to the toxicity and volatility of SO_2. Apart from its efficiency, especially under acid conditions, steam explosion offers other advantages, including moderate energy requirements, low quantities of discharges, straightforward batch implementation, low consumption of chemical reactants and good adaptation to large particles (wood chips).

For example, enzymatic hydrolysis conversion of 92% on glucans and 80% on xylans have been obtained with willow exploded after impregnation in 0.5% H_2SO_4 for 90 min [6]. The main disadvantages lie in the possible degradation of sugars and phenolic compounds into toxic products and the difficulty in extrapolating to large scale, especially for continuous processes.

d) AFEX (Ammonia Fibre Explosion) process

In this process, the lignocellulosic material is brought into contact with hot pressurised aqueous solution of ammonia (5 to 15%). The pressure is then suddenly released to explode the substrate and evaporate the ammonia. This is a variant of steam explosion conducted at lower temperature (50 to 90°C) with longer residence time (typically 30 min). No inhibiting products form. The hemicelluloses are poorly solubilised. This type of pretreatment is relatively inefficient on woody substrates and would be more suitable for herbaceous plants and agricultural residues for which good cellulose hydrolysis rates have been obtained, even with low enzyme concentrations [7]. The ammonia concentrations used have been considerably reduced (0.2-0.8 kg/kg d.m. as compared with 1 to 2 kg/kg d.m. initially [8]) and recovery and deep recycling of ammonia suggest that the process may be economically viable under certain conditions [9].

ARP (*Ammonia Recycle Percolation*) is another process based on ammonia which consists in percolating an ammonia solution (5 to 15%) on biomass at high temperature (generally 150 to 180°C), then separating and recycling the ammonia in the effluent. These conditions favour biomass delignification and allow efficient hydrolysis with low enzyme concentrations [10]. The reduction in the volumes of water required is one of the key points of the process. In a variant called SAA (*Soaking in Aqueous Ammonia*) the biomass is impregnated with an aqueous solution of ammonia [11].

e) CO_2 explosion

This is another variant of steam explosion, in which CO_2 is used as catalyst. The process has been tested and has proved efficient (pressure 5.62 MPa, 4 kg CO_2/kg d.m.) but has never been developed at large scale [2].

C. Chemical processes

a) Pretreatment with dilute acid

This type of pretreatment is preferably carried out in the presence of dilute sulphuric or hydrochloric acid in proportions of 0.5 to 3% of the substrate (dry matter). Two methods are used: a continuous one at high temperature (> 160°C), suitable for feeds with low dry matter content (5 to 10%), another discontinuous at moderate temperature (< 150°C), which can be used for dry matter concentrations of between 10 and 40%. The higher the temperature, the greater the loss in dry matter. The residence times depend on the temperature, for example 10 s at 200°C in plug flow reactor and 30 min at 120°C in percolation reactors.

This technique can be used to hydrolyse hemicelluloses into monomers and makes the cellulose much more digestible (enzymatic hydrolysis yields greater than 90% have been obtained in some cases) although the cellulose degree of crystallinity is unchanged. A part of the lignin would also be solubilised. It has been observed that above 130°C, especially under acidic conditions, lignin would disperse as droplets then deposit non-uniformly on the residual biomass, hindering the enzymatic action [12].

This technology is efficient on numerous substrates. Its cost is often estimated to be slightly higher than that of steam explosion. At high temperature, the sugars form inhibiting furanic compounds (furfural from pentoses, hydromethylfurfural – HMF – from hexoses). If this process is used, the acid residue must be neutralised with lime, which generates calcium sulphate which is difficult to find a further use. This technique is considered in a process developed by NREL (National Renewable Energy Laboratory) [13], in which the lignocellulosic material is first brought into contact with low pressure steam at 100°C. Concentrated sulphuric acid is then added to obtain a final concentration of 0.5%. Contact with the substrates takes place in a co-current downflow reactor for 10 min at 190°C, with a solid content of 22%. After cooling in flash mode, which removes 60% of furfural and HMF, the solid and liquid phases are separated. The liquid phase is neutralised with excess lime, then passed over an ion-exchange resin to eliminate the acetic acid from the hemicelluloses, the sulphuric acid and the other remaining inhibiting products. This liquid phase is then mixed with the solid phase for enzymatic saccharification.

Apart from the formation of inhibiting products, already mentioned, the main limitations of the process are the corrosiveness of the reactants and the resulting need to use expensive materials, the need to neutralise the pretreated solution before hydrolysis and the associated formation of salts, which create discharges.

b) Pretreatment at controlled pH

This pretreatment, unlike the previous one, aims to keep the pH above 4 [14]. Under these conditions (water and temperature), the treatment is similar to thermohydrolysis, but the pH is controlled to avoid hydrolysing hemicelluloses into monomers which are likely to degrade. It reduces the cellulose crystallinity and less association of lignin with cellulose is observed.

c) Pretreatment in alkaline medium

These treatments result in significant solubilisation of lignin and a part of hemicelluloses, release of acetyl groups and swelling of the cellulose fibres, making them much more accessible to enzymes. The cellulose DP and crystallinity index are sometimes reduced. Sodium hydroxide treatments (8 to 12% NaOH (w/w dry matter), 80 to 120°C, 30 to 60 min) induce a significant loss of lignocellulosic dry matter (30 to 35%), require washing and neutralisation of the solid residue as well as recycling of the sodium hydroxide. They are therefore expensive and less interesting than lime treatments. These treatments appear to be more suitable for agricultural residues, which are less lignous than wood. Corn stover pretreated with 7.5% of lime/biomass for 4 hours at 120°C loses about 40% of the initial lignin. The hydrolysis yield reaches 88% with long action times and relatively high enzyme concentrations [15].

d) Organosolv process

Derived from the paper industry, this process consists in solubilising and extracting lignin and hemicelluloses in an organic solvent (generally methanol or ethanol). A small quantity of acid catalyst (HCl or H_2SO_4) is often added. This process has been assessed on softwood using a 40-60% aqueous solution of ethanol, a liquid-wood ratio between 7 to 10, a reaction time of 30 to 60 min and a temperature of 200°C. The cellulose hydrolysis yield is greater than 90% and the lignin obtained is very pure [16]. In these processes, the solvent must be extracted (by evaporation) then recycled. The need to recycle all the solvent hinders the development of this process. It nevertheless allows recovery of nearly all the lignin which could therefore be upgraded independently. Use of supercritical fluids and ionic liquids instead of the usual organic solvents has also been proposed. In the latter case, the ability of ionic liquids such as BMIMCl (1-butyl-3-methylimidazolium chloride) to dissolve cellulose is exploited [17].

e) Pretreatment with phosphoric acid

This treatment is based on dissolution of cellulose in solutions of concentrated phosphoric acid (> 81%) at low temperature (50°C). The cellulose becomes amorphous and is therefore more reactive, while the use of moderate temperatures prevents any degradation of the sugars. After a reaction time of 30 to 60 minutes, cellulose and hemicellulose are precipitated by addition of acetone and the lignin is solubilised. After washing with water, the phosphoric acid is first precipitated by addition of calcium carbonate, then regenerated by addition of sulphuric acid. The process allows separation of the various fractions and produces a cellulose which is easy to hydrolyse [18]. Successfully tested on corn stover, hybrid poplar and *switchgrass,* the process is inefficient on Douglas pine.

f) Oxidation processes

In the chemical oxidation processes, especially ozonolysis and wet oxidation, lignin and, to a lesser extent, hemicelluloses are degraded while cellulose remains intact. Ozone has been tested on substrates such as wheat straw, sugar cane bagasses as well as pine and poplar wood with a noticeable increase in enzymatic digestibility.

These treatments offer efficient delignification, limited production of inhibiting compounds and, in case of ozonolysis, low reaction temperature. In the latter case, however, large quantities of ozone are required, making the process too expensive.

g) Biological processes

Delignification processes implementing peroxidase type enzymes or lignolytic fungi such as *Phanerochaete chrysosporium* have been proposed [19]. For the time being, they are only used at laboratory scale. The main disadvantages are the cost of the lignolytic enzymes in the first case and the slow rate of reaction in the second case.

In conclusion, there is no "ideal" universal pretreatment process [20]. *A priori*, the most attractive technologies are thermohydrolysis, pretreatment with dilute acid and steam explosion, AFEX and, more recently, pretreatment with phosphoric acid and Organosolv pretreatment, which both offer the additional advantage of allowing the lignocellulosic material to be fractionated into its constituent polymers. Thermohydrolysis seems efficient and minimises the formation of degradation products. It is developed at large scale. Pretreatment with dilute acid has been the dominant option for many years. It improves the cellulose enzymatic digestibility and allows good hemicellulose hydrolysis without excessive formation of sugar degradation products. Steam explosion, especially after soaking the lignocellulosic material in dilute sulphuric acid, is also attractive for the same reasons. Large-scale extrapolation for continuous processes nevertheless remains difficult for treatment of low-density products such as cereal straws. The choice of pretreatment (and operating conditions) depends largely on the substrate used. In addition, the techniques proposed often involve pressing and washing steps which consume energy and water. They must be optimised by trying to integrate these flows as much as possible for obvious economic and environmental reasons. A technico-economic study comparing five types of pretreatment, dilute acid, thermohydrolysis, AFEX, ARP and lime treatment carried out on the same substrate, in this case corn stover, and using the same hydrolysis, fermentation and analysis conditions demonstrated that there were only minor cost differences between these pretreatments [21]. The cost of the *a priori* least expensive pretreatments increases when catalyst recycling and ethanol recovery in very dilute medium are included. According to some estimations, in fact, pretreatment would represent approximately 20% of the ethanol cost and the other individual operations have such a high impact that an integrated approach is necessary to improve it [22].

6.1.2.2 Hydrolysis

Due to its structure, crystallinity and association with lignin and hemicelluloses still present even after pretreatment, hydrolysis of cellulose into fermentable monomers (glucose) is a difficult operation which can be carried out using two methods: chemical hydrolysis catalysed by acid or enzymatic hydrolysis.

A. Chemical hydrolysis

Dilute acid hydrolysis generally requires no pretreatment other than conditioning [3]. It takes place in two steps to optimise the release of sugars from the hemicelluloses and the cellulose. The first step, similar to dilute acid pretreatment, is used to digest the hemicelluloses and solubilise the sugars produced. After separation, the solid fraction undergoes further hydrolysis. The salt formed must be neutralised and precipitated. Various types of reactor can be used, e.g. percolation, co-current or plug flow. The NREL described the following operating conditions: 3 minutes at 190°C in the presence of 0.7% sulphuric acid for the first step, 3 minutes at 215°C in the presence of 0.4% sulphuric acid for the second step. The hydrolysis yields are 89% for mannose, 82% for galactose and 50% for glucose. This process was developed jointly with BCI (BC International Corporation). In some cases, the conditions during the second step are more severe: 240°C at 2% sulphuric acid. A flash hydrolysis technology (< 2 s, 260-280°C, 0.5-1% of H_2SO_4) developed by CIEMAT (Spanish National Research Centre for Energy, Environment and Technology) would lead to higher glucose recovery yields (70-80%).

Concentrated acid hydrolysis is generally more efficient when it combines two steps: firstly, decrystallinisation of the cellulose which has become completely amorphous due to cleavage of the inter-chain hydrogen bonds and, secondly, dilute acid hydrolysis [23]. In some cases, prior dilute acid pretreatment is carried out to separate hemicelluloses before decrystallinisation.

These hydrolysis processes were developed in the United States by Arkenol and the Masada Resource Group to produce ethanol from lignocellulosic raw material. In the Arkenol process, cellulose decrystallinisation is obtained by addition of 70-77% sulphuric acid to the substrate which has been dried to 10% moisture content. The acid is added to obtain an acid/(cellulose + hemicellulose) ratio of 1.25, while keeping the temperature below 50°C. Water is then added to dilute the acid to 20-30% and the resulting mixture is then heated to 100°C for 1 hour. After pressing, the residual solids undergo a second hydrolysis step while the sugars and acid in the soluble fraction are separated in a chromatography column. 98% of the sugars would be recovered and the loss of acid in the sweet juice is less than 3%. For this process to be economically viable, acid recycling must be maximised. In addition, corrosion-resistant reactors are necessary to withstand the concentrated acid.

B. Enzymatic hydrolysis

Enzymatic cellulose hydrolysis is often the recommended pathway to obtain fermentable sugars, for the following reasons:

- the results of most financial evaluations come down in favour of enzymatic hydrolysis rather than chemical hydrolysis,
- it generates only small volumes of effluents to be treated and no corrosion problems,
- its improvement perspectives are much more promising than chemical hydrolysis, on which numerous studies have been conducted over several decades.

In nature, cellulose is mostly biodegraded by microorganisms (eubacteria and fungi) under aerobic (e.g. on the surface of the ground) as well as anaerobic (e.g. in animal rumen) conditions [24]. The cellulolytic bacteria include bacteria belonging to the genera *Ruminococcus*, *Clostridium*, *Cellulomonas*, *Thermonospora* and *Streptomyces*. Although some bacteria, such as *Clostridium thermocellum*, produce cellulases which have a high specific activity (rate of hydrolysis per unit weight of proteins), their production capacities and growth rates are too low to consider use at industrial scale. The cellulolytic fungi belong to various genera: *Aspergillus*, *Schizophyllum*, *Penicillium*, *Trichoderma*, *Chrysosporium*, *Fusarium*, etc.

Of these fungi, the species most studied for the production of cellulases is probably *Trichoderma reesei* [24]. It is capable of secreting high concentrations of very active cellulases and has been used industrially. In addition, apart from cellulases , *T. reesei* produces hemicellulases which contribute to hydrolysis of the plant walls. During the process, the enzymes produced by this fungus in the culture medium must be first recovered and possibly conditioned before they can be implemented in hydrolysis.

Hydrolysis is carried out by simply placing the pretreated material in contact with the enzymatic solution, keeping the suspension homogeneous and maintaining the optimum conditions for *T. reesei* cellulases, i.e. a temperature of 40-50°C and a pH between 4.5 and 5. The action time depends on the quantity of enzymes added and their activity, as well as the substrate concentration. It would appear in fact that the optimum temperature and pH depend not only on the enzyme but also the substrate, due to the multi-enzymatic nature of the enzymatic mixtures used [25]. During enzymatic hydrolysis, the reducing sugars are produced first as oligomers then cellobiose and finally glucose. The soluble products resulting from enzymatic hydrolysis, mainly cellobiose and glucose, have a major inhibiting effect on the activity of the cellulolytic enzymes [26]. The sugar and lignin degradation products formed during the pretreatments, for example by steam explosion, could also have a negative effect, although limited, on the cellulolytic enzymes [27].

The cost of the utilities and reactors required for enzymatic hydrolysis is low. Recycling of enzymes is currently rarely considered since most are adsorbed on the lignin-rich solid residue. The cost of cellulases is relatively high and this stage is often considered to be the most expensive in ethanol production from lignocellulosic materials. Extensive research is therefore dedicated to understanding the enzymatic cellulolysis mechanism in order to improve it. It is in fact a complex process involving the action of soluble proteins on an insoluble and recalcitrant substrate, which is not yet fully understood.

6.1.2.3 Production of enzymes and their action mode

A. Cellulases and cellulolysis

a) Enzymatic systems

The enzymes involved in cellulose degradation, commonly called cellulases, are either secreted in the medium (mainly the case of fungi), or associated with the outer surface of the microorganisms forming complex systems called cellulosomes (case of anaerobic bacteria).

Cellulases were initially classified according to their catalytic action mode. They are now also classified according to their structure, like all enzymes acting on sugars [28]. Complete cellulose hydrolysis involves three types of complementary cellulolytic enzymatic activity:

- the endoglucanases (EG) or 1,4-β-D-glucane-4-glucanohydrolases (EC 3.2.1.4) which cut the cellulose randomly in the cellulose amorphous regions, generating new chain ends,
- the exoglucanases, including the 1,4-β-glucane glucanohydrolases (EC 3.2.1.74) or cellodextrinases and the 1,4-β-glucane cellobiohydrolases (EC 3.2.1.91) or cellobiohydrolases (CBH), which act processively on the free ends of the cellulose chains, releasing glucose for the glucanohydrolases or cellobiose for the cellobiohydrolases,
- the β-glucosidases or β-glucoside glucohydrolases (EC 3.2.1.21) which hydrolyse the soluble cellodextrins and the cellobiose into glucose.

The cellulolytic systems, multi-enzymatic systems with these three action types [1], are therefore capable of degrading the cellulose completely (Figure 6.4).

Most cellulases have a modular structure comprising several structural and functional domains. The catalytic domain – Carbohydrate Binding Module (CBM) (CBMs were previously classified as CBDs, cellulose-binding domains) is the model most frequently encountered for excreted systems. Cellulosomes have a special organisation with the existence of dockerin or fibronectin type III domains which bind to cohesin domains of the scaffolding protein (scaffoldin) (Figure 6.5) [29].

CBMs promote cellulose hydrolysis by anchoring the catalytic site on the substrate. They are essential to initiate the action of the cellobiohydrolases and for their processivity. They are located at the ends of the protein. CBMs would also splay a role in destructuring the cellulose fibrils. There is also a linker between the catalytic domain and the CBM composed of highly glycosylated fragments for the fungal cellulases and especially rich in some amino acids such as proline, threonine and serine (Figure 6.6) [30].

1. A new type of enzyme activities, Polysaccharide copper oxidases, which may cleave cellulose have very recently been discovered. Vaage-Kolstad *et al.*, Science, 2010, 330, p. 219-222.

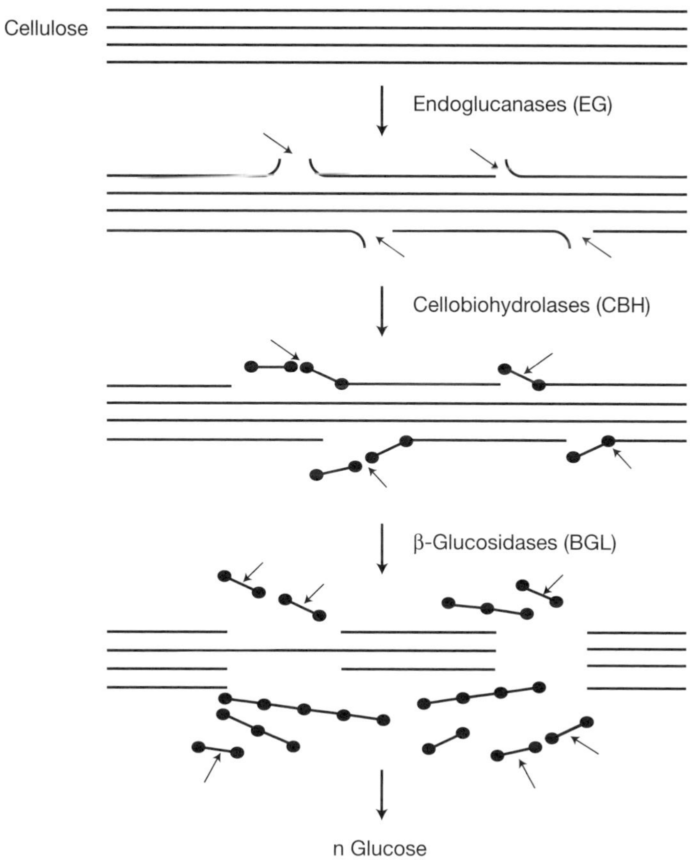

Figure 6.4

Action mode of cellulases on cellulose fibrils.

The catalytic domains have a clift configuration for the endoglucanases and tunnel configuration for the cellobiohydrolases so that the enzyme can move along the cellulose chain. In *Trichoderma reesei,* the CBH1 and CBH2 tunnels are 50 Å and 20 Å long respectively. The reaction mechanism is acid-base catalysis from two catalytic residues of the enzyme: an acid (proton donor) and a nucleophilic base. Depending on the distance between the two catalytic sites, the hydrolysis mechanism conserves or inverts the anomeric configuration of the glycosylated carbon. For example, the anomery is conserved for CBH1 and inverted for CBH2. Cellulolytic microorganisms have several endo and exo type cellulolytic enzymes in order to act on the amorphous and crystalline regions, be active towards a large range of substrates and adapt to the changes occurring to the cellulose chain during its degradation.

The cellulolytic complex secreted by *Trichoderma reesei* therefore contains in particular two CBHs, five endoglucanases and two β-glucosidases (Table 6.3).

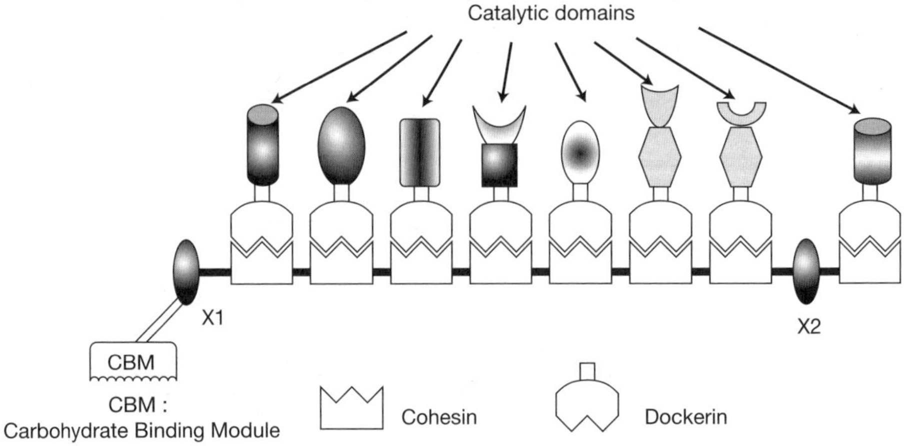

Figure 6.5

Structure of the *Clostridium cellulolyticum* cellulosome.

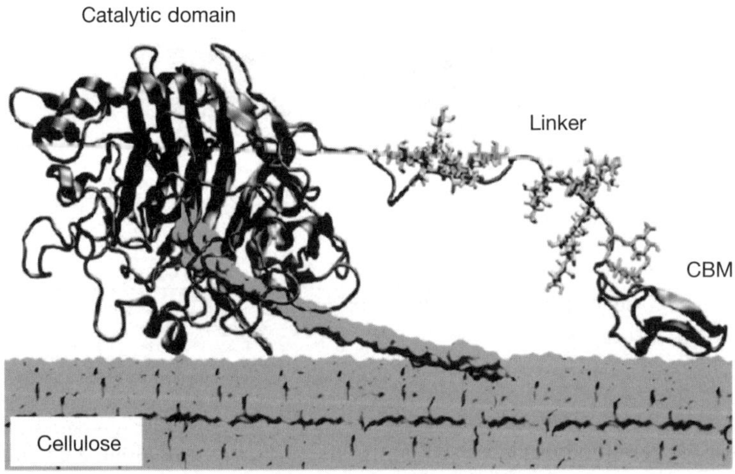

Figure 6.6

CBH1 structure and action mode on a cellulosic substrate.

Source: Beckham GT, Matthews JF, Bomble YJ, Adney WS, Himmel ME, Nimbs MR, Crowley MF. Identification of Amino Acid Responsible for Processivity in a family I Carbohydrate-Binding Module from a Fungal Cellulase. J. Phys. Chem. B, 114, p. 1447-53 (2010)-ACS Publications.

Table 6.3 Some characteristics of known cellulases of *T. reesei*.

Enzyme	Mw [a] (kDa)	Nb AA [b]	% (w/w) [c]	Domains [d]	pI [e]	Reaction mechanism [f]	Specificity	KM [g] (μM)
CBH1/Cel7A	65	513	60	CM + CBM	4.65	Conservation	Cellulose reducing end	48 (methyl β-D-cellotetraoside)
CBH2/Cel6A	58	471	20	CM + CBM	5.11	Inversion	Non-reducing end	4 (methyl β-D-cellotetraoside)
EG1/Cel7B	55	459	10 to 20	CM + CBM	4.71	Conservation	Cellulose amorphous regions	113 (cellotetraose)
EG2/Cel5A	44*	418		CM + CBM	4.97	Conservation	idem	356 (cellotetraose)
EG3/Cel12A	25	234		CD	6.69	Conservation	idem	230 (cellotetraose)
EG4/Cel61A	35*	344		CM+CBM	5.29	Unknown	idem	ND
EG5/Cel45A	23	242		CM + CBM	4.21	Inversion	idem	ND
BGL1/Cel3A	80	744	1		6.38		cellobiose	1250 (cellobiose)
BGL2/Cel1A	52*	466			5.33		cellobiose	

[a] *Molecular weight (* = Theoretical Mw).*
[b] *Number of amino acids.*
[c] *Percentage by weight with respect to the total cellulases.*
[d] *CM: Catalytic module; CBM: Cellulose Binding module.*
[e] *Theoretical isoelectric point.*
[f] *Conservation or Inversion of the anomery.*
[g] *Michaelis constant (substrate specified between parentheses).*
ND: not determined.

The nine enzymes belong to eight different structural classes. We observe the same diversity in *T. reesei* hemicellulases. A genomic study has in fact demonstrated the existence, in *T. reesei*, of a much more complex mixture of biomass degradation activities which would include 2 cellobiohydrolases, 8 endoglucanases, 7 β-glucosidases, 4 xylanases, 1 β-xylosidase, 2 acetyl xylane esterases, 2 arabinofuranosidases, 1 β-mannanase, 5 galactosidases, one swollenin and two proteins of unknown function [31]. Referring to the known cellulases (Table 6.3), we see that the CBHs alone represent 80% (by weight) of the *T. reesei* enzymatic cocktail, whereas the β-glucosidases are under-represented. This apparent imbalance may in fact reflect an optimisation of the enzymatic hydrolysis process, the weakness of some specific enzymatic activities being compensated by their high proportion in the enzymatic mixture produced. It is often suggested, however, that the low proportion of β-glucosidases is one of the main weaknesses of *T. reesei* cellulases, at least for *in vitro* saccharification, especially since *T. reesei* cellobiohydrolases are inhibited by cellobiose and β-glucosidases are inhibited by glucose, which is not the case for other β-glucosidases such as those of *Aspergillus*. The genome sequencing of the two main enzyme producing fungi, *A. niger* and *T. reesei*, has recently been published [32, 33]. Note that these 2 genomes are roughly the same size, about 34 Mb. Out of all the other fungi of known genome capable of hydrolysing plant wall polysaccharides, the *T. reesei* genome has the fewest families of genes encoding cellulases and hemicellulases. In addition, analysis of its genome does not explain its high protein secretion capacity.

b) Synergy of cellulolytic activities

Apart from the complementarity of their actions to perform complete cellulose hydrolysis, the cellulases exhibit high synergies, in other words their activity when used jointly is greater than the sum of their individual activities. Four types of synergy have been described: endo-exo synergy between endoglucanases and exoglucanases; exo-exo synergy between exoglucanases acting at the reducing ends of the cellulose chains and those acting at the non-reducing ends; synergy between the exoglucanases and the β-glucosidases which hydrolyse the cellobiose, final inhibiting product of the cellobiohydrolases; and intramolecular synergy between catalytic domains and CBM.

It is difficult to measure the activity of cellulases due to the diversity of usable substrates. Pure crystalline celluloses are available (cotton, fibre paper, Avicel) but they are often physically heterogeneous and can only be used to evaluate the activity of a complete cellulolytic system. The filter paper activity (FPU), expressed in rate of appearance of reducing sugars, is often used. The amorphous or soluble celluloses such as carboxymethylcelluloses (CMC) are also often used but they do not reflect the activities of the exoglucanases. Low-molecular weight oligomers such as the soluble cellodextrins can also be used. Chromogenic or fluorogenic substrates such as *p*-nitrophenyl-β-D-cellobioside, in which coloured or fluorescent compounds are bonded to the glucose anomeric carbon, are practical to study specific activities.

The hydrolytic action of cellulases on pure cellulose is complex and depends on several key parameters such as the crystallinity of cellulose, its degree of polymerisation and the accessible area. Numerous quantitative models aimed at describing cellulose enzymatic hydrolysis have been developed and classified as non-mechanistic, semi-mechanistic, based on the function and based on the structure [34]. Other factors are involved for pretreated lignocellulosic substrates, in particular the presence of other constituents such as hemicelluloses and lignins, which have a direct effect on enzyme accessibility to cellulose and an indirect effect (well known for lignins) on non-specific adsorption of cellulases. The important role of accessibility is also true for cellulases in complexes [35].

B. Cellulase synthesis

a) Genetics

In *T. reesei*, as with most other fungi, the genes encoding cellulases are distributed randomly in the genome, each gene having its own regulation elements. In contrast, we observe that the promoters of the *cbh1*, *cbh2*, *eg1* and *eg2* genes, which encode enzymes CBHI, CBH2, EGI, EGII, have identical regulating protein binding sites. The bacterial genes of the cellulases are chromosomic and grouped in clusters in *Cl. cellulolyticum* or distributed on the genome in *C. thermocellum*.

b) Regulation

Protein synthesis by living organisms is subject to regulation mechanisms. In *T. reesei*, production of cellulases is regulated by induction and repression processes, induction related to the presence of substrates, in this case cellulose, repression by easily assimilable carbon sources such as glucose via catabolic repression mechanisms [36]. In practice, induction is not caused by the cellulose itself which cannot enter the cell but by soluble products resulting from its hydrolysis such as cellodextrins and cellobiose. A low level of constitutive expression of cellulases therefore occurs, even in the absence of cellulose; the cellulases are then produced in abundant quantities in the presence of degradation products. In practice, the induction mechanisms seem to be complex. Sophorose, a dimer of glucose molecules linked by β-1-2 bonds which can be produced by transglycosylation catalysed by a β-glucosidase, is a very powerful inducer. In addition, the inducing role of cellobiose is complex since it inhibits the cellulases. Hemicellulases are also inducible enzymes. The production of *T. reesei* cellulases is regulated during transcription [37], i.e. when the DNA is converted to mRNA, via transcription factors (XYR1, ACEI and ACEII) which bind to the promoters of DNA. This expression is coordinated for the *cbh1*, *cbh2*, *eg1*, *eg2* and *eg5* genes since the same transcription factors are involved. However, mechanisms used by sophorose and the other regulators (inducers and repressors) to stimulate the expression of these factors have not yet been fully elucidated.

Catabolic repression also takes place during transcription in *T. reesei*, via the CRE1 protein. CRE1 synthesis is repressed in the presence of glucose. *T. reesei* strains insensitive to catabolic repression have been obtained using the traditional mutagenesis techniques, for example strains RUT C30 developed by Rutgers and CL 847 developed by CAYLA.

c) Production in reactor

For production in reactor, the complex mechanisms used to regulate cellulase synthesis require special operating conditions. Use of strains not sensitive to catabolic repression and soluble inducing substrates that can be used at large scale, such as lactose, has led to high production: 40 $g.l^{-1}$ of extracellular proteins by *T. reesei* CL847 according to a process developed by IFP Energies nouvelles [38]. In this process, fermentation is carried out in two phases, a first batch phase for the production of *T. reesei* cells, a second fed-batch phase during which the inducer (lactose) is fed at a rate avoiding its accumulation in the medium (Figure 6.7). Cellulase production rates greater than 20 000 Filter Paper Units (FPU) per litre of fermenter have been obtained in 30 m^3 reactors during 150 hours fermentation.

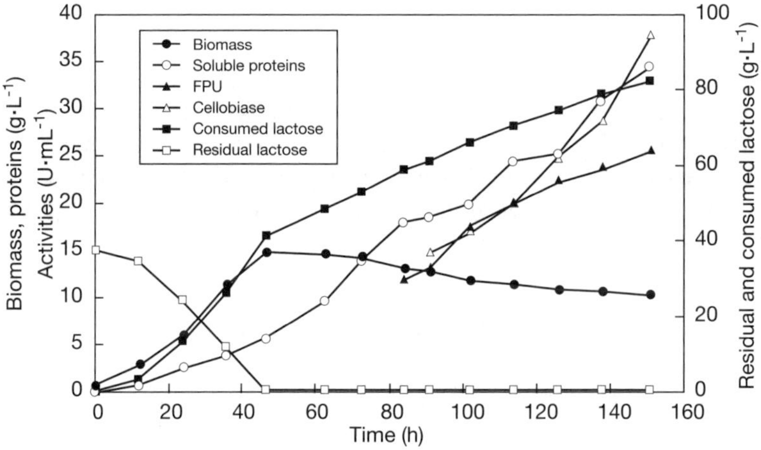

Figure 6.7

Kinetics of cellulase production by *Trichoderma reesei* CL-847 by fermentation on lactose.

d) Improvement axes

As mentioned earlier, enzymatic hydrolysis is expensive, especially due to the cost of enzymes. Currently, cellulase rates of about 10 to 15 FPU/g of cellulose allow efficient hydrolysis of pretreated substrates. Hydrolysis times are relatively long, however (48 to 72 h). Current research studies focus on reducing the cost of producing cellulases and increasing their specific activity. Production rates are already high (greater than 40 $g.l^{-1}$ of extracellular proteins) and progress would mainly concern productivities and fermentation costs. More specific studies concerning its physiology in relation to its secretion power are underway [39 ; 40]. Significant resources are dedicated to improving cellulolytic activities [41]: overexpression of often limiting β-glucosidases [42], production of more active cellulases by rational construction or directed evolution (combinatory methods) [43], to preventing inhibition by cellobiose and glucose, to obtaining cellulases that are more active and/or more stable at high temperature [44], to optimum use of complex structures such as cellulosomes [45] and to searching other cellulases (taking advantage of biodiversity). According to the NREL, the cost of enzymatic hydrolysis would be $0.119/l of ethanol [13] but press releases suggest that it has been reduced by a factor of 10, or even 20 further to the results obtained since 2000 in this field by Genencor International and Novozymes, the two largest producers of industrial enzymes, which benefit from major financing from the US DOE and USDA. The pathways which led to this progress are not yet known.

C. Other lignocellulosic material degradation enzymes

a) Hemicellulases

Depending on the pretreatment mode and its efficiency on hemicellulose hydrolysis and solubilisation, implementation of enzymatic cocktails containing hemicellulases could prove useful. Hemicellulases are as varied as their substrates: endo- and exo-xylanases, β-xylosidases, α-arabino-furanosidases, α-glucuronidases, acetyl xylane esterases, mannanases, galactosidases. These activities are often found in the enzymatic mixtures of cellulolytic microorganisms including *T. reesei* [46]. Other enzymes such as ferulate esterases or *p*-coumarate esterases are auxiliary activities required to cleave some ester bonds involved in the lignin-hemicellulose interactions, more precisely between the arabinose residues of hemicellulose side chains and the lignin phenolic acids. Although more heterogeneous than cellulose, hemicelluloses are more easily hydrolysed than cellulose since they do not form compact crystalline structures. It is nevertheless clear that the presence of hemicellulases in the enzymatic cocktail is necessary, that these enzymes, especially xylanases [47], may exhibit synergies with cellulases and that their presence will be increasingly important if pretreatments not involving acids are used.

b) Ligninases

Ligninases may also prove to be interesting enzymes. They are classified into three categories: lignin peroxidases, Mn-dependent peroxidases and laccases (monophenol oxidases). They are mainly produced by fungi such as white rot basidiomycetes and by a few actinomycetes. The main lignolytic fungi include *Phanerochaete chrysosporium*, *Phlebia radiata*, *Pycnoporus cinnabarinus*, *Coriolus versicolor*. As with hemicellulases, but apparently to a lesser extent, the ligninases could improve the cellulose hydrolysis yield with some lignocellulosic materials and some pretreatments.

6.1.2.4 Ethanol fermentation

Once the cellulose has been hydrolysed into glucose, it is fermented in the same way as the glucose produced from starch. Problems specific to the use of lignocellulosic materials as initial substrate remain, however, such as fermentation of pentoses into ethanol, presence of toxic and inhibiting compounds produced from hemicelluloses and lignin, the possibility of performing enzymatic hydrolysis and fermentation in a single step.

A. Ethanol fermentation of pentoses

a) Yeasts

While *Saccharomyces cerevisiae*, the traditional baker's yeast (of which some strains are used industrially for alcohol fermentation) is unable to use xylose as carbon source, other yeast species can convert xylose into ethanol [48]. The most efficient species are *Pichia stipitis*, *Candida shehatae* and *Pachysolen tannophilus*. These yeasts present numerous weaknesses, however: fermentative performance well below that of *S. cerevisiae* on glucose (Table 6.4), inhibition by ethanol above concentrations in the region of 3 to 5% (w/w), high sensitivity to the inhibitors present in the hydrolysates such as acetic acid.

Table 6.4 Production of ethanol from xylose.

Strain	Yield (g product/g xylose)		Productivity ($g.l^{-1}.h^{-1}$)	Reference
	Ethanol	Xylitol		
Candida shehatae				
ATCC 22 984	0.39		0.56	Wayman and Pareckh, 1985
NRRL Y 12856	0.45		0.29	Slininger *et al.*, 1985 [a]
Pachysolen tannophilus				
RL 171	0.28		0.28	Woods and Millis, 1985
NRRL Y 2460	0.25		0.13	Slininger *et al.*, 1985 [b]
Pichia stipitis				
CBS 5776	0.45		0.34	Tran and Chambers, 1986
NRRL Y 7124	0.43		0.18	Linko *et al.*, 1986
FPL-Shi-21 [a]	0.46	0	0.13	Shi *et al.*, 1999
Saccharomyces cerevisiae				
TMB 3001 [b]	0.31	0.29	0.20	Jeppsson *et al.*, 2002
TMB 3255 [c]	0.41	0.05	0.04	Jeppsson *et al.*, 2002
H158 – pXks [b]	0.20	0.03	0.03	Johansson *et al.*, 2001
1400 (pLNH32) [b]	0.30	0.08	0.33	Ho *et al.*, 1998
H2684 [d]	0.31	0.35		Verho *et al.*, 2003
FPL-YSX3P [c]	0.29	0.46	0.11	Jin *et al.*, 2002
F12	0.26	0.40		Sonderegger *et al.*, 2004 [e]
E. coli				
K011 [a]	0.53		1.3	Ohta *et al.*, 1991 [a]
FBR05 [a]	0.43			[f]
Klebsiella oxytoca M5A1 [a]	0.48		2	Ohta *et al.*, 1991 [b]
Zymomonas mobillis 8b [a]	0.47			[f]

[a] *Mutated strain.*
[b] *Recombinant strain with genes XYL1, XYL2 and XKS1 (see text).*
[c] *Recombinant strain identical to* [b], *+ mutation of the glucose-6-P-dehydrogenase gene.*
[d] *Recombinant strain identical to* [c], *+ overexpression of the glyceraldehyde-3-P-dehydrogenase gene.*
[e] *Recombinant strain having integrated XR/XDH/XK.*
[f] *[Hahn-Hägerdal and Pamment, 2004].*

These limitations and constraints represent obstacles to their industrial use. They have nevertheless been studied extensively for their potentialities and to characterise how they use xylose with a view to using their genes to construct new strains of *S. cerevisiae* capable of using this sugar.

b) Bacteria

Xylose can be converted into ethanol by several species of thermophilic bacteria such as *Thermoanaerobacter ethanolicus*, *Clostridrium thermohydrosulfuricum* and modified strains of *Bacillus stearothermophilus* [3]. These strains have been tested experimentally either alone, or in co-culture with *Clostridium thermocellum*, but their use is faced with numerous limitations: low ethanol yield, production of co-products (acetate), need for growth factors, high sensitivity to inhibitors and low ethanol tolerance. Mesophilic bacteria such as *Escherichia coli* and *Klebsiella oxytoca* or *K. planticola* are also capable of fermenting the pentoses. Good ethanol production rates have been obtained with *E. coli* strains in which the *pdc* (pyruvate dicarboxylase) and *adh* (alcohol dehydrogenase) genes of *Zymomonas mobilis* have been introduced (for example strain K011). Studies have been conducted, aiming at removing restrictions on the use of these strains, such as production of organic acids, nutritional requirements (vitamins and growth factors) and the high sensitivity to the inhibitors present in the hydrolysates and to the high ethanol contents. Significant improvements have been obtained for some of them: high expression of *pdc* and *adh* genes to direct the carbon flow towards ethanol production, creation of ethanol-resistant mutants, deletion of the fumarate reductase gene responsible for the formation of succinate, fermentation conducted in contaminated media. Productivities greater than 1 $g.l^{-1}.h^{-1}$ have been obtained on real hydrolysates with yields between 80 and 92% of the theoretical ethanol yield, but with low ethanol productions, from 27 to 35 $g.l^{-1}$.

Modified strains of *Klebsiella oxytoca* and *planticola* have also been constructed by transferring the same genes (*pdc* and *adh*) and good performance has been obtained (46 $g.l^{-1}$ of ethanol with a yield of 0.48 $g.g^{-1}$ of xylose and productivity of 2 $g.l^{-1}.h^{-1}$). *Klebsiella oxytoca* offers the advantage of being able to use a wide spectrum of substrates: disaccharides such as cellobiose, trisaccharides and tetrasaccharides. However, use of strains of the *Klebsiella* genus is faced with the existence of numerous pathogenic strains within this genus and the resulting bad reputation.

Lastly, strains of *Zymomonas mobilis*, a bacterium recognised for its good ethanol fermentation performance, have been adapted to use xylose after cloning *E. coli* genes encoding xylose isomerase, xylulokinase, transcetolase and transaldolase which are involved in the metabolism of pentoses (see Section B). The recombinant strain CP4 (p285) obtained offers good performance: 88% of the theoretical yield and productivity of 1.04 $g.l^{-1}.h^{-1}$ on corn stalk hydrolysate. In view of the properties specific to *Z. mobilis* such as good tolerance to ethanol and the inhibitors, these modified strains are highly interesting microorganisms which have been chosen by the NREL in some of their ethanol production process diagrams. Lastly, note that strains capable of using arabinose have also been constructed by genetic modification, in particular *Z. mobilis* 206c (p28301) containing several plasmids carrying the genes encoding xylose and arabinose utilisation.

B. Genetic construction of new strains of Saccharomyces

In view of the advantages related to the use of *Saccharomyces cerevisiae*, numerous studies have been, and still are, dedicated to obtaining *S. cerevisiae* strains capable of converting xylose and arabinose into ethanol [48]. The Pentose Phosphate Pathway (PPP) allows *S. cerevisiae* to assimilate xylulose. This pathway is described in Figure 6.8 with its connections with glycolysis, which is the traditional pathway for utilisation of sugars such as glucose [65].

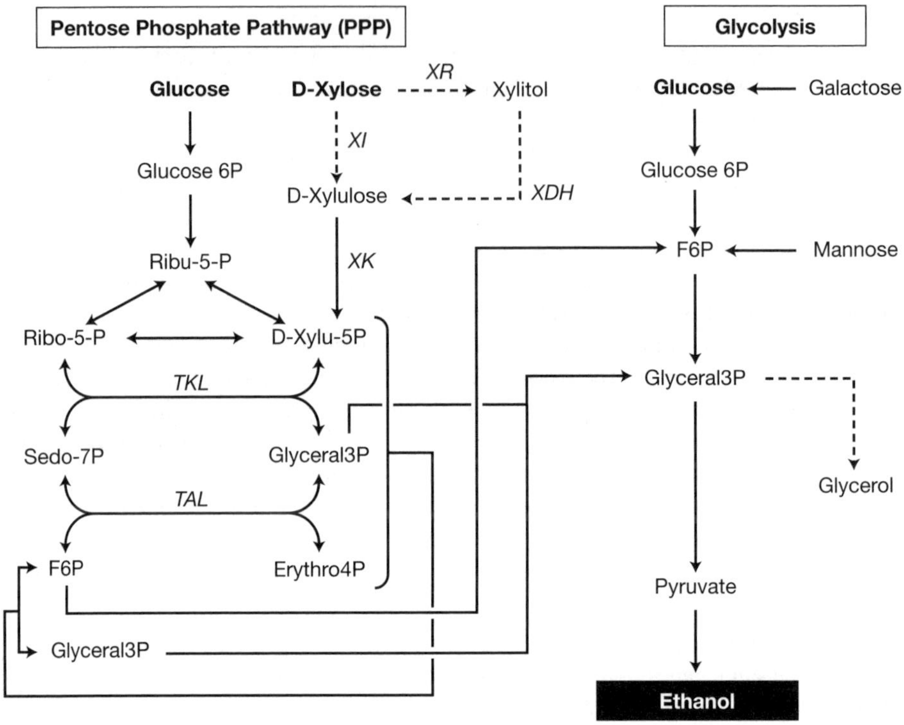

Figure 6.8

Metabolic pathways for the production of ethanol from glucose and xylose.

The first tests to insert the xylose isomerase (XI) gene in *S. cerevisiae* proved unsuccessful, a failure first attributed to the bacterial origin (*E. coli*, *Bacillus subtilis, Lactobacillus pentosus, Clostridium thermosulfurogenes*) of this gene and to unsuitable protein folding within the yeast. Heterologous expression of active bacterial XI has recently been obtained, however, using the *Thermus thermophilus* xylose isomerase. The problems due, firstly, to the optimum temperature differences between *T. thermophilus* and *S. cerevisiae* and, secondly, to inhibition by ethanol, have been attenuated by modifying the xylose isomerase gene by directed evolution. An XI gene of fungal origin has recently been found in a *Pyromyces* strain and successfully introduced in *S. cerevisiae*. This enzyme is strongly inhibited by xylitol. A haploid *S. cerevisiae* strain exhibiting good performance on xylose was first obtained then improved by evolutionary engineering, thereby producing recombinant strains capable of utilising glucose and xylose [66]. These haploid strains are not robust enough for industrial use, however. More recently, the functional expression of a new bacterial xylose isomerase produced from *Clostridium phytofermentans* in *S. cerevisiae* opens new perspectives [67].

Introduction of genes of the yeast *Pichia stipitis* encoding Xylose Reductase (XR) and Xylitol DeHydrogenase (XDH) in *S. cerevisiae* allowed growth on xylose. Ethanol production is low, however, and accumulation of xylitol is observed. Xylitol accumulation is partly explained by an imbalance in the cofactors required for each of these first two reactions (NADPH/NAD^+). The difference in the need for cofactors would lower the maximum ethanol yield from xylose from 0.51 $g.g^{-1}$ to 0.46 $g.g^{-1}$. The relative increase in XDH activity compared with that of XR reduces xylitol accumulation and increases ethanol production, which nevertheless remains low. Xylulose phosphorylation is in fact a key step in the metabolism of xylose by the pentose phosphate pathway and the xylulokinase gene must be overexpressed to favour utilisation of xylitol and, consequently, xylose. The best performance has therefore been obtained using a strain of *Saccharomyces* (*Saccharomyces* sp. 1 400 pLNH33) obtained by fusion of *S. diastaticus* and *S. uvarum*, which contains the XR and XDH genes of *P. stipitis* and in which the *S. cerevisiae* xylulokinase is overexpressed. Ethanol production of 50 $g.l^{-1}$ has been obtained in 36 h using a mixture containing 53 $g.l^{-1}$ of glucose and 56 $g.l^{-1}$ of xylose. A similar industrial strain of *S. cerevisiae* has been developed, although its performance is not as good. Overexpression of xylulokinase reduces the xylose utilisation rate, in particular under anaerobic conditions, probably due to depletion in intracellular ATP (assimilable energy source). Enzymes such as transaldolase involved in PPP have also been overexpressed. One of the main difficulties to be solved is how to obtain strains capable of growth on xylose under anaerobic conditions. The strains obtained today can convert xylose into ethanol under anaerobic conditions, but require oxygen for growth on xylose. Another point concerns xylose transport inside the cell which involves protein transporters identical to those used for glucose transport. In this case there is competition for these two substrates with a preferential affinity for glucose.

Table 6.5 lists some of the main results obtained with yeast strains modified to assimilate pentoses. Despite the number of results obtained, it appears that construction of genetically modified yeast strains capable of efficiently converting xylose and arabinose into ethanol is a difficult objective since it involves modifying the metabolic fluxes of yeast and its redox and energy balances. The main breakthroughs have been obtained incrementally but new techniques, for example transcriptome analysis using DNA chips to display the expression of several thousand genes, should lead to a better understanding of the modifications generated by each change or incorporation of a gene and thereby target the necessary improvements more accurately. Co-production of xylitol, low ethanol production rates and oxygen requirements are the main problems encountered in the recombinant strains obtained to date. In addition, priority should be given to the use of industrial strains whose genetic profile is often quite different from laboratory strains. Although the problem is clearly complex and it will be necessary to control the expression rates of numerous genes, real and substantial breakthroughs have recently been made [73], [41].

C. The problem of inhibitors

The inhibitors present in the hydrolysates come from i) degradation of sugars into furfural and hydroxymethyl furfural, themselves degradable into formic and levulinic acids, ii) groups present in the hemicelluloses (acetyls released as acetic acid), iii) lignin (aldehydes, acids, phenolic alcohols). Their presence depends on the type of initial lignocellulosic material and its pretreatment conditions. Their concentration also depends on the water recirculation ratio in the process. The effects of these inhibitors have been studied on *S. cerevisiae*, *Z. mobilis* and *E. coli*. The studies conducted on *E. coli* have shown that the toxicity is due to combined

Table 6.5 Production of ethanol from glucose-xylose mixture or on ligno-cellulosic hydrolysates.

Strain	Substrate	Ethanol yield ($g.g^{-1}$)	Ethanol ($g.l^{-1}$)	Productivity ($g.l^{-1}.h^{-1}$)	Reference
E. coli KO11	Bagasse hemicellulose hydrolysate (100 g/l xylose)	0.49		0.37 [a]	[b]
	Pine hydrolysate (20 g/l xylose, 15 g/l glucose)	0.42	31	1.53	Barbosa *et al.*, 1992
C. shehatae	Whole barley hydrolysate (54 g/l xylose, 126 g/l glucose)	0.48	84	1.16	Wayman and Pareckh, 1985
Pichia stipitis	Poplar hydrolysate (28 g/l xylose, 64 g/l glucose)	0.45	41	0.85	Pareckh *et al.*, 1986
S. cerevisiae TMB 3400	Spruce hydrolysate (6 g/l xylose, 32 g/l hexoses)	0.43		0.25 [a]	[b]
	Wheat straw hydrolysate (9% water insoluble solids)	0.36	38	0.38	Olofsson *et al.*, 2008
	Sugar cane bagasse hydrolysate (35 g/l glucose, 32 g/l xylose)	0.30	27	0.27	Rudolf *et al.*, 2008
S. cerevisiae 424A	Sugar mixture (70 g/l glucose, 40 g/l xylose)	0.43	46	1.55	Sedlak and Ho, 2004
S. cerevisiae 424A	Corn fibre hydrolysate (40 g/l xylose, 42 g/l glucose, 21 g/l arabinose)	0.44	30	0.62	Sedlak and Ho, 2004

[a] *Specific productivity ($g.g^{-1}.h^{-1}$).*

[b] *[Hahn-Hägerdal and Pamment, 2004].*

effects rather than isolated effects, that furfural inhibits the glycolysis enzymes and that the aromatic compounds are more toxic than the furans. Acids like acetic acid act on the intracellular pH of alcohol-producing microorganisms. Sensitivity to inhibitors varies considerably from one strain to another, even within the same species.

Microorganisms, in particular *S. cerevisiae*, may degrade some inhibitors. Furfural, for example, can be oxidised into furoic acid. Under anaerobic conditions, furfural, 5-hydroxymethyl furfural, vanillin, hydroxybenzaldehyde and syringaldehyde are reduced by *S. cerevisiae*. Microorganisms exhibiting low sensitivity to these inhibitors can be obtained by successive adaptations. Another possibility is to construct tolerant strains by introducing a gene, for example laccase, which oxidises some aromatic inhibitors.

Lastly, hydrolysate detoxification processes, for example by treatment over ion exchange resins, steam entrainment, excess base or enzymatic treatment, have been proposed. Since these processes incur significant additional costs, research studies are currently focusing on the development of tolerant strains.

D. Simultaneous fermentation and hydrolysis

a) SSF process

The Simultaneous Saccharification and Fermentation (SSF) process consists in performing enzymatic hydrolysis and ethanol fermentation in a single step. SSF offers the following main advantages:

- lower investments by removing the operations required for prior enzymatic hydrolysis,
- no inhibition of cellulases by glucose which is consumed by the fermentative microorganisms as soon as it appears. This often increases the hydrolysis yields and rates and the overall ethanol productivities. In addition, the risks of microbial contamination of the glucose-rich hydrolysate are reduced. This type of process can be used both with yeasts and bacteria such as *Z. mobilis*.

The main disadvantages of SSF are the difference in optimum temperatures of enzymatic hydrolysis (50°C) and ethanol fermentation (30 to 35°C) and the fact that the yeasts cannot be recycled, due to the presence of insoluble lignin.

Some *Saccharomyces ceverisiae* strains can grow at 37°C and some *S. pombe* strains at 41°C. Yeast strains belonging to the *Kluyveromyces* genus can ferment glucose at temperatures between 32°C and 45°C [74]. Their performance and properties are very similar to those of *S. cerevisiae*.

It appears that the benefits of SSF, especially financial, justify studies on this type of process [75]. In addition, SSF seems to be suitable for *S. cerevisiae* stains capable of converting pentoses. Some aspects must be optimised, however, especially the initial dry matter concentration, which is still too low to obtain high ethanol concentrations [76].

b) Direct microbial conversion or consolidated bioprocessing

This type of process aims at using the same microorganism for production of enzymes, and therefore cellulose hydrolysis, and for conversion of the sugars obtained – hexoses and pentoses – into ethanol[77].

A first possibility is to use cellulolytic microorganisms and favour the conversion of sugars into ethanol. *Clostridium thermocellum*, for example, is a thermophilic cellulolytic anaerobic bacterium capable of producing ethanol from glucose. It would form fewer co-products than the other thermophilic bacteria and would be active at relatively high substrate concentrations. For the time being, however, its performance and the difficulties involved in its large-scale culture in medium containing no growth factors or vitamins have restricted its applications. Other potential obstacles include the existence of powerful genetic tools required to prevent the co-product formation pathways and the related metabolic energy imbalances. *Thermoanaerobium saccharolyticum* is another thermophilic anaerobic microorganism considered. Its modified strain ALK2 has led to productions of 37 $g.l^{-1}$ of ethanol on a glucose-xylose mixture [78]. This hemicellulolytic strain could be used in combination with *C. thermocellum* in a thermophilic process.

Another possibility is to clone genes encoding cellulases in alcohol-producing microorganisms, *Z. mobilis*, *E. coli* or *S. cerevisiae*. To obtain an active cellulolytic mixture, at least two cellobiohydrolases are generally required, one attacking the reducing ends of the cellulose, the other attacking the non-reducing ends, an endoglucanase and a β-glucosidase and at least one of the CBHs must have a CBM. Heterologous expression of cellulases in *Z. mobilis* has rarely been described, apart from that of an *Erwinia chrysanthemi* endoglucanase and an *Acetobacter xylinum* cellulase. More promising results have been obtained with expression of the same cellulases in *E. coli* or *K. oxytoca*. These studies are still far from the industrial stage. Numerous genes of cellulases, whether endoglucanases, cellobiohydrolases or β-glucosidases of bacterial or fungal origin have been cloned and expressed in *S. cerevisiae*. The proteins expressed are often hyperglycosylated and of heterogeneous size and therefore relatively inactive. Furthermore, the heterologous protein secretion rates represent an obstacle to the use of *S. cerevisiae* as host-yeast to clone foreign genes. These two points are currently being investigated [79]. The DMC pathway is very attractive which has made significant progress over the last few years. Some companies are considering industrialisation in the medium term (Mascoma).

6.1.2.5 Use of lignin

In most pathways for the production of ethanol from lignocellulosic biomass, lignin is recovered then treated and converted into steam to be used as energy source. The volumes of lignin recovered are largely sufficient to supply the heat required for the various steam- and energy-consuming steps, in particular pretreatment and distillation. The possibility of using lignin makes these processes self-sufficient and improves the CO_2 balance of the process.

With some raw materials, such as wood, excess quantities of lignin can be obtained. Serious consideration is being given to their use as energy source (electricity or heat) for distribution to local consumers

Lignin can obviously be upgraded in other ways which are, *a priori* more profitable, mainly after conversion (resins, glues, raw materials for fine chemistry). These conversion pathways are far from being able to absorb all the quantities of lignin associated with the production of ethanol fuel and, for the time being, remain marginal.

6.1.2.6 The various production diagrams

Based on the generic process diagram (Figure 6.3) and the various possibilities we have mentioned, a number of process diagrams have been considered (Figure 6.9). The choices often depend on the substrate to be treated. Only processes which have really been developed will be mentioned.

The process most frequently proposed by the NREL combines co-current pretreatment with dilute acid, detoxification treatment then Simultaneous Saccharification and Co-Fermentation (SSCF) by a *Zymomonas mobilis* strain modified to utilise pentoses. This process has been described in detail and the units have been designed, especially when using corn cobs and poplar wood as raw material.

The BCI processes implements two-step acid hydrolysis, described previously. Xylose fermentation is carried out by a modified *E. coli* strain and is independent of glucose fermentation.

The process recommended by IFP Energies nouvelles and Iogen, especially from wheat straw, involves pretreatment (steam explosion or cooking) under dilute acid conditions, separate enzymatic hydrolysis, followed by separation of lignin before fermentation.

The process described by Etek considers wood as raw material. It differs from the previous diagram by dilute acid pretreatment without steam explosion and an SSF.

These processes are currently under development. They are generally based on efficient pentose fermentation, which is not yet the case, even with the modified strains available. Apart from this point, these processes have all been tested or are operating at pilot stage, for example 1 t/day for the NREL pilot, 3.5 t/day for the Iogen pilot, 2 t/day for the Etek pilot, the pilot formerly used by IFP Energies nouvelles being capable of treating 2 t/h.

Over the last few years, larger projects have emerged and several configurations are considered in the numerous operations in progress, especially in the United States thanks to the determined efforts made by the DOE.

Without being exhaustive, the main American projects are those of Dupont-Danisco, Verenium, POET and Abengoa. In Europe, in addition to the processes already indicated, we must mention Biogasol in Denmark, Abengoa at Salamanca, the Dong Inbicon process which includes LHW pretreatment to operate at high dry matter contents, the Futurol project in France Other countries are also highly involved, such as Brazil and China (Novozymes project, for example).

6.1.2.7 Conclusion

Use of lignocellulosic biomass for the production of ethanol would offer numerous environmental advantages (CO_2 emission balance more favourable than ethanol from sugar or amylaceous plants, upgrading of co-products and waste) and socio-economic advantages (no competition with agricultural land dedicated to food or feed use, lower cost of the raw material). It is undoubtedly the most long-term solution to an extension of the use of bioethanol. Various process diagrams and books are available, based on experiments conducted at representative scales. Some steps are complex, however, and their improvement involves a

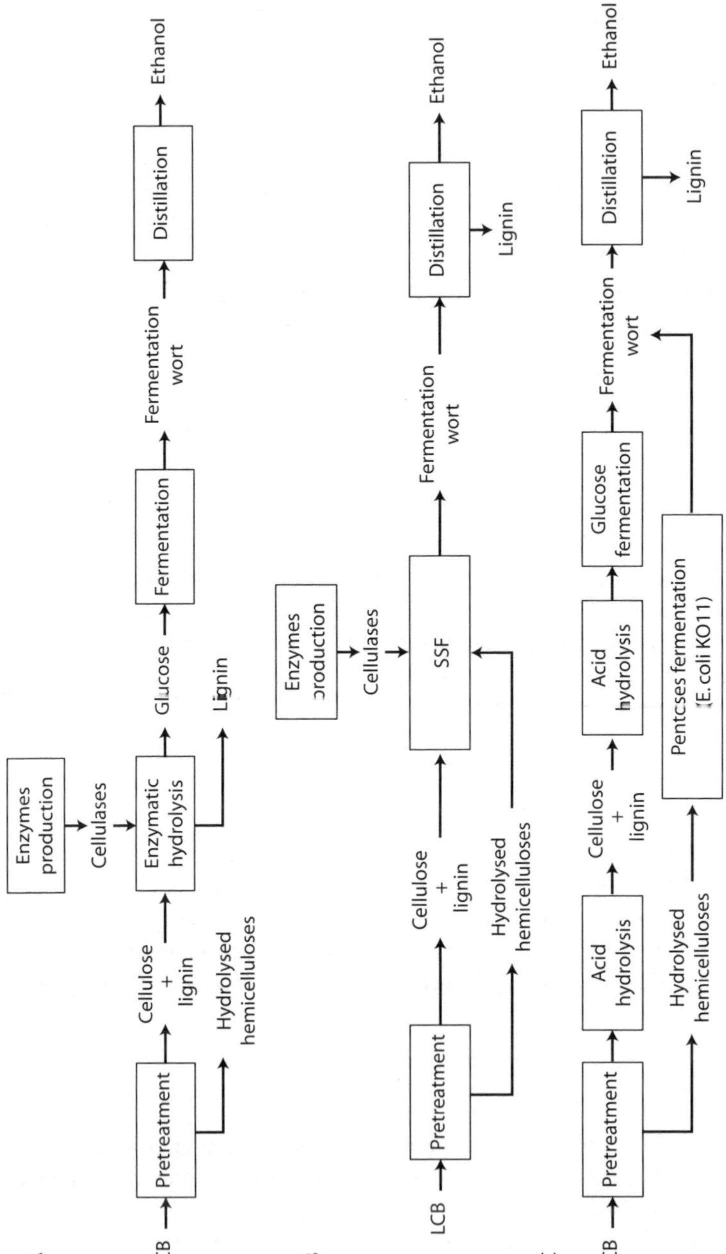

Figure 6.9
The various diagrams for the production of ethanol from lignocellulosic biomass.

greater understanding of a number of aspects, in particular the effect of pretreatment, the enzymology of cellulolysis and the physiology of yeasts. The construction of industrial production units is a major stake whose future largely depends on the progress made on the outcomes of biotechnologies.

6.1.3 Environmental and energy aspects

In terms of greenhouse gas effect, an important gain is obtained when converting lignocellulosic materials into biofuels (especially ethanol) and using them instead of fossil origin fuels. The JRC/EUCAR/CONCAWE study [80] demonstrates that this gain can reach 78% compared with the petroleum fuel reference, i.e. an emission of 36 g of CO_2/km travelled, versus a discharge of 164 g of CO_2/km travelled with gasolines.

To convert these lignocellulosic materials, some of the initial biomass feed is used to supply the utilities required for the conversion process, heat in particular, thereby giving substantial gains in terms of greenhouse gas emissions. This explains that the best results are achieved with sugar cane, whose bagasse is used to supply the electricity and heat required for the distillation step and evaporation of effluents.

Considerable gains are achieved in terms of fossil energy consumptions, as indicated in the above-mentioned study. They amount to about 76%, with a consumption of 0.51 MJ/km travelled, compared with a value of 2.16 MJ/km travelled with gasolines. It is generally recommended to install an energy production unit running on biomass feedstocks, but also solid woody effluents and liquid effluents concentrated after evaporation [81a]. In the latter case, the energy recovered from these residues would largely compensate for the thermal and electrical energy expenditures incurred in the manufacture of ethanol from biomass, estimated at 16.4 MJ and 1 kW per litre of ethanol. The steps consuming the most energy in the process are the distillation and stillage concentration steps, due to the low alcohol contents obtained in fermentation [81b].

6.1.4 Economic aspects

Several studies conducted in the United States and Sweden are described in this paragraph. While there are marked differences between them, they nevertheless give an idea of the investment and operating costs, as well as the costs of manufacturing ethanol from lignocellulosic materials, even though they are rather targets than actual current costs.

The first [82a] gives an accurate picture of the financial calculations performed on a unit with a potential production of 150 Mgal/year gasoline equivalent, from different substrates. The prices of the materials taken into account are \$2.12/bushel of corn and \$50/t for the cellulosic biomass. The CAPEX, estimated in \$/barrel/day, would amount to 13 000 for corn (grain) and 76 000 for the lignocellulosic material, while the operating costs would be respectively \$1.22 and \$1.76 per gallon of gasoline equivalent.

Other data sources are available. A study conducted by the NREL [82b] indicates that, for a unit of 75 000 t/year of ethanol manufactured from corn stover, the investments amount to \$136 M, with a production cost of \$1.5/gal. As a comparison, the investments amount to $28 M for a unit of the same size manufacturing ethanol from corn grains, with a cost of $0.88/gal. of ethanol produced.

Another NREL study [82c], conducted jointly with DOE, explains the costs obtained for a unit designed for processing 2 000 t/day of biomass (at a price of $35/t) and whose ethanol production capacity would be 205 000 t/year. The total investment costs would amount to $198 M and the manufacturing costs $1.07/gal. ($0.28/l).

A Swedish study [81a] indicates investment costs of $100 M for a unit with a capacity of 45 000 t/year of ethanol from spruce and $55 M for a unit of the same capacity manufacturing ethanol from corn starch. The total production costs would amount to $0.6 $/l. of ethanol including $0.2/l for the raw material, $0.19/l for the investments, $0.16/l for energy and $0.09 $/l for the enzymes. The total costs of manufacturing ethanol from corn starch would be $0.57/l.

6.1.5 First and second generation ethanol production on the same industrial site

Ethanol producers, in the United States, Brazil and Europe, are showing increasing interest in the development of second-generation processes. They are also considering the possibility of using some of the facilities on a unit producing first-generation ethanol to manufacture ethanol from lignocellulosic biomass on the same site.

Being on the same site should reduce the investment and product manufacturing costs, since integration of the two processes is fully optimised, and the same equipment being used in the distillation, centrifugation, filtration and fermentation steps. Figure 6.10 shows the possible integration pathways for enzymatic hydrolysis, fermentation and distillation.

It is also generally accepted that processes using lignocellulosic materials produce an energy surplus, due to conversion of the lignin. This surplus can be used in the first-generation process, which is always energy deficient.

Moreover, by combining the fermentation worts before distillation, to obtain an average ethanol content greater than that achieved with lignocellulosic biomass, significant energy savings can be made during the distillation step and when treating the distillation effluents.

R&D and modelling studies as well as energy and financial calculations are currently being conducted. A project is underway as part of the Enerbio Dedicated Fund [83].

6.2 PRODUCTION OF ETHANOL FROM CO + H_2

With the ethanol production processes described in the previous chapters, only sugars and/or polysaccharides are converted into ethanol, irrespective of the raw materials used, whether starchy, sacchariferous or lignocellulosic.

With lignocellulosic biomass, lignin is primarily used to supply energy (electricity and heat) while only the other two main constituents, cellulose and hemicelluloses, produce

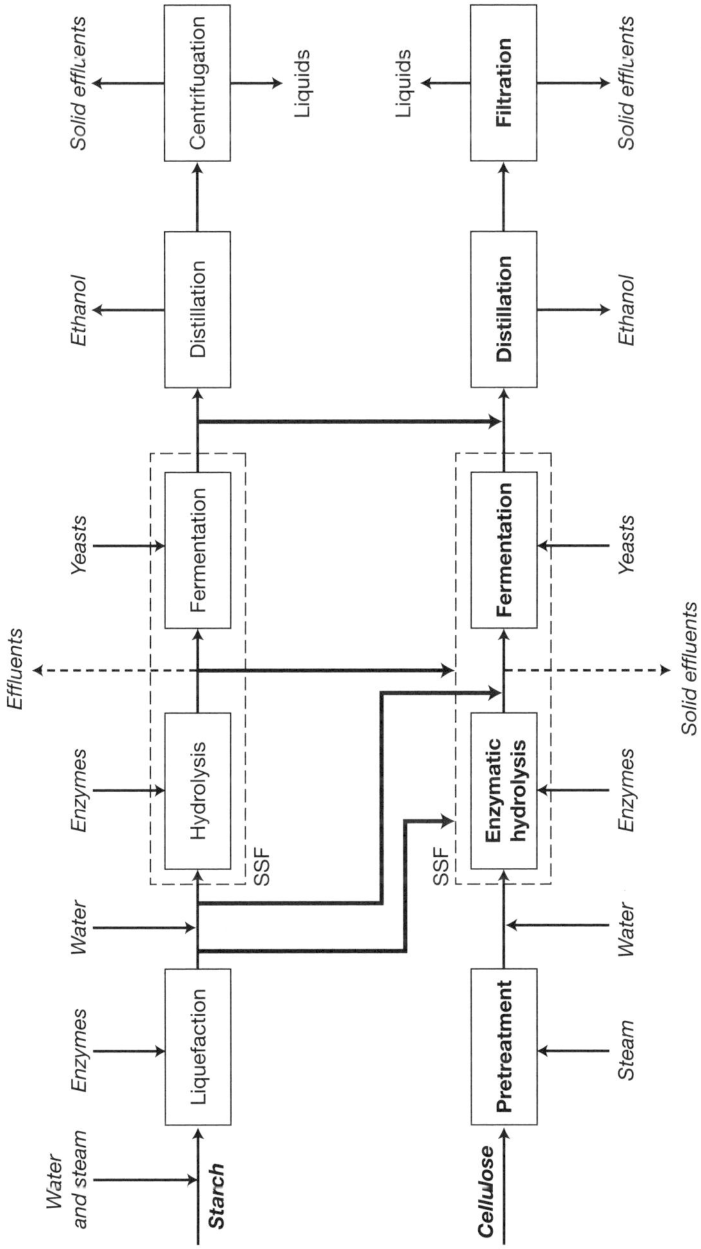

Figure 6.10
Integration of first and second generation ethanol production on the same industrial site.

ethanol-fermentable sugars, after enzymatic or chemical hydrolysis. Consequently, some of the biomass carbon is not used directly to manufacture liquid biofuels.

By thermochemical pathway, syngas, an intermediate in the production of synthetic liquid fuels (BTL, Biomass-to-Liquids pathway), is obtained by converting all the organic matter, including lignin.

To increase the yields obtained by biological conversion of lignocellulosic biomass into ethanol, teams of researchers, including those from Oklahoma University (United States) and the VTT Technical Research Centre (Finland), as well as companies such as Coskata, Ineos Bio and LanzaTech, are trying to convert syngas, mostly composed of CO, $CO_2 + H_2$, into liquid products including ethanol, by fermentation.

As early as the start of the 1990s, scientists demonstrated that anaerobic bacteria such as *Clostridium ljungdahlii* and *Clostridium autoethanogenum* have the metabolic ability to synthesise ethanol from CO, CO_2, H_2 [84], [85], [86], [87].

The stoichiometric equations, representing the conversion of syngas into ethanol, are as follows:

$$6\,CO + 3\,H_2O \longrightarrow C_2H_5OH + 4\,CO_2$$

$$6\,H_2 + 2\,CO_2 \longrightarrow C_2H_5OH + 3\,H_2O$$

With carbon monoxide as the only source of carbon substrate, just one third of the carbon from the biomass and coming from the CO is converted into alcohol.

With an equimolar H_2/CO mixture, two thirds of the carbon from the CO are theoretically converted into ethanol. The theoretical yield will depend largely on the syngas composition:

$$6\,CO + 6\,H_2 \longrightarrow 2\,C_2H_5OH + 2\,CO_2$$

Other compounds may be co-produced, including acetic acid and butanol, as described in the following equations:

$$4\,CO + 4\,H_2 \longrightarrow 2\,CH_3 - COOH$$

$$12\,CO + 12\,H_2 \longrightarrow 2\,CH_3 - CH_2 - CH_2 - CH_2OH + 2\,H_2O + 4\,CO_2$$

Another central ideal in these studies was that the syngas specifications concerning, in particular, the type and concentrations of the by-products (CH_4, tars, etc.) had to be less severe for the biological conversion pathway than for catalytic conversion into biofuels, for example via the Fischer-Tropsch synthesis. This point resulted in simplification of all syngas purification steps. Furthermore, dilution of syngas by nitrogen, when the gasification step is carried out in the presence of air, does not represent an obstacle for the subsequent fermentation. Residual traces of oxygen may be a handicap, however.

One of the main constraints related to conversion by biological pathway is due to the metabolism of the selected microorganisms which, being strictly anaerobic, do not tolerate any traces of oxygen.

Two other difficulties occur during fermentation, firstly the problems caused by placing a gaseous substrate in contact with microorganisms suspended in a liquid medium. With conversion rates for each pass far from 100%, management of the gaseous flows is one of the key points to be optimised.

The second difficulty concerns the absolute necessity to work in strictly anaerobic atmosphere and under sterile conditions in order to protect the active microflora, whose growth rates are very low, against the risks of contamination by other microorganisms.

Results of laboratory experiments conducted with strain P7 [88], [89] have recently been published. The main points identified are as follows:

- stabilisation of the continuous fermentation process, over short periods, not exceeding about ten days,
- low concentrations of ethanol, butanol and acetic acid, respectively 0.16%, 0.11% and 0.03%,
- conversion rates reaching 60 to 70% of the theoretical yields,
- cellular growth rate of about 0.03 h^{-1},
- pronounced sensitivity of the biological conversion to the impurities contained in the syngas, which remains after partial elimination of the tars and particulates in suspension.

Based on these preliminary results, it is clear that significant progress is still required before this biological pathway for conversion of syngas into ethanol can be extrapolated to pilot or industrial scale.

6.3 ACETONOBUTYLIC FERMENTATION

6.3.1 Principle

Acetonobutylic fermentation (ABE fermentation) [1] which generates butanol as main product, as well as acetone and ethanol in smaller quantities, is carried out by "solventogenic" bacteria belonging to the *Clostridium* genus and, more specifically, to the group of butylics such as the main species *Clostridium acetobutylicum* and *Clostridium beijerinckii*.

Bacteria of this species form straight rods of length 2.5 to 5 µm, equipped with peritrich cilia allowing them to move. Their metabolism is strictly anaerobic and their optimum growth temperature lies between 35 and 37°C.

Acetonobutylic fermentation of sugars can be illustrated by the following three equations which describe the conversion of glucose into butanol, acetone and ethanol. Release of CO_2 and H_2 decreases the solvent yield.

$$\underset{\text{Glucose}}{C_6H_{12}O_6} \longrightarrow \underset{n\text{-Butanol}}{CH_3-CH_2-CH_2-CH_2OH} + 2\,CO_2 + H_2O$$

$$C_6H_{12}O_6 + H_2O \longrightarrow \underset{\text{Acetone}}{CH_3-CO-CH_3} + 3\,CO_2 + 4\,H_2$$

$$C_6H_{12}O_6 \longrightarrow 2\,\underset{\text{Ethanol}}{CH_3-CH_2OH} + 2\,CO_2$$

1. ABE: Acetone-Butanol-Ethanol.

Unlike ethanol fermentation, generally carried out on industrial scale with yeasts such as *Saccharomyces cerevisiae*, acetonobutylic fermentation, carried out by bacteria of species *Clostridium acetobutylicum*, must be conducted under septic conditions and under strictly anaerobic atmosphere.

6.3.2 History

It was Pasteur who, in 1862, detected the presence of butanol for the first time, while fermenting lime lactate [90].

From 1910, Fernbach found the solution to the problem faced by the chemists who had just managed to synthesise synthetic rubber by condensing the butadiene obtained by *n*-butanol dehydration and who were looking for sources of this primary alcohol.

He developed a first process for the production of butanol by fermentative pathway. At the same time, he discovered the concomitant formation of acetone with the butylic bacteria, in particular with *Clostridium acetobutylicum*. The weight distribution of acetone and butanol is 1 to 2.

This type of fermentation was to develop rapidly during World War I, further to the very high demand for acetone used to manufacture cordite, an explosive. In 1915, upon request from the English government, Weizmann [1] managed to start the first production on industrial scale.

The production of acetone and butanol by fermentative pathway was subsequently extended to other countries, especially France, Japan and the United States. For example, the Peoria plant in Illinois (United States) at the peak of its production in the 1930s had an installed capacity of 96 fermenters of at least 200 m^3 each [91].

The substrates used were mainly corn meals and sugar cane molasses. The average solvent concentrations were in the region of 18 to 20 g/l. The fermentation worts were distilled in a first column which enriched the solvent content of the phlegm recovered up to about 50%. A second column produced the acetone and butanol fractions by discontinuous distillation of the phlegm.

By the start of the 1960s, the fermentative pathway had been almost completely replaced by petrochemical processes. Only a few countries such as China and South Africa continued using acetonobutylic fermentation until the start of the 1980s.

Several units are still currently operating in China [92]. The physiological problems induced by ABE fermentation have been detailed in recent review articles [93], [94], [95], [96], [97], [98].

Interest was revived in the 1980s, then in 2000, due to the potential of this type of fermentation to convert all the component sugars of the cellulose and hemicelluloses contained in lignocellulosic biomass into energy products such as biofuels [99], [100], [101].

---

1. When Llyod George, head of the Munitions Committee in the British Cabinet of the time, asked Weizmann what he would like as a reward for his efforts and results, Weizmann replied that his dearest wish would be to see a Jewish state created in his lifetime. Chaim Weizmann was the first President of Israel from 1949 to 1952.

6.3.3 Raw materials

Clostridium acetobutylicum can ferment a large variety of monomer sugars: glucose, lactose, saccharose, xylose, arabinose, ribose, and polymers such as starch, inulin, dextrins and xylans.

The most frequently used industrial substrates have been sugar cane and beet molasses as well as cereal meals.

Pentoses and hexoses produced by hydrolysis of lignocellulosic biomass may represent interesting carbon sources.

6.3.4 The process

A. Operating parameters

The inoculum transfer chain has a major impact on the reaction kinetics and yield. The optimum ratio between the volumes of the series reactors, prefermenters and production fermenter lies in the region of 5%. For example, the inoculation chain consists of 3 series reactors, respectively 25 l, 500 l and 10 m^3 to obtain a production fermenter of 200 m^3 (Photo 6.1)

The empty fermenter is sterilised for 1 h at 120°C and placed under inert atmosphere by introducing nitrogen during the cooling phase.

The fermentation medium, with the carbon source, is sterilised in line at 140°C for 30 seconds. The initial sugar concentration lies between 60 and 70 $g.l^{-1}$.

Photo 6.1

ABE production fermenters on the Soustons platform.

The pH varies between 5.5 and 6.2. The pH is important to trigger the production of ABE after the initial production of acetic and butyric acids (Figure 6.11). The temperature is kept at 35°C. The medium is stirred with a propeller rotating at about 100 r.p.m.

B. Production kinetics

Figure 6.11 gives an example of the kinetics for the production of ABE from corn stalk hydrolysates [102].

Fermentation is almost complete in about 50 hours. Xylose is converted more slowly than glucose which is fully consumed in less than 40 hours. Pentose assimilation improves if the initial sugar concentration is low. The acetic acid initially present is partly consumed and used for the production of solvents.

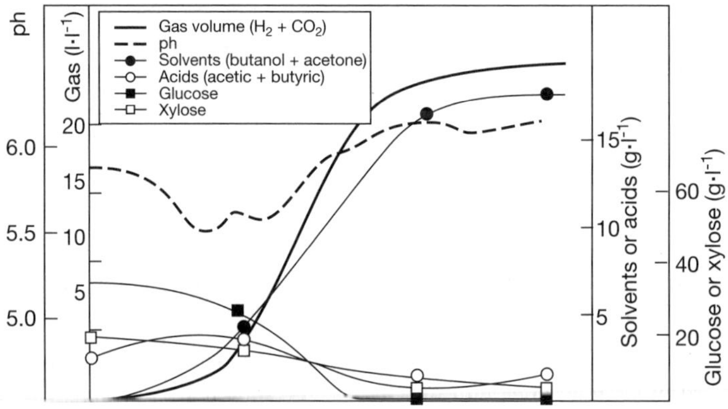

Figure 6.11

Production of ABE from corn stalk hydrolysates.

The corn stalks, pretreated with steam in acid medium, have been hydrolysated by *Trichoderma reesei* cellulases. The hydrolysate has been neutralised with lime, sterilised and supplemented with 3 g.l^{-1} of $(NH_4)_2$ SO_4 and 3 g.l^{-1} of corn steep liquor.

C. Mass and energy balances

Table 6.6 shows an example of mass balance. It was produced on a 48 m^3 fermenter with a corn cob hydrolysate as carbon source which contains 2.9 g.l^{-1} of acetate and sugars at an initial concentration of 66.5 g.l^{-1} [103].

Fermentation lasts 46 h, but the glucose is fully consumed after just 28 hours. At end of reaction, 20% of the xylose has not yet been assimilated. The concentration of ABE produced is 20.5 g.l^{-1}, with a yield by weight of 31%, calculated from the sugars consumed. The ABE synthesis productivity is 0.45 g.l^{-1}.h^{-1}.

ABE is separated from the fermentation juices by distillation; the schematic diagram is shown on Figure 6.12. The first column produces solvent-enriched phlegms and stillages. The phlegms are sent to a second column which separates the acetone at the top, still

Table 6.6 Balance of ABE fermentation from corn cobs.

Type of products	Final concentration
Acetone ($g.l^{-1}$)	7.8
Butanol ($g.l^{-1}$)	12.4
Ethanol ($g.l^{-1}$)	0.3
ABE ($g.l^{-1}$)	20.5
Acetate ($g.l^{-1}$)	1.2
Butyrate ($g.l^{-1}$)	0.8
Gases produced ($CO_2 + H_2$) ($l.l^{-1}$)	31.1
Sugars consumed ($g.l^{-1}$)	59.0
Residual pentoses ($g.l^{-1}$)	7.5

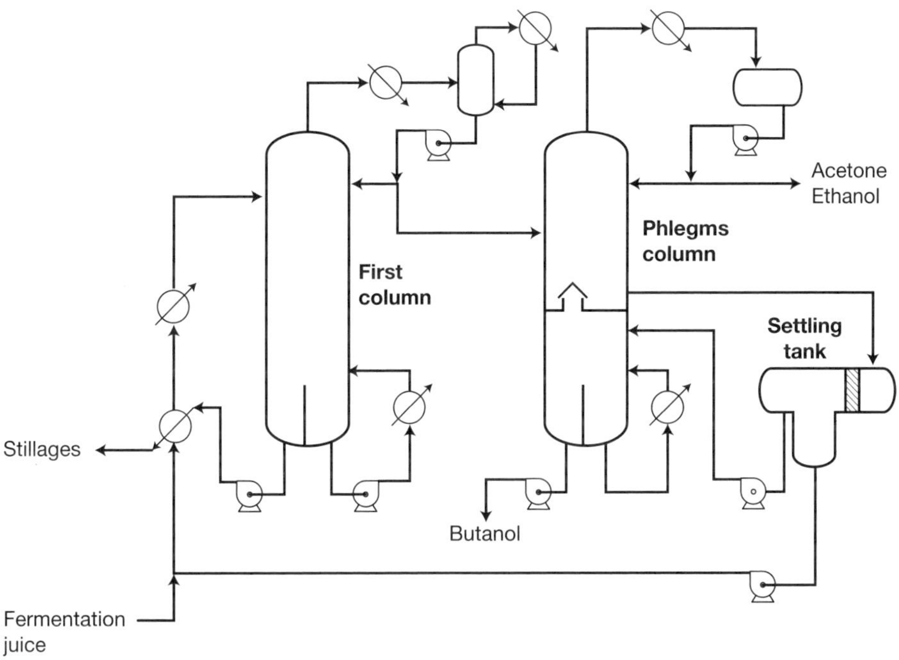

Figure 6.12

Schematic diagram of ABE distillation.

containing approximately 1% water and ethanol, if any, and butanol at the bottom. A side-stream and a settling tank on the side of this column are used to recycle a phase very rich in water to the first column and the light phase, rich in acetone and butanol, to the second column [104].

Fermenter sterilisation and distillation are the main energy consumers. The energy requirements per m^3 of ABE are 253 kWh of electricity and 3.3 tonnes of steam.

D. Product characteristics

The ABE mixture has a high octane number. Mixed with gasoline, it creates no stability problems in the presence of traces of water and, under these conditions, no demixing occurs, which is sometimes the case with ethanol.

Globally, there are no difficulties regarding its use provided that the resistance of the materials has been checked, since problems could appear due to the presence of acetone. It may also act as methanol cosolvent in gasoline.

Its calorific value (31.9 MJ/kg) is significantly greater than that of ethanol (26.8 MJ/kg).

It is miscible with Diesel. The cetane number of a 90/10 Diesel-ABE mixture (by volume) is 49.

E. Financial data

A study [104] shows that on selecting the Brazilian conditions, with sugar cane juice as carbon substrate, the investment for an ABE unit is about 15% greater than that of an ethanol unit, processing the same quantity of sugar cane, i.e. 1 600 t/day. Setting the same internal rate of return of 10.6% after tax on the profits, the selling price of ABE is roughly 50% higher than that of ethanol.

F. Improvement axes

The main handicap with ABE fermentation lies in the low concentration of products at end of fermentation (2% by weight as compared with 7 to 10% by weight for ethanol fermentation). This limit is due to inhibition of *Clostridium acetobutylicum* metabolism by *n*-butanol.

Several studies show that this inhibition can be removed by *in situ* gas stripping of the solvents as they are synthesised [105] or by selective extraction across a silicone-based membrane, with oleic alcohol as extraction solvent [106]. In the latter case, 87.2 g of solvent are produced per litre of installed fermentation capacity from 277 g of lactose with a productivity of 0.2 $g.l^{-1}\ h^{-1}$. Progress is still required to improve the membrane selectivity and prevent oleic alcohol from passing into the fermentative medium.

REFERENCES

[1] Lapierre C (1993) Application of New Methods for the Investigation of Lignin Structure, *In: Forage Cell Wall Structure and Digestibility*, HG Jung, DR Bruxton, RD Hatfield, J. Ralph Eds, ASA-CSSA-SSSA, Madison, WI, pp 133-166.

[2] Sun Y and Cheng J (2002) Hydrolysis of Lignocellulosic Materials for Ethanol Production: a review. Bioresource Technol., 83, pp 1-11.

[3] Ogier JC, Ballerini D, Leygue JP, Rigal L, Pourquié J (1999) Production d'éthanol à partir de biomasse lignocellulosique. Oil & Gas Science and Technology – Revue de l'IFP, **54**, 1, pp 67-94.

[4] Thomsen MH, Thygesen A and Thomsen AB (2008) Hydrothermal Treatment of Wheat Straw at Pilot Plant Scale using a Three-step System Aiming at High Hemicellulose Recovery, High Cellulose Digestibility and Low Lignin Hydrolysis, Bioresource Technol., 99, pp 4221-4228.

[5] Ballerini D, Desmarquest JP, Pourquié J, Nativel F and Rebeller M (1994) Ethanol Production from Lignocellulosics: Large Scale Experimentation and Economics. Bioresource Technol., 50, pp 17-23.

[6] Sassner P, Galbe M and Zacchi G (2006) Bioethanol Production Based on Simultaneous Saccharification and Sermentation of Steam-pretreated Salix at High Dry-matter Content, Enzyme Microb. Technol., 39, pp 756-762.

[7] Teymouri F, Laureano-Perez L, Alizadeh H and Dale B (2005) Optimization of the Anmmonia Fiber Explosion (AFEX) Treatment Parameters for Enzymatic Hydrolysis of Corn Stover. Bioresource Technol., 96, pp 2014-2018.

[8] Schell DJ, Torget R, Power A, Walter PJ, Grohmann K and Hinmann ND (1991) A Technical End Economic Analysis of Acid-Catalysed Steam Explosion and Dilute Sulfuric Acid Pretreatment Using Wheat Straw or Aspen Wood Chips. Appl. Biochem. Biotechnol. 28/29, pp 87-97.

[9] Sendich E, Laser M, Kim S, Alizadeh H, Laureano-Perez L, Dale B and Lynd L (2008) Recent Process Improvements for the Ammonia Fiber Expansion (AFEX) Process and Resulting Reductions in Minimum Ethanol Selling Price. Bioresource Technol., 99, pp 8429-8435.

[10] Yang B and Wyman CE (2008) Pretreatment: the Key to Unlocking Low-cost Cellulosic Ethanol, Biofuels Bioproducts Bioref., 2, pp 26-40.

[11] Kim TH, Taylor F and Hicks KB (2008) Bioethanol Production from Barley Hull Using SAA (Soaking in Aqueous Ammonia) Pretreatment. Bioresource Technol., 99, pp 5694-5702.

[12] Selig MJ, Viamajala S, Decker SR, Tucker MP, Himmel ME and Vinzant TB (2007) Deposition of Lignin Droplets Produced During Acid Pretreatment of Maize Stems Retards Enzymatic Hydrolysis of Cellulose, Biotechnol. Progr. 23, pp 1333-1339.

[13] Wooley R, Ruth M, Glassner D and Sheehan J (1999) Process Design and Costing of Bioethanol Technology: a Tool for Determining the Status and Direction of Research and Development, Biotechnol. Prog. 15, pp 794-803.

[14] Well J, Brewer M, Hendrickson R, Sarikaya A and Ladisch MR (1998) Continuous pH Monitoring During Pretreatment of Yellow Poplar Wood Sawdust by Pressure Cooking in Water, Appl. Biochem. Biotechnol. 70-72, pp 91-111.

[15] Kaar WE and Holtzapple MT (2000) Using Lime Pretreatment to Facilitate the Enzymic Hydrolysis of Corn Stover. Biomass Bioenergy, 18, pp 189-199.

[16] Pan X, Arato C, Gilkes N, Gregg D, Mabee W, Pye K, Xiao Z, Zhang X and Saddler J (2005) Biorefining of Softwoods Using Ethanol Organosolv Pulping: Preliminary Evaluation of Process Streams for Manufacture of Fuel-grade Ethanol and Co-products, Biotechnol. Bioeng., 90, pp 373-481.

[17] Zhu S, Wu Y, Chen Q, Yu Z, Wang C, Jin S, Ding Y and Wu G (2006) Dissolution of Cellulose with Ionic Liquids and its Application: a Mini-review, Green Chem., 8, pp 325-327.

[18] Zhang Y-HP, Ding S-Y, Mielenz JR, Cui JB, Elander RT, Laser M, Himmel ME, McMillan JR and Lynd LR (2007) Fractionating Recalcitrant Lignocellulose at Modest Reaction Conditions, Biotechnol. Bioeng., 97, pp 214-223.

[19] Lee J (1997) Biological Conversion of Lignocellulosic Biomass to Ethanol. J. Biotechnol., 56, pp 1-24.

[20] Mosier N, Wyman C, Dale B, Elander R, Lee YY, Holtzapple M and Ladisch M (2005) Features of Promising Technologies for Pretreatment of Lignocellulosic Biomass, Bioresource Technol., 96, pp 673-686.

[21] Eggeman T and Elander RT (2005) Process and Economic Analysis of Pretreatment Technologies, Bioresource Technol. 96, pp 2019-2025.

[22] Yang B and Wyman CE (2008) Pretreatment: the Key to Unlocking Low-cost Cellulosic Ethanol, Biofuels Bioproducts Bioref., 2, pp 26-40.

[23] Ackerson M, Ziobro M and Gaddy GL (1981) Two-Stage Acid Hydrolysis of Biomass. Biotechnol. Bioeng. Symp. 11, pp 103-112.

[24] Lynd LR, Weimer PJ, van Zyl WH and Pretorius IS (2002) Microbial Cellulose Utilization: Fundamentals and Biotechnology. Microbiol. Molec. Biol. Rev., **66**, 3, pp 506-577.

[25] Ortega N, Busto MD and Perez-Mateos M (2001) Kinetics of Cellulose Saccharification by Trichoderma Reesei Cellulases, Int. Biodeter. Biodeg., 47, pp 7-14.

[26] Peri S, Karra S, Lee YY and Karim MN (2007) Modeling Intrinsic Kinetics of Enzymatic Cellulose Hydrolysis, Biotech. Progr., 23, pp 626-637.

[27] Palmqvist E, Hahn-Hegerdal B, Galbe M and Zacchi G (1996) The Effect of Water-soluble Inhibitors from Steam-pretreated Willow on Enzymatic Hydrolysis and Ethanol Fermentation. Enzyme Microb. Technol., 19, pp 470-476.

[28] Coutinho RM and Henrissat B (1999) Carbohydrate – Active Enzymes Server at URL: http: //afb.cnrs-mrs.fr/cazy/index.html.

[29] Schwarz WH (2001) The Cellulosome and Cellulose Degradation by Anaerobic Bacteria. Appl. Microbiol. Biotechnol, 56, pp 634-649.

[30] Beckham GT, Matthews JF, Bomble YJ, Bu L, Adney WS, Himmel ME, Nimlos MR and Crowley MF (2010) Identification of Amino Acids Responsible for Processivity in a Family 1 Carbohydrate-Binding Module from a Fungal Cellulase. J. Phys. Chem. B. 114, pp 1447-53.

[31] Foreman PK, Brown D, Dankmeyer L, Dean R, Diener S, Dunn-Coleman NS *et al.* (2003) Transcriptional Regulation of Biomass-degrading Enzymes in the Filamentous Fungus *Trichoderma reesei*. *J Biol Chem* 278, pp 31988-31997.

[32] Pel HJ, de Winde JH, Archer DB *et al.* (2007) Genome Sequencing and Analysis of the Versatile cell Factory *Aspergillus niger* CBS 513.88, Nature Biotech. 25, pp 221-231.

[33] Martinez D, Berka RM, Henrissat B *et al.* (2008) Genome Sequencing and Analysis of the Biomass-degrading Fungus *Trichoderma reesei* (syn. *Hypocrea jecorina*), Nature Biotech. 26, pp 553-560.

[34] Zhang H-YP and Lynd LR (2004) Toward an Aggregated Understanding of Enzymatic Hydrolysis of Cellulose: Noncomlexed Cellulase Systems. Biotechnol Bioeng., 88, pp 797-824.

[35] Fierobe HP, Bayer EA, Tardif C, Czjek M, Mechaly A, Belaich A, Lamed R, Shoham Y and Belaich JP (2002) Degradation of Cellulose Substrates by Cellulosome Chimeras: Substrate Targeting Versus Proximity of Enzyme Engineering, J. Biol. Chem. 277, pp 49621-49630.

[36] Suto M and Tomita F (2001) Induction and Catabolite Repression Mechanisms of Cellulase in Fungi. J. Biosc. Bioeng., 92, pp 305-311.

[37] Aro N, Pakula T and Penttilä M (2005) Transcriptional Regulation of Plant Cell Wall Degradation by Filamentous Fungi, FEMS Microbiol. Rev., 29, pp 719-739.

[38] Warzywoda M, Larbre E and Pourquié J (1992) Production and Characterization of Cellulolytic Enzymes from *Trichoderma reesei* Grown on Various Carbon Sources. Bioresource Technol., 39, pp 125-130.

[39] Sticker AR, Mach RL and de Graaf LH (2008) Regulation of Transcription of Cellulases- and Hemicellulases-encoding Genes in *Aspergillus niger* and *Hypocrea jecorina* (*Tricherderma reesei*). Appl. Microbiol. Biotechnol., 78, pp 211-220.

[40] Le Crom S, Schackwitzd W, Pennacchio L, Magnuson JK, Culley DE, Collett JR, Martin J, Druzhinina IS, Mathis H, Monot F, Seiboth B, Cherry B, Rey M, Berka R, Kubicek CP, Baker SE and Margeot A (2009) Tracking the Roots of Cellulase Hyperproduction by the Fungus Trichoderma Reesei using Massively Parallel DNA Sequencing Proc. Nat. Acd. Sci. United States.

[41] Margeot A, Hahn-Hägerdal D, Edlund M, Slade R and Monot F (2009) New Improvements for Lignocellulosic Ethanol, Curr. Op. Biotechnol. 20, pp 372-380.

[42] Kovacs K, Megyeri, Szakacs G, Kubicek CP, Galbe M and Zacchi G (2008) Trichoderma Atroviride Mutants with Enhanced Production of Cellulase and β-glucosidase on Pretreated Willow. Enzyme Microbiol. Technol., 43, pp 48-55.

[43] Zhang Y-HP, Himmel ME and Mielenz JR (2006) Outlook for Cellulase Improvement: Screening and Selection Strategies. Biotechnol. Adv., 24, pp 452-481.

[44] Blumer-Schuette SE, Kataeva I, Westpheling J, Adams MWW and Kelly RM (2008) Extremely Thermophilic Microorganisms for Biomass Conversion: Status and Prospects. Curr. Op. Biotechnol., 19, pp 210-217.

[45] Bayer EA, Lames R and Himmel ME (2007) The Potential of Cellulases and Cellulosomes for Cellulosic Waste Management, Curr. Op. Biotechnol., 18, pp 237-245.

[46] Margolles-Clark E, Ilmen M and Penttila M (1997) Expression Patterns of Ten Hemicellulases of the Filamentous Fungus Trichoderma Reesei on Variuos Sources, J. Biotechnol., 57, pp 167-179.

[47] Fierobe HP, Mingardon F, Mechaly A, Belaich A, Rincon MT, Lamed R, Tardif C, Belaich JP and Bayer EA (2005) Action of Designer Cellulosomes Versus Complex Substrates: Controlled Incorporation of Three Distinct Enzymes Into a Defined Tri-functional Scaffoldin. J. Biol. Chem., 280, pp 16325-16334.

[48] Jeffries TW and Jin YS (2004) Metabolic Engineering for Improved Fermentation of Pentoses by Yeasts. Appl. Microbiol. Biotechnol., 63, pp 495-509.

[49] Wayman M and Pareckh SR (1985) Ethanol and Sugar Tolerance of *Candida shehatae*. Biotechnol. Lett., 7, pp 909-912.

[50] Slininger SP, Bothast RJ, Okos MR and Ladish MR (1985) Comparative Evaluation of Ethanol Production by Xylose-Fermenting Yeasts Presented High Xylose Concentration. Biotechnol. Lett., 7, pp 431-436.

[51] Woods MA and Millis NF (1985) Effect of Slow Feedings of Xylose on Ethanol Yield by *Pachysolen tannophilus*. Biotechnol. Lett., 7, p. 679.

[52] Slininger SP, Bothast RJ, Black LT and McGhee JE (1985) Continuous Conversion of D-xylose to Ethanol by Immobilized *Pachysolen tannophilus*. Biotechnol. Bioeng., 24, pp 2241-2251.

[53] Tran AV and Chambers RP (1986) Ethanol Fermentation of Red Oak Acid Prehydrolysate by the Yeast *Pichia Stipitis* CBS 5576. Enz. Microbiol. Technol., 8, pp 439-444.

[54] Linko YY, Hautola H, Uiotila S and Linko P (1986) Alcoholic Fermentation of D-xylose by Immobilized *Pichia stipitis* Yeast. Biotechnol. Lett., 8, pp 47-52.

[55] Shi NQ, Davis B, Sherman F, Cruz J and Jeffries TW (1999) Disruption of the Cytochrome c Gene in Xylose-utilizing Yeast *Pichia stipitis* Leads to Higher Ethanol Production. Yeast 15, pp 1021-1030.

[56] Jeppsson M, Johansson B, Hahn-Hägerdal B and Gorwa-Grauslund MF (2002) Reduced Oxidative Pentose Phosphate Pathway Flux in Recombinant Xylose-utilizing *Saccharomyces cerevisiae* Strains Improves the Ethanol Yield from Xylose. Appl. Environ. Microbiol., 68, pp 1604-1609.

[57] Johansson B, Christensson C, Hobley T and Hahn-Hägerdal B (2001) Xylulokinase Overexpression in Two Strains of *Saccharomyces cerevisiae* also Expressing Xylose Reductase and Xylitol Dehydrogenase and its Effect on Fermentation of Xylose and Lignocellulosic Hydrolysate. Appl. Environ. Microbiol., 67, pp 4249-4255.

[58] Ho NW, Chen Z and Brainard AP (1998) Genetically Engineered *Saccharomyces* Yeast Capable of Effective Co-fermentation of Glucose and Xylose. Appl. Environ. Microbiol., 64, pp 1852-1859.

[59] Verho R, Londesborough J, Penttilä M and Richard P (2003) Engineering Redow Cofactor Regeneration for Improved Pentose Fermentation in *Saccharomyces cerevisiae*. Appl. Environ. Microbiol., 69, pp 5892-5897.

[60] Jin YS, Jones S, Shi NQ and Jeffries TW (2002) Molecular Cloning of XYL3 (D-xylulokinase) from *Pichia stipitis* and Characterization of its Physiological Function. Appl. Environ. Microbiol., 68, pp 1232-1239.

[61] Sonderegger M, Jeppsson M, Larsson C, Gorwa-Grauslund MF, Boles E, Olsson L, Spencer-Martins I, Hahn-Hägerdal B and Sauer U (2004) Fermentation Performance of Engineered and Evolved Xylose-fermenting *Saccharomyces cerevisiae* Strains. Biotechnol. Bioeng., 87, pp 90-98.

[62] Ohta K, Beall DS, Meija JP, Shanmugam KT and Ingram LO (1991) Genetic Improvement of *E. coli* for Ethanol Production: Chromosal Integration of *Zymomonas mobilis* Genes Encoding Pyruvate Decarboxylase and Alcohol Dehydrogenase II. Appl. Environ. Microbiol., 57, pp 893-900.

[63] Hahn-Hägerdal B and Pamment N (2004) Microbial Pentose Metabolism. Appl. Biochem. Biotechnol., Vol. 113-116, pp 1207-1209.

[64] Ohta K, Beall DS, Meija JP, Shanmugam KT and Ingram LO (1991) Metabolic Engineering of *Klebsiella oxytoca* M5A1 for Ethanol Production from Xylose and Glucose. Appl. Environ. Microbiol., 57, pp 2810-2815.

[65a] Zaldivar J, Nielsen J and Olsson L (2001) Fuel Ethanol Production from Lignocellulose: a Challenge for Metabolic Engineering and Process Integration. Appl. Microbiol. Biotechnol, 56, pp 17-34.

[65b] Zaldivar J, Borges A, Johansson B, Smits HP, Villas-Bôas SG, Nielsen J and Olsson L (2002) Fermentation Performance and Intracellular Metabolite Patterns in Laboratory and Industrial Xylose-fermenting *Saccharomyces cerevisiae*. Appl Microbiol Biotechnol. 59, pp 436-442.

[66] Kuyper M, Hartog MMP, Toirkens MJ, Almerig MJH, Winkler AA, Van Dijken JP and Pronk JT (2005) Metabolic Engineering of a Xylose-isomerase-expressing *Saccharomyces cerevisiae* Strain for Rapid Anaerobic Xylose Fermentation, FEMS Yeast Res., 5, pp 399-409.

[67] Brat D, Boles E and Wiedemann B (2009) Functional Expression of a Bacterial Xylose Isomerase in *Saccharomyces cerevisiae*. Appl Environ Microbiol, 75, pp 2304-2311.

[68] Barbosa MFS, Beck MJ, Fein JE, Potts D and Ingram LO (1992) Efficient Fermentation of *Pinus sp*. Acid Hydrolysates by an Ethanologenic Strain of *E. coli*. Appl. Environ. Microbiol., 58, pp 1382-1384.

[69] Parekh SR, Yu S and Wayman M (1986) Adaptation of *Candida shehatae* and *Pichia stipitis* to Wood Hydrolysates for Increased Ethanol Production. Appl. Microbiol. Biotechnol., 25, pp 300-304.

[70] Olofsson K, Rudolf A and Liden G (2008) Designing Simultaneous Saccharification and Fermentation for Improved Xylose Conversion by a Recombinant Strain of *Saccharomyces cerevisiae*. J. Biotechnol., 134, pp 112-120.

[71] Rudolf A, Baudel H, Zacchi G, Hahn-Hägerdal B and Liden G (2008) Simultaneous Saccharification and Fermentation of Steam-pretreated Bagasse Using *Saccharomyces cerevisiae* TMB3400 and *Pichia stipitis* CBS6054. Biotechnol. Bioeng., 99, pp 783-790.

[72] Sedlak M and Ho NWY (2004) Production of Ethanol from Cellulosic Biomass Hydrolysates Using Genetically Engineered *Saccharomyces* Yeast Capable of Cofermenting Glucose and Xylose. Appl. Biochem. Biotechnol., 113-116, pp 403-416.

[73] Hahn-Hägerdal B, Karhumaa K, Fonseca C, Spencer-Martins I and Gorwa-Grauslund MF (2007) Towards Industrial Pentose-fermenting Yeast Strains. Appl Microbiol Biotechnol 74, pp.937-53.

[74] Tomas-Pejo E, Oliva JM, Gonzalez A, Ballesteros I and Ballesteros M (2009) Bioethanol Production from wheat Straw by Thermotolerant yeast *Kluyveromyces marxianus* CECT 10875 in a Simultaneous Saccharification and Fermentation Fed-batch Process. Fuel, 88, pp 2142-2147.

[75] Wingren A, Galbe M and Zacchi G (2003) Techno-Economic Evaluation of Producing Ethanol from Softwood: Comparison of SSF and SHF and Identification of Bottlenecks. Biotechnol Prog., 19, pp 1109-1117.

[76] Olofsson K and Liden G (2008) A Short Review on SSF – an Interesting Process Option for Ethanol Production from Lignocellulosic Feedstocks. Biotechnol Biofuels, 1, p 7.

[77] Lynd LR, van Zyl WH, McBride JE and Laser M (2005) Consolidated Bioprocessing of Cellulosic Biomass: an Update. Curr. Op Biotechnol. 16, pp 577-583.

[78] Shaw JA, Podkaminer KK, Desai SG, Bardsley JS, Rogers SR, Thorne PG, Hogsett DA and Lynd LR (2008) Metabolic Engineering of a Thermophilic Bacterium to Produce Ethanol at High Yield. Proc. Nat. Acad. Sci. United States, 105, pp 13679-13774.

[79] van Zyl WH, Lynd LR, den Haan R and McBride JE (2007) Consolidated Bioprocessing for Bioethanol Production using *Saccharomyces cerevisiae*. Adv. Biochem. Engin./Biotechnol. 108, pp 205-235.

[80] http://ies.jrc.ec.europa.eu/WTW, Well-to-Wheels Analysis of Future Automotive Fuels and Powertrains in the European Context. Concawe, Eucar, JRC, Well-to-Wheels report, March 2007.

[81a] Wingren A (2005) Ethanol from Softwood. Technico-economic Evaluation for Development of the Enzymatic Process. Department of Chemical Engineering, Lund University, Sweden.

[81b] Wingren A, Galbe M and Zacchi G (2003) Technico-economic Evaluation of Producing Ethanol from Softwood: Comparison of SSF and Identification of Bottlenecks. Biotechnol. Prog., 19, pp 1109-1117.

[82a] Wright MM and Brown RC (2007) Biofuels Bioprod. Bioref., 1, pp 49-56.

[82b] McAloon A, Taylor F, Yee W, Ibsen K and Wooley R (2000) Determining the Cost of Producing Ethanol from Corn Starch and Lignocellulosic Feedstocks. Technical Report NREL/TP-580-28893, October 2000.

[82c] Aden A *et al.* (2002) Process Design Report for Stover Feedstock: Lignocellulosic Biomass to Ethanol Process Design and Economics Utilizing Co-current Dilute Acid Prehydrolysis and Enzymatic Hydrolysis for Corn Stover. Technical Report NREL/TP-510-32438, June 2002.

[83] Zacchi G and Wahlberg O (2009-2011) Technico-economic Evaluation of Combined First and Second Generation Bio-ethanol Production. Projet Enerbio.

[84] Abrini A, Naveau H and Nyns EJ (1994). *Clostridium autoethanogenum,* sp. nov., an Anaerobic Bacterium that Produces Ethanol from Carbon Monoxide. Archives of Microbiology, 161, pp 345-351.

[85] Klasson KT, Ackerson MD, Clausen EC and Gaddy JL (1992). Bioconversion of Synthesis Gas Into Liquid or Gaseous Fuels. Enzyme Microbiology and Technology, 14, pp 602-608.

[86] Tanner RS (2008) Production of Ethanol from Synthesis Gas. *In:* Wall JD, Harwood CS, Demain A (eds) Bioenergy. ASM Press, Washington DC, pp 147-151.

[87] Henstra AM, Sipma J, Rinzema A and Stams AJM (2007) Microbiology of Synthesis Gas Fermentation for Biofuel Production. Current Opinion in Biotechnology, 18, pp 200-206.

[88] Rajagopalan S, Datar RP and Lewis RS (2002) Formation of Ethanol from Carbon Monoxide via a New Microbial Catalyst. Biomass and Bioenergy, 23, pp 487-493.

[89] Datar RP, Shenkman RM, Cateni BG, Huhnke RL and Lewis RS (2004). Fermentation of Biomass-Generated Producer Gas to Ethanol. Biotechnology and Bioengineering, **86**, 5, pp 587-594.

[90] Brechot P (1982) Fermentation acétonobutylique depuis Pasteur, Fernbach et ses collaborateurs jusqu'au carburol, Colloque Soc. Fr. Microbiol., IFP Energies nouvelles, Rueil-Malmaison, pp 111-122.

[91] Vandecasteele JP et Marchal R (1993) Les bactéries solvantogènes du genre *Clostridium*, pp 489-537, in Microbiologie Industrielle. Les micro-organismes d'intérêt industriel, Ed. J.Y. Leveau, M. Bouix, Collection Sciences et Techniques Agro-Alimentaires – Technique et Documentation Lavoisier, Paris.

[92] Ni Y and Sun ZH (2009) Recent Progress on Industrial Fermentative Production of Acetone-butanol-ethanol by *Clostridium acetobutylicum* in China. Applied Microbiology and Biotechnology 83, pp 415-423.

[93] Dürre P (2008) Fermentative Butanol Production – Bulk Chemical and Biofuel. Annals of the New-York Academy of Sciences, 1125, pp 353-362.

[94] Rogers P, Chen J-S and Zidwick M (2006) Organic Acid and Solvent Production. Part III; Butanol, Acetone, Isopropanol;1,3- and 1,2-propanediol Production; and 2,3-butanediol Production. *In:* Dworkin M (ed) The Prokaryotes. A Handbook on the Biology of Bacteria. 3rd Edition. Springer, New York, United States, pp 672-755.

[95] Paredes CJ, Jones SW, Senger RS, Borden JR and Sillers R (2008) Molecular Aspects of Butanol Fermentation. *In:* Bioenergy. ASM Press, Washington DC, pp 323-334.

[96] Zheng Y-N *et al.* (2009) Problems with the Microbial Production of Butanol. Journal of Industrial Microbiology and Biotechnology, 36, pp 1127-1138.

[97] Gheshlaghi R, Scharer JM, Moo-Young M and Chou CP (2009) Metabolic Pathways of *Clostridia* for Producing Butanol. Biotechnology Advances, 27, pp 764-781.

[98] Ezeji TC and Blaschek HP (2008) Practical Aspects of Butanol Production. *In:* Wall JD, Harwood CS, Demain A (eds) Bioenergy. ASM Press, Washington DC, pp 335-346.

[99] Zverlov VV, Berezina O, Velikodvorskaya GA and Schwarz WH (2006) Bacterial Acetone and Butanol Production by Industrial Fermentation in the Soviet Union: Use of Hydrolyzed Agricultural Waste for Biorefinery. Applied Microbiology and Biotechnology, 71, pp 587-597.

[100] Sanchez OJ and Cardona CA (2008) Trends in Biotechnological Production of Fuel Ethanol from Different Feedstocks. Bioresource Technology, 99, pp 5270-5295.

[101] Nativel F, Pourquié J, Ballerini D, Vandecasteele JP and Renault Ph (1992) The Biotechnology Facilities at Soustons for Biomass Conversion. Int. J. Solar Energy, 11, pp 219-229.

[102] Pourquié J et Vandecasteele JP (1998) Conversion des composés lignocellulosiques par hydrolyse enzymatique et fermentation acétone-butanol. *In :* Biotechnologie. R. Scriban coordonnateur – Editions Technique et Documentation – Lavoisier, Paris, pp 773-792.

[103] Marchal R, Ropars M, Pourquié J, Fayolle F and Vandecasteele JP (1992) Large-Scale Enzymatic Hydrolysis of Agricultural Lignocellulosic Biomass. Part 2: Conversion into Acetone-Butanol. Bioresource Technology, 42, pp 205-207.

[104] Arlie JP (1985) Etude préliminaire des possibilités d'utilisation pour le Brésil des produits de fermentation acétonobutylique en mélange avec le gas-oil. Rapport IFP Energies nouvelles n° 33252.

[105] Maddox IS, Qureshi N and Roberts-Thomson K (1995) Production of Acetone-Butanol-Ethanol from Concentrated Substrates using *Clostridium acetobutylicum* in an Integrated Fermentation Product Removal Process. Process Biochem., 30, pp 209-215.

[106] Qureshi N and Maddox IS (2003) Production of Butanol from Concentrated Lactose/Whey Permeate Using *Clostridium acetobutylicum* and Removal by Perstraction. Poster Presentation 3-10, 25th Symposium on Biotechnology for Fuels and Chemicals, Breckenridge, Colorado, May 4-7, 2003.

7

Other biomass-to-energy biochemical pathways

D. Ballerini

Several biomass-to-energy biochemical pathways exist or are under development. The most advanced concerns the production of biomethane which, depending on the countries, is developing on industrial scale at a fast pace. Methane, produced by anaerobic fermentation of organic matter, is used after purification as energy source and as fuel after compression to 200 bar. Biomethane is considered to be a renewable energy.

Two other pathways are under development, i.e. production of algae and/or microalgae for energy purposes, while the synthesis of hydrogen by biological pathway is still at research phase, with an increasing number of scientific teams working in this field.

These three pathways are described below.

7.1 METHANE FERMENTATION

7.1.1 History

Methane or "marsh gas" was discovered in 1776 by Volta. Unaware of this discovery, two other famous men, not especially recognised as scientists, George Washington, the future first president of the United States, and Thomas Paine, an American propagandist of British origin and future member of the National Convention in France, made the same type of observation as Volta in 1783 [1].

In 1787, Lavoisier named this flammable gas "gas hidrogenium carbonatrum".

It was to take almost another century before the formation of methane was recognised as being the result of a microbiological process, further to the studies of Bechamp, a student of Pasteur, in 1868.

In 1906, Omelianski and Söhngen proved that the products resulting from the degradation of cellulose such as formate, acetate, butyrate, ethanol and even hydrogen and carbon dioxide represented carbon substrates for methanogenic bacteria. The first installations producing methane at small scale date back to this period.

In 1920, continuous methane production was carried out for the first time on large scale by Imhoff, before progressively being adapted to different raw materials and optimised. The situation concerning the production of biogas from biomass in Germany and Brazil has been described recently [2].

7.1.2 Principle

Microbiological decomposition of organic matter into methane is a complex anaerobic process, requiring the presence of several species of microorganisms, and occurring in several successive phases [3]. A simplified diagram shown on Figure 7.1 illustrates the methanisation of organic matter.

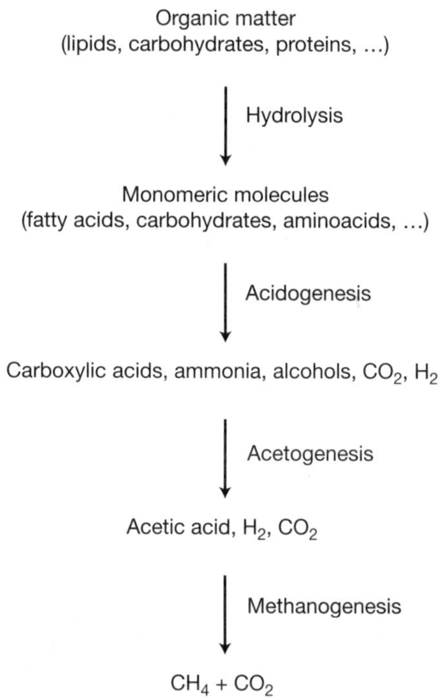

Figure 7.1

Methanisation of organic matter.

The first step consists in hydrolysis of organic polymers (lipids, cellulose, proteins, etc.) which are converted into monomers by the action of hydrolytic bacteria. This is often the limiting step in the overall methanisation process.

During the next phase, acidogenesis, fatty acids (acetic, propionic, butyric acids, etc.) and alcohols such as methanol and ethanol, generally accompanied by carbon dioxide and hydrogen, are synthesised from the monomers produced by hydrolysis of the organic matter. Acidogenesis is carried out by mixed cultures of anaerobic bacteria.

In the next step, the products obtained during acidogenesis are converted by acetogenic bacteria into methane synthesis precursors such as acetic acid, hydrogen and carbon dioxide.

Acetogenic bacteria generally live in symbiosis with methanogenic bacteria and, like them, are strict anaerobic microorganisms. Methanogenic bacteria produce methane during the last step of the process, according to the following two equations:

$$CH_3COOH \longrightarrow CH_4 + CO_2$$

$$CO_2 + 4\,H_2 \longrightarrow CH_4 + 2\,H_2O$$

Using the anaerobic digestion technique, the most biodegradable organic products are eliminated and converted into methane and carbon dioxide. The non-biodegraded solid residue can be used for agricultural purposes as soil conditioner.

Although methane fermentation is used in a few cases and a few countries to produce gaseous biofuels, it is above all a method aimed at converting biomass and waste into energy (cogeneration of electricity and heat).

Methane is used as biofuel in the United States, New Zealand and Sweden. In Sweden, about 10 Mm^3 of biogas are used in engines adapted to run on natural gas, in fleets of municipal vehicles, buses and taxis.

7.1.3 Raw materials

Methanisation is suitable for the treatment of wet waste and liquid effluents loaded with organic matter (e.g. domestic wastewater).

Anaerobic digestibility of the substrates is generally assessed by measuring the quantity of methane synthesised in 50 days from a known mass of waste.

Examples of raw materials include by-products from animal breeding (manure, pig slurry, etc.), agribusiness industries (abattoirs, dairies, breweries, fisheries, etc.), the paper industry, etc. Sludge discharged from urban wastewater purification stations could also be used.

Industrial methanisation units generally process several types of raw materials, either successively or mixed together.

The raw materials are also characterised by the extreme variability of the organic matter load which, evaluated in terms of COD (Chemical Oxygen Demand), ranges between 1 500 and 65 000 mg/l.

Methane production can also be optimised and controlled from urban garbage dumps. The creation of new environmental regulations on domestic wastes management has led to the development of this new biogas production sector.

Lastly, lignocellulosic biomass will undoubtedly be used in the longer term.

7.1.4 Processes

7.1.4.1 Operating parameters

The reaction temperature is an important parameter which must be kept constant, even though methane fermentation take place over a very broad temperature interval ranging from 10 to 70°C. It is currently difficult to identify a clear trend in the choice between digesters operating under thermophilic conditions and those operating under mesophilic conditions, both options having their advocates. By operating at a temperature of between 50 and 55°C, however, the risks of pathogenic microflora appearing and developing are significantly reduced. In addition, with thermophilic bacteria, selected under these conditions, the good methanogenic activities obtained allow operation with short raw material residence times in the fermenters and low volumes of residual solid matter are produced.

The pH, which is almost neutral, is optimum for methanogenesis.

Accumulation of volatile acids during the acidogenesis phase lowers the pH which inhibits methanogenesis. This is one of the reasons why, in some industrial units, acidogenesis and methanogenesis are carried out in two separate reactors.

The presence of high concentrations of heavy metals in the raw materials also inhibits the biochemical processes.

By favouring contact between the substrate and the microorganisms, stirring the reaction medium (either mechanical stirring or recirculation of some of the biogas produced) increases the hydrolysis and anaerobic digestion rates, thereby reducing the residence time and the reactor volumes. Stirring prevents stratification phenomena by efficiently keeping the solid matter in suspension and avoiding the formation of a crust on the surface.

The residence times can be reduced even further by increasing the concentration of microorganisms by inoculation with microflora and by recycling the solid sludge enriched with active microorganisms.

7.1.4.2 Technologies

Batch fermentation is very rare, apart from small installations operating at farm scale.

Industrial installations operate on the continuous fermentation technology. As mentioned earlier, stirred, temperature-regulated reactors are used.

Several systems exist, including either a single fermenter, a fermenter equipped with an internal subdivision into two parts – the first being mainly dedicated to predigestion of the organic matter – or two fermenters in series, methanogenesis being carried out in the second.

At fermentation outlet, the effluents are transferred to a settling tank where the undigested residual sludge enriched with microorganisms is separated from the lixiviates. Some of the sludge is recycled back to the inlet of the installation. Internal settling systems, fitted in some installations, retain some of the active sludge. Since the residence time here is longer than that of the feed to be degraded, the sludge becomes enriched with microflora and biodegradation rates increase.

For waste of dry matter contents less than 25-30%, no pretreatment is required. Otherwise preliminary grinding is carried out which, at the same time, homogenises the substrate.

When the dry matter content is greater than 30%, the feed often requires preconditioning. This generally consists of grinding-mixing-preheating with some of the recycled lixiviate.

Digester volumes vary considerably, from 100 m^3 to several thousand m^3.

Residence times vary between 10 and 25 days, depending on the organic matter feed.

Residence times are lower, about ten days, when treating polluted liquid effluents.

7.1.4.3 Performance [4]

Raw material feeds (reduced to 100% dry matter) vary between 1.5 and 2 $kg.m^{-3}.d^{-1}$ for high capacity continuous reactors. Methane production varies between 0.5 and 1 m^3 per kg of dry matter of waste treated

The waste volume and COD are reduced by about 60%. In addition, digested solid sludge is easier to dehydrate, before it can be used as a fertiliser, subject to acceptance due to its heavy metal composition. Liquid effluents must be purified, in some cases, by ultrafiltration or reverse osmosis, before they can be discharged.

Table 7.1 shows the average composition of biogas. H_2S and NH_3 are often detected in addition to CH_4 and CO_2, the main constituents. Generally saturated with water vapour, the gas also contains dust and traces of H_2, N_2 and CO.

Table 7.1 Average dry biogas composition

Constituent	% volume
CH_4	55-75
CO_2	25-45
H_2S	0-1.5
NH_3	0.05

Its calorific value varies between 20 and 30 MJ/Nm^3, depending on the methane content.

Up to 50% of the biogas produced can be consumed to supply the energy required to run the methanisation plant. This percentage drops when the degree of dryness of the waste treated increases, and may drop down to 20%.

7.1.4.4 Purification of biogas

So that it can be used in engines and gas turbines, biogas must be dried to avoid corrosion problems and the H_2S removed to reach a final concentration of less than 1 000 ppm.

To be used on the gas network, several purification operations are required:

- elimination of solid particulates and dust,
- dehydration by adsorption of water on activated alumina, silica gel or molecular sieve,

- elimination of sulphur compounds and halogenated compounds by adsorption on activated carbon, for example,
- elimination of CO_2 by absorption techniques, separation on membrane, pressure swing adsorption (PSA) or by washing with water or solvent.

Two types of gas are obtained, one with low methane content and low calorific value, recycled in the installation, the other with high methane content and high calorific value, which is compressed before being injected in the gas network.

For fuel applications, the specifications are as follows:

- CH_4 > 96-97% volume
- H_2S < 23 mg/Nm^3
- CO_2 < 3% volume
- H_2O < 32 mg/Nm^3

The gas is then compressed to about 250 bar.

7.1.4.5 Financial aspects

The main data, which are averages, are taken from the article published by De Mes *et al.* [4].

In Denmark, the average investment for waste treatment units of capacity in the region of 100 000 t/year amounts to about €65/t. This figure includes the transport and storage of raw materials, the digesters and cogeneration units. All units built must have a solid waste disposal network.

In the Netherlands, the costs of biogas produced in large columns amount to €9/GJ with treatment costs of €70-80 per tonne of waste.

The costs of electricity manufactured using biogas from anaerobic digesters vary between €0.1 and €0.22/kWh, dropping to €0.04-0.07/kWh for gas obtained from refuse tips.

While small units, at farm scale, may be economically viable, this is not true for large installations which require financial grants to operate.

In France, the cost of producing purified biogas, from waste, is less than 7 c€/kWh, and between 5 and 15 c€ when methane is manufactured from pig slurry. The price of natural gas at the pump is 8 c€/kWh [5].

7.1.4.6 Industrial sites in Europe

It has been estimated that, by 2015-2020, the biogas production potential in France [5] would reach 1.4 Mtoe obtained with installed methanisers, together with 0.9 Mtoe produced by the recovery of methane generated by all the refuse tips.

Table 7.2 gives an overview of the sites of annual treatment capacity greater than 2 500 tonnes/year in Europe. As can be seen, the development of methanisation technologies varies depending on the country. The largest industrial units are found in Germany, followed by Denmark and Austria. De Baere's article [6] provides further details concerning the development of industrial methanisation equipment in Europe. In Sweden, the demand for biomethane as biofuel increased from 0.2 Mm^3/year to 5 Mm^3/year between 2005 and 2009. The distribution capacity in service stations is increasing and the fleet of vehicles adapted to run on biogas, combined or not with another fuel, is expanding.

One of the reasons explaining the differences between European countries is probably the extent of the financial incentives granted, for example on the buy-back price of electricity generated from methanisation units.

Table 7.2 Main industrial methanisation installations in Europe (processing capacity > 2 500 t/year).

Country	Number of units	Capacities (tonnes of waste/year)
Austria	15	2 500-160 000
Germany	46	2 500-120 000
Denmark	25	10 000-190 000
Sweden	9	3 000-105 000
The Netherlands	5	10 000-85 000
Italy	4	2 500-300 000
France	1	85 000

The buy-back prices are respectively €0.12/kWh in Austria, €0.10/kWh in Germany and €0.08/kWh in Denmark.

7.1.4.7 Advantages – Disadvantages – Improvement axes

Waste methanisation is competing more and more with composting treatment. Despite higher investment costs, methanisation, which accounts for only 5% of the total waste composting capacity in Europe, is developing and, in some countries, has reached 25%.

Greenhouse gas emissions would be reduced by about 80% for the use of methane produced by anaerobic fermentation of waste, substituted for gasoline.

The main advantages with this type of process include:

- rustic installations,
- possibility of stopping and starting the installations with no major constraints,
- low energy consumption,
- average investments for methanisation, although significantly increased with the gas purification and compression section,
- concerning the actual treatment of the waste, anaerobic processes reduce the volume of sludge while simplifying subsequent dehydration. They eliminate the problem of odours and reduce the content in organic compounds and mineral elements such as N, P and K.

Disadvantages include:

- the biogas production costs, which are still very high,
- very large reactors are required (several thousand m^3) due to low productivities,

- production of a gas with high carbon dioxide content and impurities which must be separated,
- production of solid and liquid effluents not completely depolluted, for which it is essential to find applications.

All improvement axes aimed at increasing the rate and efficiency of the anaerobic digestion process must be taken into account, especially those related to pretreatment of the raw material and hydrolysis of the organic polymers (cellulose, proteins, lipids, etc.) which, as already mentioned, represent the limiting step of biochemical conversion processes.

7.2 PRODUCTION OF ALGAE FOR ENERGY USES

Algae: energy of the future, third generation biofuels.

For many people this is a statement, whereas for others it is above all a question for which numerous technico-economic points must still be clarified in order to reply as seriously as possible. Studies have been, and are still being, conducted in the fields of energy and chemical conversions of algae.

Two types of alga are considered: marine macroalgae and microalgae.

7.2.1 Marine macroalgae

Marine macroalgae, or seaweeds, represent a non-negligible renewable plant resource. They are harvested in numerous countries throughout the world. In Europe, France and Norway are the two main countries harvesting macroalgae, primarily as a source of alginates. In the Far East, they are also harvested along the coast, and more and more frequently cultivated on large scale in marine farms, for food purposes. Along the coast, they represent a way of reducing surplus fertilisers from land crops and decreasing carbon dioxide emissions.

The main species are *Laminaria* and *Ulva*. About 1 Mt/year (wet matter) is harvested along the coast, mainly in China and Chile. In 2006, harvesting in the marine farms reached 15 Mt/year (wet matter) in China, the Philippines, Korea, Indonesia and Japan. In China, the estimated productivity of *Laminaria japonica* is approximately 25 tonnes dry matter.ha^{-1}.year^{-1} [7].

Pretreatment consists in sand removal and pressing to obtain dry matter contents of between 20 and 30%. The costs of producing and harvesting macroalgae lie in the region of 200 to 265 euros/tonne of dry matter in France, compared with only 165 euros in the Philippines.

The average composition of species *Laminaria* is as follows: proteins 12%, sugar compounds 60%, lipids 2% and salts 26%.

The feasibility of producing methane by fermentation of algae has already been investigated. Projects combining offshore wind turbine installations and macroalgae farms are proposed in Germany, the Netherlands and the United States [8].

It is generally recognised that brown algae are more fermentable than green and red algae.

Anaerobic fermentation of one tonne of wet *Laminaria* can produce 22 to 28 m^3 of methane. Methane production involves optimising the pretreatment of algae and selecting the best microorganisms capable of withstanding the toxicity problems generated by high phenol, salt and sulphur contents. During methanisation, some macroalgae may generate large quantities of hydrogen sulphide, which may limit the production of biogas and create safety problems. Evacuation of residues and effluents, both at sea and in refuse tips, raises problems which must be solved.

Many of these algae contain sugar compounds (mannitol, laminarin, cellulose, etc.) which must be hydrolysed chemically or by enzymatic pathway, before they can be converted into ethanol by selected microorganisms resistant to inhibitors such as those mentioned previously. New hydrolysis and ethanol fermentation processes need to be developed.

In conclusion, macroalgae-to-energy pathways exist, provided that the culture and harvesting costs can be significantly reduced. Methane fermentation is the most feasible application, before ethanol fermentation for which fermentable sugars must be obtained.

7.2.2 Microalgae

Microalgae can produce lipids, sugars, hydrogen and hydrocarbons which can be converted into biofuels. In addition, they can undergo alcohol fermentation (in case of carbohydrate-rich algae) or methane fermentation. They can also be pyrolysed and gasified.

They can be grown in different types of water (fresh or salt), in ponds or photoreactors installed on unproductive land which is not used or only rarely used for agriculture of forestry.

Algae use solar energy for growth, via photosynthesis, to convert carbon dioxide in an aqueous medium into biomass, in the presence of nitrogenated and phosphorated mineral elements. Some species can develop by consuming organic matter such as sugars.

Numerous studies have already been conducted on the manufacture, by algal pathway, of high added-value products, for human food, with protein-rich species exhibiting good amino acid distribution, such as *Spirulina maxima*. Investigations are currently focusing on how to obtain pigments from algae cultivated in photobioreactors and a product rich in polyunsaturated fatty acids. Company NBT in Israel manufactures β-carotene for food purposes, as well as for cosmetic and pharmaceutical applications. β-carotene is produced in open ponds measuring 0.3 to 0.4 hectare over a total area of 7 ha, using species *Dunaliella* which is capable of growth in salt water with salt content higher than that of seawater.

Numerous studies have also been conducted mainly at laboratory scale, and some in small pilot units, to isolate various species of algae from diversified aquatic environments, in order to select the best, in terms of growth and capacity to produce oils for energy purposes.

7.2.2.1 Algal strains

Between 1979 and 1996, the American Aquatic Species Program (ASP) was funded by the DOE (Department of Energy) and coordinated by NREL (National Renewable Energy Laboratory) to select algal strains. This program focused mainly on the production of biofuels from lipid-rich algae [9]. Some 3000 strains were selected, including 300 species composed mainly of diatoms and green algae. This strain collection is kept at Hawaii University. The strains are available for use by scientists.

Diatoms (measuring from 5 microns to 5 mm) contain energy reserve lipids. The cell wall has a high silica content; silica deficiencies favour the intracellular accumulation of lipids.

Green algae or *Chlorophytes* can accumulate lipids when their growth is limited by a nitrate deficiency. The genera used most frequently to produce biofuels from lipids are *Dunaliella, Chlorella* and *Scenedesum*. The other classes of algae selected include golden algae, haptophytes and cyanobacteria.

One of the main lessons to be learned from this project concerns the difficulties in maintaining the growth in pilot installations with open ponds of microorganisms that have been isolated in the laboratory. Species which seem promising in the laboratory are not robust enough to adapt to exterior conditions and withstand contaminations by algae present on the site. It would appear that the best approach is to develop the algae already present.

7.2.2.2 Yield and productivity

Photosynthetic organisms are capable of capturing 60% of the photons in Photosynthetically Available Radiation (PAR) which represents only 50% of the total radiation. With an average biosynthesis efficiency of 25%, the maximum energy that can be stored as algal biomass would be 770 $kJ.m^{-2}.day^{-1}$. Assuming that 18 kJ are required to manufacture one gram of dry matter of biomass, the maximum productivity would be 43 $g.m^{-2}.day^{-1}$, achieved only under optimum algal growth conditions. An average productivity of 20 $g.m^{-2}.day^{-1}$ represents a realistic objective over one year of culture, i.e. 70 tonnes of algae (100% dry matter) per hectare per year. Intracellular oil contents are often announced with highly variable and sometimes unrealistic values. A content of 25% by weight, constant over time, would be an acceptable figure. Under these conditions, algal oil productivity could reach 17.5 $t.ha^{-1}.year^{-1}$, or approximately 20 $m^3.ha^{-1}.year^{-1}$ (5 300 $gallons.ha^{-1}.year^{-1}$).

Carbon dioxide consumption is estimated at between 3 and 4 g/g of algae.

7.2.2.3 Culture of microalgae

Microalgae can be grown in open ponds or in closed environment in reactors. Growth is continuous, with a permanent supply of nutrients and continuous removal of some of the algae cultured.

A. Culture in open ponds

Very large, shallow ponds are generally used. They are installed in excavations covered with waterproof plastic membranes, or built with low concrete walls, covered or not by waterproof, easy to clean membranes. In order to optimise the photosynthetic efficiency, it is important to ensure that each algal cell has the optimum quantity of light. The water is therefore always shallow, about 15 to 20 cm, so that light can reach the bottom of the pond. It is recommended to stir the culture medium gently to make the mixture more homogeneous and ensure the algae receive a good supply of light and CO_2. Algae concentrations are generally low, between 1 and 2 g (dry matter)/l, which allows the diffusion of light in the pond.

Ponds are either rectangular or circular. Several types of stirring are used:

- by supplying CO_2 in small wells inside the pond, which will cause the liquid to circulate via a horizontal air-lift system [10],
- by mechanical systems such as blades mounted on a shaft partly submerged in the liquid.

This open system offers several advantages:

- easy extrapolation to large areas,
- low investment, operating and maintenance costs,
- very easy cleaning,
- available technology,

and a number of disadvantages:

- low cell concentration,
- need for large areas of land,
- high risk of contamination,
- water evaporation,
- susceptibility to weather conditions (storms, rain, etc.) and temperature variations.

B. Culture in photobioreactors

These reactors have various shapes and are generally tubular with small diameters and variable inclinations. These systems offer good control over several parameters such as temperature, homogenisation of the medium by continuous circulation in the reactor, supply of CO_2 and mineral nutrients. This technology is still under development but nevertheless offers several advantages compared with open ponds:

- higher cell concentration,
- lower risks of contamination (but more complicated cleaning operations), probably offering the possibility of maintaining a single algal species in culture,
- higher rate of productivity per unit volume (1.5 $g.l^{-1}.day^{-1}$ versus 0.12 $g.l^{-1}.day^{-1}$ in open pond), with however productivities which depend on the area of land occupied, so the overall result is more or less the same,
- average dependency on weather conditions.

The main disadvantage is financial, since the investment, operating and maintenance costs are higher. Energy expenditures are high, to obtain good circulation of the culture medium in the series of tubes forming the reactor. Temperature regulation, required during period of strong sunshine, is expensive due to the large volumes of water which must circulate over the reactor walls. Generally speaking, the financial aspects are a handicap, difficult to overcome when the energy yield of the algae is low, to remain within reasonable limits with respect to fossil energy.

C. Culture in fermenters

Companies such as Solazyme in the United States and Fermentalg in France are trying to develop productions of heterotrophic algae, capable of growth on carbon substrates such as sugars. Using this technology, the cell concentrations are high and investment and operating costs moderate. Risks of contamination are limited, under aseptic production conditions. Oxygen is the main limiting factor. Solazyme has already supplied biofuels based on algal oils to Mercedes.

D. Culture in lagoons

The algae are produced to purify polluted liquid effluents. Companies in New Zealand use this technology in large lagoons covering several dozen hectares, into which polluted urban effluents are discharged. Heterotrophic algae consume the organic pollution and the mineral elements present in the effluents. This technique is applied by company Aquaflow and NIWA (National Institute of Water and Atmospheric Research) [11].

7.2.2.4 The overall process for production of algae and oils

After the culture, which it is absolutely essential to stabilise over time, the algal suspensions must be harvested and concentrated then the oils extracted and converted into biofuels.

A. Concentration

The algal content in the culture medium is always low: 1 to 2 g/l in open ponds, 3 to 4 g/l in photobioreactor. Due to the very small size of the algae and their low concentration, recovery and concentration represent a major problem to be solved from both the technological and financial aspects, since this item may reach 30% of the algae production cost. This process is often broken down in two steps: preconcentration by flocculation or flotation up to dry matter contents of 1 to 5%, prior to further concentration par filtration or centrifugation to reach contents of 15 to 20% of dry matter, the extracellular water being largely eliminated. Flocculation by chemical pathway has already been studied. Addition of mineral compounds is only efficient in the presence of high concentrations of salts such as $Al_2(SO_4)_3$, $Fe_2(SO4)_3$, $FeCl_3$ and $Ca(OH)_2$. One conclusion from the studies conducted by the ASP (Aquatic Species Program) was that this technique should not be recommended, being too expensive and likely to cause problems when using the products or co-products for food purposes and when evacuating the effluents due to the large quantities of metal salts present. Flocculation with organic polyelectrolytes seems to be more realistic, but at non negligible cost. Studies are also being conducted on bioflocculation of some algal species which, for example under limiting nitrogen or oxygen conditions, form aggregates and settle. Other techniques, such as electroflocculation and flotation, which can be used to recover the surface flocs, are being investigated.

As a second step, centrifugation would be a suitable technique, despite considerable investment and operating costs (high energy expenditures), which are prohibitive for the algae-to-energy pathway. Filtration would seem to be a better technique, less expensive than centrifugation. Problems caused by membrane clogging have not yet been fully solved, however.

B. Oil production

Should the algae be dried before extracting the oils? If yes, the energy expenditures and hence the costs will increase considerably. The best solution would be to extract the oils from algal suspensions that are still wet, for example with 20% dry matter. Another solution could consist in thermal treatment of this suspension in order to obtain a crude oil.

Algal oils have a high content in polyunsaturated fatty acids, with four or more double bonds [12], making them sensitive to oxidation. These oils must be hydrogenated, cracked and hydroisomerised to methyl paraffins, before they can be incorporated in Diesel (see Chapter 3) if used in kerosene as jet fuel. These treatments will stabilise them during storage.

7.2.2.5 Financial aspects

At the current time, the costs published are only estimations. They vary considerably. They are based on data from small installations and can only be validated when all steps in the process are defined and the fate and conversions of the co-products (remember that oils represent only 20 to 30% of the algal biomass produced) and the effluents are determined.

The estimated costs [13] would be \$0.5 to \$0.6/kg of algae for a production of 10 000 t/year of biomass and \$0.34/kg for a 10 hectare open pond installation (Seambiotic data). Calculated with respect to oil production, assuming a content of 25% lipids in the algae, the production cost would be estimated between \$2.4 and \$1.4/l of oil. The expenses related to obtaining and converting the oil, treating the co-products and effluents must be added to these estimations, and the potential revenue generated by upgrading them (proteins, polysaccharides, etc.) must be taken into account. It is nevertheless clear that the expenses related to the supply of CO_2, freshwater or seawater and the nutrients must be as low as possible. Being located near a power plant to obtain free supplies of CO_2-rich gaseous emissions and near the sea for the culture of algae in saline environment are advantages when deciding where to install an algae-to-energy plant.

7.2.2.6 Pilot operations

Several projects were conducted during the ASP:

- HRP (High Rate Pond) program in California. This is a continuous algae production process, in open ponds of area 100 to 200 m^2, and depth between 15 and 30 cm. Amongst the main results obtained, the average productivity varies between 15 and 25 $g.m^{-2}.day^{-1}$ the algal species selected depends highly on the ambient temperature, and bicarbonate ions are required to keep the pH below 8. In addition, a high contamination rate was observed,
- ARPS (Algal Raceway Production System) project in 50 m^2 ponds. This project demonstrated that homogenisation of the culture medium is an important factor and that the ponds must be shallow.

Other projects are underway, e.g. the Seambiotic pilot plant in Israel which feeds a small proportion of the desulphurised flue gases from the nearby Ashkelon power plant into the ponds. The average productivity is 20 $g.m^{-2}.day^{-1}$, with various algal species, including

Nannochloropsis, cultivated in seawater. To produce high lipid concentrations, the algae are mainly cultivated from November to early May when the surrounding conditions are not favourable to the development of contaminating strains with lower lipid concentrations, whose growth rates are higher than those of the targeted species. To build up their intracellular lipids, which act as energy reserves, the algae consume a considerable amount of energy, at the expense of development and growth rates, much lower than for algae with low lipid contents. The latter algae, containing more carbohydrates, could be cultivated during the other hotter and sunnier periods of the year, by focusing on other energy conversion pathways such as the production of biogas or ethanol, or other applications such as food, for example.

7.2.2.7 Main companies involved in the production of microalgae for energy purposes

The main companies involved in the production of microalgae for energy purposes are given below [11] (non-exhaustive list):

- Solazyme,
- General Atomics,
- Solix Biofuels,
- Seambiotic,
- Petroalgae,
- Valcent-Vertigro,
- Cellana,
- BioFuel Systems,
- Kent Bioenergy,
- Aquaflow Bionomic.

One important point concerns the increasing role played by the oil companies in developing the production of algal oils. This is the case of Exxon associated with Synthetic Genomics, Chevron with NREL, Shell with Cellana on a project at Hawaii, ConocoPhillips with C2 B2 (Colorado Center for Biorefining and Biofuels). ENI in Italy has been interested in the production of algae for many years.

Aeronautical companies are also involved in assessing some of the fuels produced by algae, used in mixtures with kerosene.

7.2.2.8 R&D programmes

In France, several R&D projects have been financed by the Agence Nationale de la Recherche (National Research Agency) as part of the national research programme on bioenergies, for example the Shamash project.

In the United States, microalgal biofuels are considered as “advanced biofuels” by the RFS (Renewable Fuels Standard). Several programmes are funded by the DOE and the USDA, for a total amount of $200 M. In addition, the DARPA (Defense Advanced Research Projects Agency) is an active partner in R&D programmes aimed at producing algal biofuels, which would partially replace fossil origin fuels. The DARPA objective is to obtain a biofuel costing less than $2/gallon. The DARPA has invested $35 M in two projects, one with General Atomics, the other with the Californian company SAIC, and it will launch an invitation to tender for the construction of a unit capable of producing about 20 000 m^3 of jet fuel.

7.2.2.9 Conclusions

Testifying to the current interest in this new sector, numerous studies are underway on the production of energy and biofuels from microalgae for the automotive and aircraft industries, and various oil, aeronautical and automotive companies are involved in this field.

Biofuels can be obtained via several pathways, the main one being the production of lipid-rich algae for conversion into biodiesel or biokerosene. Other pathways exist, which may be easier to extrapolate to industrial scale. This is the case, for example, with the conversion of organic compounds from algae with low lipid contents into methane or of intracellular sugars into ethanol.

The manufacture of algal Diesel or kerosene nevertheless remains an objective generating considerable interest.

Numerous aspects still need to be solved or improved, however. The main ones are as follows:

- lower production costs,
- choice of culture mode, either in open ponds, photobioreactors or fermenters. Implementation of a dual system is also being considered, with open ponds to grow algae, then photobioreactors to increase the intracellular lipid content,
- improved production stability, during the microorganism growth stage, better average lipid concentrations (20 to 30%), over long periods, several months of the year, with efficient protection against contamination by other strains,
- better algae recovery and concentration technologies,
- extraction of lipids, if possible without a prior drying step which is very expensive in terms of energy,
- conversion of these lipids into crude oils or into hydrogenated, cracked and/or isomerised hydrocarbons.

It is also extremely important to find profitable outlets for the remaining 70% of the biomass and for the effluents. Life cycle assessments will only be performed when all steps in the process have been fully defined.

Lastly, the culture of heterotrophic strains in fermenter, or in lagoons for the depollution of polluted effluents, represents another centre of interest, offering more potential for obtaining process stability, as well as cell concentrations and productivities higher than those measured during photosynthetic development of algae.

7.3 HYDROGEN PRODUCTION

Hydrogen is considered as a future source of renewable energy. It can be produced by natural gas reforming, gasification of coal, gas and biomass and by electrolysis of water. Much R&D effort is dedicated to synthesis by biological pathway.

Hydrogen can be produced by biological pathway according to several processes:

- anaerobic fermentation of organic compounds used by the microorganisms as carbon and energy source, known as heterotrophic hydrogen production,
- photobiological processes, either photoautotrophic by photolysis of water in the presence of light energy, or photoheterotrophic by conversion of the organic matter with light as energy source,
- a combination of these various pathways.

7.3.1 Production by fermentation [14]

In an anaerobic environment, compounds other than oxygen act as acceptors of the electrons released during degradation of organic carbon substrates. This is the case of protons which are reduced into molecular hydrogen in the presence of specific enzymatic systems such as the hydrogenases of hydrogen-producing microorganisms.

Genesis of hydrogen in natural anoxic environments is ubiquitous but difficult to detect. Hydrogen is in fact involved in numerous syntrophic reactions: as soon as it is produced during anaerobic fermentation of an organic substrate, it is consumed during other biological processes such as reduction of sulphates and especially methanogenesis. Hydrogen is an acceptable electron donor, especially for methanogenic bacteria, according to the reaction:

$$4\,H_2 + CO_2 \longrightarrow CH_4 + 2\,H_2O$$

The equation for hydrogen production from glucose as carbon and energy source is written:

$$C_6\,H_{12}\,O_6 + 4\,H_2O \longrightarrow 2\,CH_3\,COO^- + 2\,H\,CO_3^{\,-} + 4H^+ + 4H_2$$

During hydrogen production, on part of the substrate is mineralised into CO_2, while the other part is oxidized into acetate. In less favourable cases, for example when the partial pressure of hydrogen in the gaseous atmosphere exceeds 10 kPa, the metabolism is biased towards the synthesis of more reduced products such as ethanol and lactate, instead of acetate. In addition, it is essential to block the action of the methanogenic bacteria which are likely to use the hydrogen produced, by adopting fermentation conditions that reduce the methanogenic activity.

The theoretical efficiency for the conversion of glucose to hydrogen is 4 moles H_2/mole of glucose consumed, i.e. a weight efficiency of just 4.4%.

A wide range of microorganisms produce hydrogen by fermentation [15]. They belong to various categories:

- strict anaerobic bacteria (*Clostridia*),
- thermophilic bacteria,

– methanogens,
– facultative anaerobic bacteria, such as those belonging to the genus *Enterobacter,*
– mixed cultures of *Clostridium – Enterobacter.*

Most fermentation substrates are sugars, either pure (hexoses and pentoses) or mixed monomers, or oligomers (xylans). Fermentation represents a means of producing hydrogen from renewable resources, using the products of lignocellulosic biomass enzymatic hydrolysis. It can also be used to produce carbon monoxide.

The best conversion efficiencies, close to theoretical, have been obtained with glucose, using *Thermotoga maritima*, a thermophilic anaerobic bacterial strain whose growth is optimum at about 80°C [16].

Other hydrogen production results have been obtained, using a mixed culture of bacteria, by continuous anaerobic fermentation of a sugar refinery liquid effluent [17].

This study indicates that higher sugar-to-hydrogen conversion efficiencies and hydrogen synthesis productivities are obtained with shorter residence times, reaching respectively 2.9% (expressed as yield per unit weight of sugars consumed) and 0.4 $g.l^{-1}.d^{-1}$ (Table 7.3).

Table 7.3 Continuous anaerobic fermentation of a sugar refinery effluent.

Residence time (d)	Sugar supply ($g.l^{-1}.d^{-1}$)	Sugars consumed (%)	Fermentation products ($g.\ l^{-1}.\ day^{-1}$)					
			H_2	CO_2	CH_4	Acetate	Butyrate	Cellular biomass
0.5	19.7	70.0	0.40	4.98	Traces	2.26	4.20	0.79
1	9.8	79.7	0.22	2.50	Traces	1.30	2.38	0.46
2	4.9	93.0	0.10	1.24	Traces	0.87	1.36	0.24
3	3.3	97.4	0.07	0.87	Traces	0.78	1.06	0.17

However, the depletion rate of the sugars present in the effluents treated reaches only 70% under these operating conditions, and nearly 100% when the residence times are significantly increased to 3 days.

Some types of bacteria are also capable of producing hydrogen from CO as carbon and energy source, via the gas-shift reaction which produces more hydrogen.

$$CO + H_2O \longrightarrow H_2 + CO_2$$

This is the case of a chemo-heterotrophic bacterium of species *Citrobacter*, which develops under anaerobic conditions (generation time of one hour) and which produces hydrogen via the gas shift reaction, by anaerobic fermentation, in the absence of light, under CO atmosphere. Its H_2 productivity is 66 $mg.g^{-1}cell/h^{-1}$ and the CO to H_2 conversion efficiency is close to 100% [18].

7.3.2 Photobiological production

There are several types of production: photoautotrophic, photoheterotrophic, or a combination of the two.

7.3.2.1 Photoautotrophic production

Water biophotolysis is based on the ability of phototrophic microorganisms to convert water into hydrogen using light energy, according to the reaction:

$$2\,H_2O + \text{Light energy} \longrightarrow O_2 + 2\,H_2$$

Phototrophic microorganisms are grouped mainly into the families of microalgae, "purple" bacteria and cyanobacteria. Cyanobacteria are oxyphototrophs which tolerate the presence of oxygen, unlike anoxyphototrophs such as purple bacteria. It is highly likely that cyanobacteria are the main living organisms responsible for the transition of the terrestrial atmosphere from anoxic to oxic. These microorganisms use the visible light spectrum.

The enzymes catalysing the water photolysis reaction are hydrogenases, whose action is unfortunately inhibited by the presence of oxygen in low concentrations [19].

While the conversion efficiency of light energy to H_2 does not exceed 1.5%, in the presence of oxygen, it could reach 3 to 10%, if oxygen was eliminated as it is produced. Until solutions are found to solve the problem of oxygen inhibition, it will be impossible to extrapolate this hydrogen photobiological production process.

7.3.2.2 Photoheterotrophic production

Hydrogen production by this pathway is carried out by photoheterotrophic microorganisms, such as the purple bacteria, which have the ability to use light as energy source and organic compounds such as organic acids and alcohols as carbon source [20].

A reaction example is described below:

$$2\,CH_3COOH + 4\,H_2O \xrightarrow{\text{Light energy}} 4\,CO_2 + 8H_2$$

Nitrogenases are enzymes catalysing this type of reaction, especially in the absence of N_2, since they are primarily catalysts which activate fixation of molecular nitrogen in the same photoheterotrophic bacteria. Nitrogenases are oxygen-sensitive, but in this case the problem does not arise since the reaction is carried out under anaerobic conditions.

One problem with this type of fermentation is related to the light energy utilisation efficiency. Experiments are in progress using tubular reactors or reactors made from vertical panels; it is essential to distribute the light as efficiently as possible inside the reactors so that it reaches all the microorganism cells.

Conversion efficiencies of light energy to H_2 (energy equivalent) can reach 7%.

The conversion efficiency of acids to hydrogen is about 70% of the theoretical value and the hydrogen synthesis productivity reaches $0.15\ g.l^{-1}.d^{-1}$.

Photoheterotrophic bacteria, such as *Rhodopseudomonas palustris P4*, develop in the presence of light energy (generation time of about 2 h) to produce hydrogen from carbon monoxide, according to the gas-shift reaction. CO pressures of up to one bar do not inhibit the metabolic processes of this microbial species [21].

7.3.3 Combined productions by fermentative and photobiological pathways

The photoheterotrophic hydrogen production processes are mainly considered as a second step after converting the organic matter by fermentation [22].

It represents a good way of converting more organic matter into H_2, by completing the oxidation of volatile acids (acetate, butyrate) to CO_2. The hydrogen balance is therefore theoretically multiplied by a factor of 3, as shown by the following reaction equations:

$$C_6H_{12}O_6 + 2\,H_2O \longrightarrow 2\,CH_3COOH + 2\,CO_2 + 4H_2 \quad (1)$$

$$2\,CH_3COOH + 4\,H_2O \xrightarrow{\text{Light energy}} 4\,CO_2 + 8H_2 \quad (2)$$

$$C_6H_{12}O_6 + 6H_2O \longrightarrow 6\,CO_2 + 12\,H_2 \quad (1) + (2)$$

Reaction (1) is a fermentative process, whereas reaction (2) is carried out by photoheterotrophic microorganisms which use light as energy source and organic compounds as carbon source.

The costs of producing hydrogen from lignocellulosic biomass have been estimated at about €0.21/m^3H_2, based on a combination of the two pathways. This results does not include the cost of raw material and its treatment or the labour costs [23].

It also represents a means of avoiding oxygen inhibition of hydrogen production during water photolysis, by separating the H_2 and O_2 syntheses in time and space. A first photosynthesis reaction which can be carried out in open environment by microscopic algae leads to the production of carbohydrates. The carbohydrates are then decomposed into H_2 and CO_2, in closed reactor, by two anaerobic microbiological processes, first fermentative, then photobiological. A limitation in elementary sulphur, imposed during the first two reactions, favours hydrogen production.

$$6H_2O + 6\,CO_2 \xrightarrow{\text{Light energy}} C_6H_{12}O_6 + 6\,O_2 \quad (1)$$

$$C_6H_{12}O_6 + 2\,H_2O \longrightarrow 2\,CH_3COOH + 2CO_2 + 4H_2 \quad (2)$$

$$2CH_3COOH + 4H_2O \xrightarrow{\text{Light energy}} 4CO_2 + 8H_2 \quad (3)$$

$$12\,H_2O \xrightarrow{\text{Light energy}} 12\,H_2 + 6O_2 \quad (1) + (2) + (3)$$

7.3.4 Conclusions and perspectives

Numerous teams of researchers are working, throughout the world, on improving these potentially interesting processes for the production of hydrogen by biological pathway [24], [25], [26]. These studies are conducted in the laboratory. In order to consider extrapolation to pilot and then industrial scale, numerous aspects must be optimised and several problems solved, including:

- concerning the fermentative pathway, higher productivity (greater cell density, use of reactors with immobilised cells) and better resistance of the microorganisms to the partial pressure of hydrogen, removal of hydrogen inhibition of metabolic processes by eliminating it as it is synthesised (use of selective membranes),
- concerning the direct photobiological pathway, removal of oxygen inhibition of hydrogenases (genetic engineering studies) and, from the technological standpoint, optimisation of the light energy transfer in the reactors,
- optimised combination of the fermentative and photobiological mechanisms.

REFERENCES

[1] Wolfe RS (1993) An Historical Overwiew of Methanogenesis in Methanogenesis, Edited by James G. Ferry, Chapman et Hall Inc., New York, pp 1-32.

[2] Weiland P, Verstraete W and Van Haandel A (2009) Biomass Digestion to Methane in Agriculture: a Successul Pathway for the Energy Production and Waste Treatment Worldwide. *In:* Soetaert W, Vandamme EJ (eds) Biofuels. Wiley, Chichester UK, pp 171-195.

[3] Kubiak C et Dubuis T (1998) Valorisation des déchets et des effluents agricoles, industriels et domestiques – in Biotechnologies. R. Scriban coordonnateur. Ed. Technique et Documentation – Lavoisier – Paris, pp 731-771.

[4] De Mes TZD, Stams AJM, Reith JH and Zeeman G (2003) Methane Production by Anaerobic Digestion of Wastewater and Solid Wastes, in Bio-methane and Bio-hydrogen. Status and Perspectives of Biological Methane and Hydrogen Production. Edited by J.H. Reith, R.H. Wijfells and H. Barten, c/o Energy Research Centre of the Netherlands ECN, Unit Biomass, 1755 ZG Petten, The Netherlands, pp 58-102.

[5] Étude ADEME, AFGNV, ATEE Club Biogaz, GDF Suez et IFP Energies nouvelles, MEEDAT (2008) Synthèse : État des lieux et potentiel du méthane carburant, Novembre.

[6] De Baere L (2000) Anaerobic Digestion of Solid Waste: State-of-the-art. Water Science and Technology, **41**, 3, pp 283-290.

[7] Kelly M and Dworjanyn S (2008) The Potential of Marine Biomass for Anaerobic Biogas Production: A Feasibility Study with Recommendations for Further Research. The Crown Estate 2008.

[8] Hagerman G and McKay LB (2007) Marine Biogas from Offshore-culture Seaweds. Virginia Industry Energy Symposium.

[9] Sheehan J, Dunahay T, Benemann J and Roessler P (1998) A Look Back at the US Department of Energy's Aquatic Species Program-Biodiesel from Algae, Prepared by NREL for USDOE (Office of Fuels Development), NREL/TP-580-24190.

[10] Clément G, Rebeller M et Trambouze P (1968) Utilisation massive de gaz carbonique dans la culture d'une nouvelle algue alimentaire. Revue de l'Institut Français du Pétrole, 23, pp 702-711.

[11] Whitwham M (2009) Production de biocarburants à partir de micro-algues. Étude de la société Philgea pour le fonds dédié Enerbio, dans le cadre de la Fondation Tuck, enerbio@fondation–tuck.fr

[12] Belarbi EH, Molina E and Chisti Y (2000) A Process for High Yield and Scaleable Recovery of High Purity Eicosapentaenoic Acid Esters from Microalgae and Fish. Enzyme Microbiology Biotechnology, 26, pp 516-529.

[13] Chisti Y (2007) Biodiesel from Microalgae. Biotechnology Advances, 25, pp 294-306.

[14] De Vrije T and Claassen PAM (2003) Dark Hydrogen Fermentations. *In:* Bio-methane and Bio-hydrogen – Status and Perpectives of Biological Methane and Hydrogen Production. Edited by J.H. Reith, R.H. Wijffels and H. Barten, c/o Energy Research Centre of the Netherlands ECN, Unit Biomass, 1755 ZG Petten, The Netherlands, pp 103-123.

[15] Kengen SWM, Goorissen HP, Verhaart M, Stams AJM and Claassen PAM (2009) Biological Hydrogen Production by Anaerobic Microorganisms. *In:* Soetaert W, Vandamme EJ (eds) Bio-fuels. John Wiley & Sons Ltd, Chichester UK, pp 197-221.

[16] Schröder C, Selig M and Schönheit P (1994) Glucose Fermentation to Acetate, CO_2 and H_2 in the Anaerobic Hyperthermophilic Eubacterium *Thermotoga maritima*: Involvement of the Embden – Meyerhof Pathway. Arch. Microbiol., 161, pp 460-470.

[17] Ueno Y, Otsuka S and Morimoto M (1996) Hydrogen Production from Industrial Wastewater by Anaerobic Microflora in Chemostat Culture. Journal of Fermentation and Bioengineering, **82**, 2, pp 194-197.

[18] Jung GY, Kim JR, Jung HO, Park J-Y and Park S (1999) A New Chemoheterotrophic Bacterium Catalyzing Water-gas Shift Reaction. Biotechnol Lett., 21, pp 869-873.

[19] Seibert M, King PW, Posewitz MC, Melis A and Ghirardi ML (2008) Photosynthetic Water Splitting for Hydrogen Production. *In:* Wall JD, Harwood CS, Demain A (eds) Bioenergy. ASM Press, Washington DC, pp 273-291.

[20] Harwood CS (2008) Nitrogenase-catalyzed Hydrogen Production by Purple Nonsulfur Photosynthetic Bacteria. *In:* Wall JD, Harwood CS, Demain A (eds) Bioenergy. ASM Press, Washington, DC, pp 259-271.

[21] Jung GY, Jung HO, Kim JR, Ahn Y and Park S (1999) Isolation and Characterization of *Rhodopseudomonas palustris* P4 which Utilizes CO with the Production of H_2. Biotechnol. Lett., 21, pp 525-529.

[22] Manish S and Banerjee R (2008) Comparison of Biohydrogen Production Processes. International Journal of Hydrogen Energy 33, pp 279-286.

[23] Claassen PAM, van Groenestijn JW, Janssen AJH, van Niel EWJ and Wijffels RH (2000) Feasibility of Biological Hydrogen Production from Biomass for Utilization in Fuels Cells. *In:* Proceedings of the 1st World Conference on Biomass for Energy and Industry, Sevilla, Spain 5-9 June 2000, ed: James and James (Science Publishers) Ltd, London, UK, pp 1665-1667.

[24] Hallenbeck PC and Ghosh D (2009) Advances in Fermentative Biohydrogen Production: the Way Forward? Trends in Biotechnology 27, pp 287-297.

[25] Rupprecht J, Hankamer B, Mussgnug JH, Ananyev G, Dismukes C and Kruse O (2006) Perspectives and Advances of Biological H2 Production in Microorganisms. Applied Microbiology and Biotechnology 72, pp 442-449.

[26] Liu XM, Ren NQ, Song FN, Yang CP and Wang AJ (2008) Recent Advances in Fermentative Biohydrogen Production. Progress in Natural Science 18, pp 253-258.

Appendix 1

Environmental balance of biofuels

To draw up an environmental balance of biofuels that will be as detailed and as realistic as possible, appropriate tools and methods had to be developed to measure this environmental performance. This concept, based on analysis of a product or process throughout its life cycle, "from cradle to grave", is known by the acronym LCA: Life Cycle Assessment.

A1.1 METHODOLOGY

LCA was launched in the 1960s. Initially, it was mainly used in the selection of packaging materials (glass, steel, aluminium, etc.). Use of the method was then extended to other products, especially in the field of energy, production of electricity and fuels.

The method is based on analysis of all elementary steps involved in the creation of a product, from the supply of raw materials up to end of life (recycling, burning, landfill). Each sector leading to the same end product is therefore broken down into elementary steps, which are then linked together in a coherent chain. The global analysis therefore avoids transfer of pollution. With electric vehicles, for example, no pollution will be counted when using the vehicle; most pollutant discharges in the atmosphere will in fact be taken into account during the electricity production phase. Consequently, the environmental benefit of the electric vehicle will depend on the electricity generation mode.

For a given product or process, therefore, LCAs are carried out to:

- estimate the consumptions of energy and raw materials, together with the associated emissions into water, soils and the air,
- quantify the environmental impact related to these consumptions of energy and raw materials and to the various discharges into the environment,
- analyse these results to simplify the choice between various options.

In practice, LCAs represent a method which, according to standard ISO 14040, includes 4 steps:

- definition of the goals and scope,
- inventory analysis,
- impact assessment,
- interpretation and, if necessary, search for improvements.

A1.1.1 Definition of the goals and scope

This step consists in clearly defining the intended application of the study, the reasons for carrying out the LCA and the intended audience of the study. The scope of the study must also be clearly defined and the following points explained:

- what is the functional unit? The functional unit is the basis on which the various possible options are compared. In case of fuels, for example, the balances of pollutant and CO_2 emissions per km travelled when a vehicle drives on NEDC cycle must be determined.
- what are the system boundaries?
- what are the allocation rules if co-products are generated? (this point will be discussed later).
- what are the impact categories, how will the impacts be assessed?
- what are the data quality requirements?

The scope must be sufficiently well defined so that the breadth, depth and level of detail of the study are compatible with its goal.

A1.1.2 Inventory analysis

Process during which all inputs and outputs of the system studied are quantified (Figure A1.1). The process includes in particular:

- energy consumed,
- raw materials consumed,
- emissions into the air,
- emissions into water and soils,
- solid waste generated.

This inventory is often presented according to the key steps in the life cycle of the product studied. It can be performed in several ways. For example, the method recommended by the US EPA is described briefly below. It includes the following steps:

- develop a flow diagram of the process being evaluated,
- develop a data collection plan,
- collect data,
- evaluate and write the LCA final report.

The life cycle of a given product/process analysed is therefore broken down into elementary steps. Each step is characterised by incoming flows (raw materials, energy) and outgoing flows (product, co-products, emissions into the air/water, wastes).

All these elementary steps are organised to give a comprehensive picture of the product life cycle. Depending on the complexity of the chains of products analysed, the LCA may quickly require considerable quantities of data. Consequently, it is often recommended to clearly determine the acceptable level of uncertainty before starting the project. In particular, the use of data specific to production sites which could lead to long collection and compilation times should be avoided whenever possible. Data from international databases often

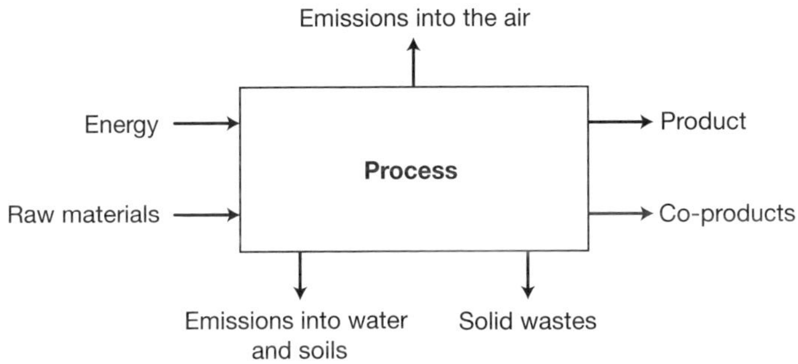

Figure A1.1

Flow diagram.

Source: World Energy Council, 2004, "Comparison of energy systems using life cycle assessment"

contained in software, which provide information on a large number of standard industrial processes should be used instead. However, when presenting the final results of an LCA, it is important to always accurately describe the methodology used, the boundaries of the system studied, the origin of the data used and the assumptions made concerning the various processes.

A1.1.3 Impact assessment

In this section of the study, the impacts of the previously quantified emissions are evaluated. The impacts usually quantified can be grouped into the following broad families:

- impact on human health,
- impact on the environment,
- impact on depletion of finite natural resources.

For example, the approaches implemented can be used to obtain partial answers to the following questions:

- what are the impacts of 9 000 tonnes of carbon dioxide or 500 tonnes of methane emissions being released into the atmosphere?
- what are their potential impacts in terms of acid rain or greenhouse effect?

This inventory phase represents a systematic approach used to classify and characterise the various effects. Table A1.1 provides examples of these effects.

A1.1.4 Interpretation

The LCA interpretation phase is the final section where all results are compiled, analysed, checked and interpreted. This section also includes the recommendations to improve the product/process environmental impact.

Table A1.1 Examples of impacts and measurement factors.

Impact categories	Scale	Data involved	Measurement factor	Description of measurement factor
Greenhouse effect	Global	Emission into the air of CO_2, N_2O, CH_4, CFCs, HCFCs, CH_3Br	Global warming potential measured in CO_2 equivalent	Impact measured in CO_2 equivalent in 50, 100 or 500 years
Stratospheric ozone depletion	Global	Emission into the air of CFCs, HCFCs, CH_3Br, Halons	Ozone depletion potential	Impact measured in CFC-11 equivalent
Acidification	Regional	Emission into the air of SO_x, NO_x, HCl, HF, NH_3	Acidification potential	Impact measured in H+ equivalent
Eutrophication	Local	Emission into the air of PO_4, NO, NO_2, Nitrates, NH_3	Eutrophication potential	Impact measured in PO_4 equivalent
Photochemical smog	Local	Emission into the air of non-methane organic compound	Photo-oxidant formation potential	Impact measured in ethane equivalent
Depletion of natural resources	Global	Quantity of raw materials used	Resource depletion potential	

A1.1.5 Main limitations of LCAs

The main limitations concerning the use of LCAs are as follows:

- this type of study may be very long and require large quantities of data. Before starting the study, it is therefore important to clearly determine the level of uncertainty required for decision-making.
- traditionally, the LCA approach is limited to assessing the environmental aspects only, which are never the only decision criteria at industrial scale. Frequently therefore, the LCA study must be combined with other financial or technical studies, for example.
- lastly, if co-products are generated by a process, the question of how the various emissions or consumptions of raw materials are allocated to the products must be answered. This question is extremely sensitive with biofuels since the quantities of co-products generated are larger than the quantities of biofuels finally obtained (for one tonne of biofuel, about 2 tonnes of co-products are generated). Aware of this difficulty for which, to date, no fully satisfactory solution has been found, the authors of standard ISO 14040 do not recommend a solution but simply propose three options: either extend the system boundaries so that there are no co-products, allocate the emissions and consumption of raw materials on a weight/energy/monetary prorata basis, etc., or allocate all emissions to the biofuel alone. The second option is generally chosen on a weight prorata basis. This choice is not based on any theoretical demonstrations,

however, and for the time being remains purely subjective. Although the results obtained are representative of a certain situation from a qualitative point of view, they will always call for a certain degree of criticism from a quantitative point of view.

A1.2 LCAs CONDUCTED ON BIOFUELS

Numerous studies have been conducted to measure the impact of biofuel production and use on the environment. These "well to wheel" studies compare the greenhouse gas emissions and energy consumption over the entire fuel life cycle between the biofuel solutions and those using oil fuels. Far fewer studies are conducted on the other impacts mentioned above since the uncertainties are greater and these impacts are highly dependent on local conditions where the pollutants are discharged.

Two studies are often considered:

- the study conducted in France, under the technical coordination of the ADEME [Life cycle assessments applied to first generation biofuels used in France – Summary – February 2010]. It provides an update which is as exhaustive as possible of the energy balances and greenhouse gas emissions related to biofuels, and an assessment of three impact indicators (eutrophication potential, photochemical oxidation potential and human toxicity potential). It does not take into account the effect of land use change. These studies nevertheless emphasise this impact, which may be discriminating when the development of energy crops leads, either directly or indirectly, to the disappearance of primary forests (like the Amazonian forest which could be converted into soya plantations, and the primary forests in Indonesia and Malaysia into palm tree plantations, etc.), or the disappearance of grasslands and wetlands. Studies are currently being conducted to increase the understanding of these land use changes which, until now, have not been accurately evaluated,
- the European study conducted jointly by the JRC (European Commission Joint Research Centre), the CONCAWE (the oil companies' European association for environment), and EUCAR (which coordinates the R&D actions of the European Automobile Manufacturers Association).

Investigations are also in progress to assess nitrous oxide emissions more precisely. The global warming potential of nitrous oxide is 300 times greater than that of carbon dioxide. No methodology has yet been validated to quantify N_2O emissions. The IPCC nevertheless recommends the use of the following empirical factors:

- for direct emissions: 1% of the nitrogen supplied by fertilisers and of that contained in the waste of the crop cultivated is emitted as N_2O,
- for indirect emissions: 0.75% of the leached nitrogen is re-emitted as N_2O, as well as 1% of the nitrogen evaporated as ammonia.

The results of these two studies concerning greenhouse gas emissions are discussed in Chapters 2-6.

REFERENCES

World Energy Council, Comparison of Energy Systems Using Life Cycle Assessment (2004) London, UK, July 2004.

Normes ISO 14040, 14041, 14042, 14043.

Écobilan de l'ester méthylique de colza. Écobilan, ONIDOL 1993, 1996 et 1999.

Écobilan de l'ETBE de betterave en comparaison avec le MTBE, pour la CGB et le SNPAA, Écobilan 1996.

Wang MQ and Huang HS (1999) A Full-Fuel-Cycle Analysis of Energy and Emissions Impacts of Transportation Fuels Produced from Natural Gas.

Wang MQ, Saricks C and Santini D (1999) Effects of Fuel Ethanol Use on Fuel-Cycle Energy and Greenhouse Gas Emissions.

Wang MQ, Sarcks C and Wu M (1997) Fuel-Cycle Fossil Energy Use and Greenhouse Gas Emissions of Fuel-Ethanol Produced from US Midwest Corn.

Well-to-Wheel Energy Use and Greenhouse Gas Emissions of Advanced Fuel/Vehicle Systems. North American Analysis. General Motors Corporation, BP, ExxonMobil, Shell, June 30, 2001.

Well-to-Wheel Analysis of Energy Use and Greenhouse Gas Emissions of Advanced Fuel/ Vehicle Systems. A European Study. GM, LBST, BP, ExxonMobil, Shell, TotalFinaElf, September 22, 2002.

Ahlvik P and Brandberg A (2001) Well-to-Wheel Efficiency for Alternative Fuels from Natural Gas and Biomass. Ecotraffic R&D AB.

JRC, EUCAR, CONCAWE (2007) Well-to-Wheel Analysis of Future Automotive Fuels and Powertrains in the European Context.

Appendix 2

Crude oil refining

Crude oil is separated into different fractions by distillation at atmospheric pressure and distillation at reduced pressure, also known as vacuum distillation.

Apart from the products used as fuels, the other fractions are used for various industrial purposes: naphthas as feedstock in petrochemistry, paraffins to manufacture lubricants, bitumens, etc.

Together with the separation operations, a number of conversion methods are available to improve the quality of the products while adjusting the quantities obtained to market requirements:

- catalytic reforming leads to the production of high octane gasolines and hydrogen from a feedstock composed mainly of low octane gasolines,
- Fluid Catalytic Cracking (FCC) produces FCC gasolines and light olefins from a vacuum distillate, converted by an alkylation process into cuts of high octane number, or used in petrochemistry,
- *straight run* Diesel, produced directly from atmospheric distillation, is hydrotreated to reduce its sulphur content,
- FCC produces LCO *(Light Cycle Oil)* which, after hydrodesulphurisation, is blended in Diesel, while thermal cracking by visbreaking produces another Diesel fraction,
- hydrocracking treatments under high hydrogen pressure complete the production of Diesel, kerosene and gasolines.

A2.1 GASOLINES (UNLEADED PREMIUM-GRADE RON 95)

A2.1.1 Average composition (% by weight)

n-paraffins:5

isoparaffins:32

naphthenes:4

aromatics:43

olefins:16

On average, the C_5-C_{10} cut represents 97% by weight of a gasoline.

A2.1.2 Some specifications

Vapour pressure: 45-90 kPa, depending on the season

Distillation: – 46-71% (v/v) evaporated at 100°C

– 75% (v/v) evaporated at 150°C

– Final distillation point: 210°C

A2.2 DIESEL

A2.2.1 Average composition (% by weight)

C_7-C_{16} *n*-paraffins:20

C_8-C_{16} isoparaffins:17

C_7-C_{15} naphthenes:13

C_7-C_{14} aromatics:15

C_{15+} cut:35

On average, the C_{10}-C_{18} cut represents 75% of Diesel.

A2.2.2 Some specifications

Distillation: – less than 65% (v/v) recovered at 250°C,

– minimum 85% (v/v) recovered at 350°C,

– 95% (v/v) recovered at 360°C.

PAHs: maximum 11% by weight

Flash point: greater than 55°C

Cetane number measured: minimum 51.0

Cetane number calculated: minimum 46.0

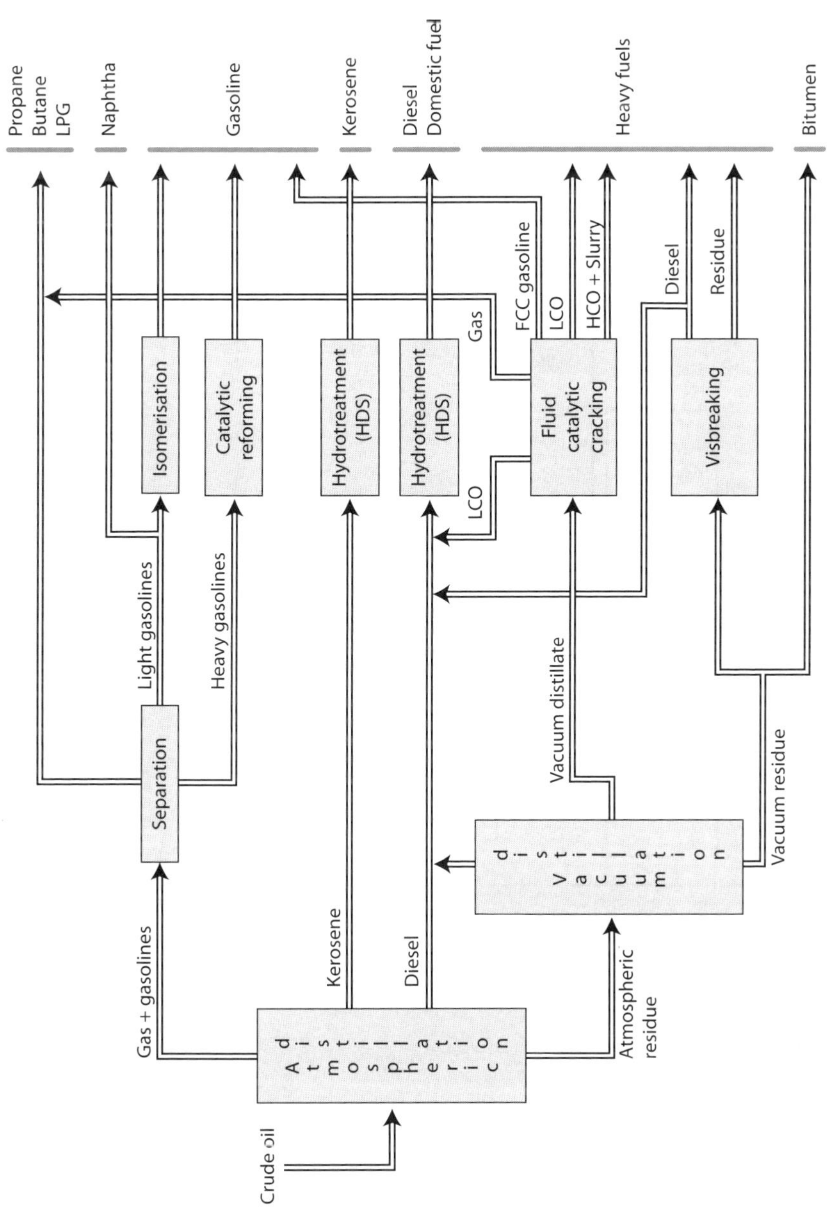

Figure A2.1
Simplified refining diagram.

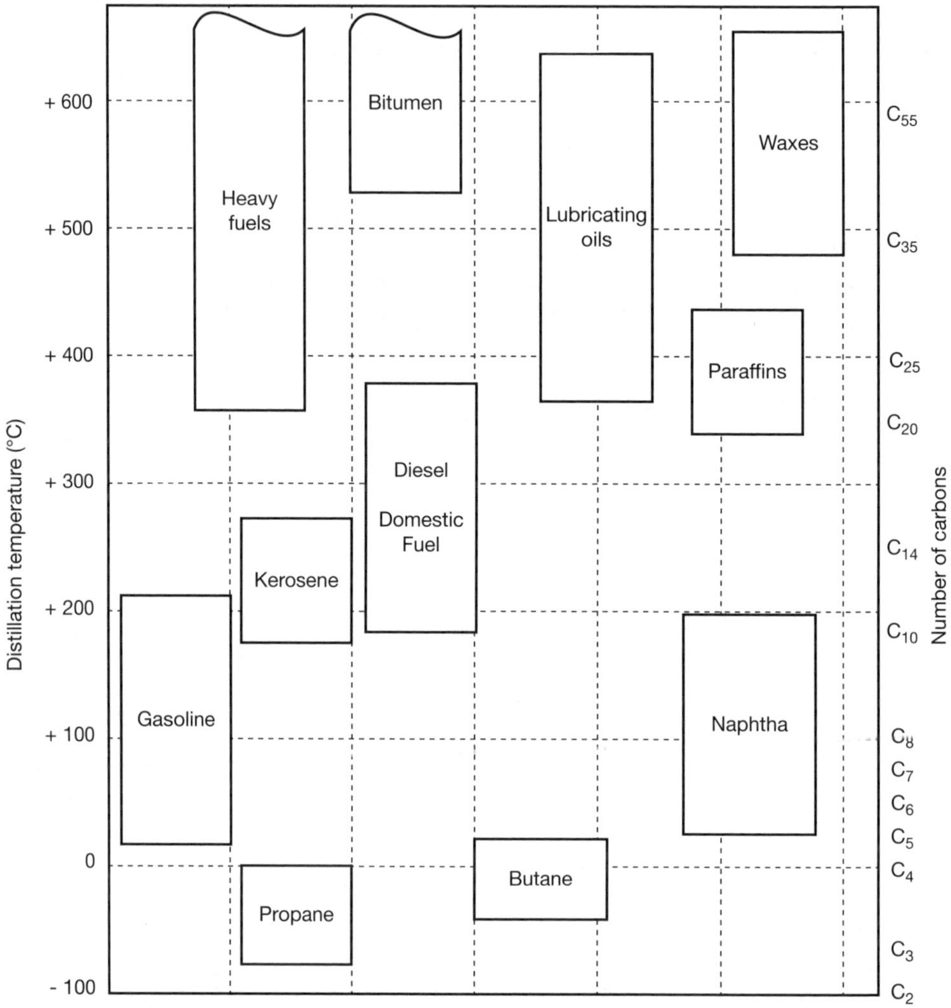

Figure A2.2

Oil fractionation by distillation.

Index

Acetogenesis 320
- acetogenic bacteria 321

Acetonobutylic fermentation 307
- acetone 307
- butanol 307
- *Clostridium acetobutylicum* 307
- mass and energy balances 310
- the process 309

Acidogenesis 321
AFEX process 281
Algae 319
- financial aspects 331
- marine macroalgae 326
- microalgae 327
- overall process for production of algae and oils 330
- pilot operations 331
- production of algae for energy uses 326
- productivity 328
- R&D programmes 332
- yield 328

Alternative fuels 31
Animal fats 139
- properties of the esters obtained 139

Bagasse 46
Barley 53
Benzol 3
Biodiesel 1
- characteristics 156
- energy balances 167
- environmental aspects 166
- greenhouse gas emissions 166
- impact on emissions 165
- main industrial players 145
- manufacture 117
- performance as fuel 163
- positive substitution effect 37
- probiodiesel programme 47
- production 146
- production costs 168
- risks 15

Biodiversity 24
BTL 32
- cetane number 263
- characteristics 263
- environmental aspects 249
- existing demonstration units 249
- technico-economic aspects 247

Butanol 105
- acetonobutylic fermentation 307
- conversion of ethanol 105

Cake 123
- composition of cakes 123

Carburol plan 4
Cellulases 286
- action mode 287
- β-glucosidases 286
- endoglucanases 286
- exoglucanases 286
- production in reactor 291
- synergy of cellulolytic activities 290
- synthesis 291
- *Trichoderma reesei* 285

Cellulose 276
- cellulolytic bacteria 285
- cellulolytic fungi 285

- chemical hydrolysis 284
- chemical structure 276
- enzymatic hydrolysis 285
- pyrolysis 202

Clean Air Act Amendments 4
Common Agricultural Policy 5
Competition with food agricultural crops 23
Corn 53
CTL 252

Direct liquefaction 253
- hydrogenolysis 253
- operating conditions 253
- principle 253
- solvolysis 253

DME 260
- catalyst 258
- characteristics 260
- compatibility with materials 261
- impact on combustion 261
- impact on emissions 261
- synthesis 257

Energy Policy Act 4
Energy supply and demand 6
Energy Tax Act 4
ETBE 29
- characteristics 95
- compatibility with materials 95
- emissions by evaporation 95
- polluting emissions 95
- production cost 112
- tolerance to water 95
- vapour pressure 95

Ethanol 51
- alcohol-producing potential of amylaceous plants 52
- alcohol-producing potential of sugar plants 52
- azeotropes 88
- azeotropic distillation 59
- cold starting 90
- compatibility of materials 90
- conversion into butanol 105
- corn gluten feed 60
- corn gluten meal 60
- DDGS 61
- dehydration on molecular sieves 60
- direct oligomerisation 102
- distillation 59
- dry milling 60
- E85 84
- effluents and co-products 60
- energy balance 64
- environmental aspects 106
- fermentation 54
- fermentation parameters 55
- hydrogen production 101
- in Diesel engines 93
- in spark ignition engines 91
- main industrial players 83
- octane numbers 87
- physical properties 86
- production cost in Brazil 110
- production cost in Europe 112
- production cost in the United States 109
- production from amylaceous plants 62
- production from sugar plants 61
- production of hydrogen 96
- raw materials for alcohol production 51
- *Saccharomyces* 54
- stability in the presence of water 89
- steam reforming 96
- vapour pressure 88
- volatility 88
- wet milling 60
- world production 83
- *Zymomonas mobilis* 55

Ethanol from CO + H2 304
Ethanol production from lignocellulosic raw materials 275
- cellulases 286
- chemical hydrolysis 284
- economic aspects 303
- environmental and energy aspects 303
- enzymatic hydrolysis 285
- ethanol fermentation of pentoses 293
- hemicellulases 293
- hydrolysis 284
- ligninases 293
- new strains of *Saccharomyces* 295
- pretreatment 278
- production of enzymes and their action mode 286
- simultaneous fermentation and hydrolysis 299
- substrate composition 275
- use of lignin 300
- various production diagrams 301

Ether fuels 67
- catalysts 74
- DIPE 67
- DPE 67
- ETAE 67
- ETBE 67
- etherification reaction 71

– MTBE 67
– processes 77
– TAME 67
Ethylene 103
– dehydration of ethanol 103
– oligomerisation by heterogeneous catalysis 104
– oligomerisation by organometallic catalysis 104
– Ziegler catalysis 104
Euro standards 41
Evolution of soil carbon stock 26

Fallow land 5
First-generation biofuel 1
Fischer-Tropsch synthesis 262
Fuel Flexible Vehicles (FFV) 46

Gasification 230
– ash 242
– carbon yield 244
– corrosion 242
– environmental aspects 249
– existing demonstration units 249
– hydrogen yield 244
– industrial processes 236
– LHVs 240
– main reactions 232
– plant energy yield 244
– primary energy yield 244
– syngas 230
– syngas compositions 240
– tars 241
– technico-economic aspects 247
– weight yield 244
Gasohol 4
Glycerine 117
– applications 148
– glycerine market 147
– production and consumption 150
– purification 149
– synthesis of GTBE 149
– world production 147
Glycerolysis of the fatty acids 133
Greenhouse effect 10
Greenhouse gas emissions 2
Greenhouse gases 10
GTL 31
Guerbet reaction 105

Hemicellulases 293
Hemicelluloses 276
Hydrogen 39, 96, 334
– combined productions by fermentative and photobiological pathway 337
– gasification 235
– methanation 263
– photobiological production 336
– pressure swing adsorption 265
– production 263
– production by ethanol steam reforming 96
– production by partial oxidation 100
– purification 99, 264

Isobutene synthesis pathways 70

Jatropha curcas 118

Kyoto Protocol 10

Land use changes 26
Life Cycle Assessment 341
– energy consumption 345
– greenhouse gas emissions 345
– impact assessment 343
– main limitations of LCAs 344
– methodology 341
Ligninases 293
Lignocellulosic biomass 1
– acetonobutylic fermentation 275
– agricultural residues 181
– biochemical conversion 185
– bioelectricity 22
– biological conversion 275
– biorefining 16
– characteristics 185
– composition 182
– dedicated crops 182
– distribution of main types 176
– energy purposes 195
– ethanol 20, 275
– forest resource 177
– gasification 199
– hydrothermal conversion 199
– potential for the production 190
– pretreatment 278
– pyrolysis 199
– resources 173
– source and current uses 173
substrate composition 275
– synthetic fuels 20
– thermal use 21
– thermochemical conversion 185, 199
– thermochemical pathway 20
– torrefaction 199
LPG 32

Manioc 52
Methane 319
Methane fermentation 319
– average biogas composition 323
– batch fermentation 322
– COD 323
– continuous fermentation 322
– financial aspects 324
– purification of biogas 323
Methanogenesis 320
– methanogenic bacteria 321
Methanol 69, 255
– catalysts 256
– characteristics 259
– high octane numbers 259
– M85 259
– M90 259
– main chemical applications 129
– main licence vendors 256
– main players 257
– production processes 129
– synthesis 129, 255
– transesterification reaction 129
MTBE 4
– characteristics 260
– conversion of isobutene 72
– octane number 259
– phasing out 4
– production 71

Oil 8
– oil demand 8
OPEC 8
Organosolv process 282

PAD 44
Palm oil 118
– production 121
– world producer 119
PEM cells 100
Plant chemistry 22
Potato 52
Proalcool Program 4
Pyrolysis 199
– applications for pyrolysis oils 218
– cellulose 202
– fast 201
– hemicelluloses 202
– lignins 203
– lignocellulosic biomass 203
– main parameters 204
– main reactor technologies 207
– mass and energy balances 215
– processes 206
– pyrolysis oils 215
– slow 200

Rapeseed 118
– composition of vegetable oils 122
– cultivated area 120
– deoiled cakes 123
– food and energy applications 126
– main producers 121
– oil extraction rates 121
– oil production yield 121
– production 122
– requirements for land dedicated to oilseeds 126
– seeds 120
– trituration 123
– yields 121
Road transport 8, 33
– characteristics of conventional fuels 40
– emissions 36
– energy consumption 9
– in the world 8
– reduction of CO_2 emissions 37

Safety stakes 12
Second generation biofuels 2
SOFC cells 100
Soya 118
– agricultural yields 121
– cakes 126
– main soybean producers 119
Splash blending 28
Starch 52
– α-amylase 53
– acid hydrolysis 53
– amyloglucosidase 53
– amylopectin 52
– amylose 52
– enzymatic hydrolysis 53
Sugar beet 52
– production of ethanol 64
Sugar cane 52
– energy potential 45
– ethanol 61
Sunflower 118
– cultivated area 120
– food and energy applications 126
– production of seeds 120
– requirements for land dedicated to oilseeds 126
– yields 121
Sweet sorghum 51, 52

Tall oil 140
- synfuel 140

Torrefaction 222
- mechanisms 223
- processes 225
- stakes 222

Transesterification reaction 129
- continuous heterogeneous catalysis processes 135
- enzyme catalysis 130
- homogeneous catalysis processes 131
- microwave heating 130

Trend in fuel characteristics 43
- Diesel 43
- gasoline 43

Use of biofuels in air transport 166

Vegetable oils 117
- agricultural yields 121
- applications 126
- castor oil 139
- characteristics 151
- cold filter-plugging point (CFPP) 153
- composition 122
- composition in fatty acids 122
- cooking oils 139
- corn and sunflower oils 125
- degumming 124
- direct use in Diesel engines 151
- european production 119
- extraction 123
- hydroisomerisation 143
- hydrotreatment 140
- impact on combustion 151
- impact on emissions 153
- manufacture 122
- oxidation reactions 153
- palm oil 125
- processes 144
- production of ethyl esters 138
- production of rapeseed oil 122
- production outside Europe 119
- refining 124
- soya oils 125
- special case of agricultural tractors 154
- trend in prices 168
- triglycerides 122
- trituration 123
- world production 118

VNG 31

VOEE 138

VOME 117
- catalysts 131, 135
- cetane number 161
- characteristics 156
- continuous heterogeneous catalysis processes 135
- continuous processes 133
- discontinuous processes 132
- economic aspects 168
- energy balances 167
- environmental aspects 166
- greenhouse gas emissions 166
- homogeneous catalysis processes 131
- impact on emissions 165
- lubricating power 161
- main industrial players 145
- performance as fuel 163
- production 146
- properties 161
- toxicity and biodegradability 160
- use in Diesel engines 155
- viscosity 161
- VOME100 164
- VOME30 163
- VOME7 163

Water-gas shift reaction 99

Wheat 53

Achevé d'imprimer en juillet 2012 par EMD S.A.S. – 53110 Lassay-les-Châteaux

N° d'impression : 26886 – Dépôt légal : juillet 2012

Imprimé en France